Responses to the Challenge

Keys to Improved Instruction

by Teaching Assistants and Part-Time Instructors

Edited by

Bettye Anne Case

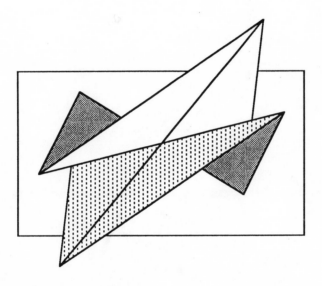

The Committee on Teaching Assistants and Part-Time Instructors

Thomas F. Banchoff, Brown University
Bettye Anne Case, Florida State University, Chair
John Philip Huneke, Ohio State University
David Kraines, Duke University
Research Associate: Annette Blackwelder

A Joint Committee of The American Mathematical Society
and The Mathematical Association of America

A grant from the Fund for the Improvement of Postsecondary Education of the U.S. Department of Education assisted the committee and supported the production costs of this publication. Despite any copyright notice on this page, permission is granted to any Federal agency or to investigators working on Federal grants to reproduce or borrow freely from all original material in this volume, and from materials reproduced here through the courtesy of their copyright holders unless the contrary is noted.

The Committee logo used on the cover of this book was devised by Tom Banchoff. It depicts a polyhedral monkey saddle which is related to, but different from, the symbol for the centenary of the American Mathematical Society. For each of these polyhedral surfaces the horizontal plane through the vertex meets the disc in six segments as in the case of the smooth monkey saddle, the graph of $x^2 - 3 \times y^2$. For this *exotic* polyhedral disc however there is no plane into which it projects in a one-to-one manner.

ISBN-0-88385-061-3

CONTENTS

PREFACE . . . . *v*

1 INTRODUCTION 1

General Remarks from the Committee 1

Panel Discussion: Teaching Assistant and Adjunct Instructor Competency
(Anaheim Joint Meetings, January, 1985.) 4
Felix Browder   Bettye Anne Case   Joseph Fiedler
Wendell Fleming   Leon Henkin   Phil Huneke   David Kraines

Views from the Committee Members 8
Tom Banchoff   Bettye Anne Case   Phil Huneke   David Kraines

References and Resources 11

2 SURVEY RESPONSES 13

Demographic Information 13

Some Observations 14

List of Tables and Figures 15

Teaching Personnel 16

Class Formats (for Lower Level, Calculus, Upper Level) 19

Class Size 27

TA Work Load 29

International Graduate Students 31

Orientation, Supervision and Evaluation 36

Prohibitions 39

References and Resources 40

3 MODELS OF PROGRAMS 41

Historical Perspective 41

Characteristics of the Models 43

Materials 45

Doctoral Program Models 47
University of California, Berkeley   University of Montana
Cornell University   Ohio State University
Duke University   Syracuse University
Florida State University   Vanderbilt University
University of Michigan   University of Wisconsin

Master's Program Models 60
Miami University, Ohio   University of Southern Mississippi
Western Washington University

References and Resources 64

4 INTERNATIONAL TEACHING ASSISTANTS 65

Introduction 65

English as a Second Language (ESL) 66

Surveying the Ground 66

Departmental and Individual Activities 68

Some Common Language Tests 69
    TOEFL   TSE   SPEAK

A Verbal Snapshot: The Classroom Observation 71

A Verbal Snapshop: Examples 71

References and Resources 75

5 PART-TIME AND TEMPORARY INSTRUCTORS 77

Who Are These Teachers? 77

Is Teaching by Nontraditional Faculty Widespread?
    Is the Practice Increasing? 78

Practices and Materials in Two-Year College Departments 79

Related Questions About Ethics and Academic Integrity 81

References and Resources 84

6 PERCEPTION SURVEYS 85

The TA Survey 85

The Survey of Student Perceptions 87

APPENDIX 91

The Table of Contents for the **Appendix** begins on page 92. Much of the material may be copied directly from this volume for departmental use. Included are guides for TAs, part-time and temporary instructors; course guidelines for the professor of a TA workshop; orientation schedules and sample guides for specific courses; different types of classroom observation forms and student evaluation forms; a TA Workshop evaluation form; guides and projects for international TAs; and an example of a helpful recommendation letter as a TA enters the job market. The 1985 and 1987 survey forms are included.

PREFACE

In many departments, mathematicians are concerned about and recognize the importance of effective orientation and careful assignments for new teachers. As those new teachers progress to teaching assignments with more independent responsibility, there is attention to continued support and supervision.

In 1984, when I asked in titling an American Mathematical Society (AMS) abstract "How Can We Help Graduate Students Develop Teaching Competency?", there was not a channel for the sharing of information. Ensuing discussions with colleagues led to a panel discussion and the establishment of the Committee on Teaching Assistants and Part-Time Instructors as a subcommittee of the Mathematical Association of America (MAA) Committee on the Teaching of Undergraduate Mathematics.

After the anecdotal information from the panel, the Committee felt a strong need for a reliable data base. A survey was distributed to the departments with graduate programs; preliminary results were presented in a 1986 panel discussion. An interim committee report was published in 1987 as an unnumbered volume of the MAA *Notes* under the title Teaching Assistants and Part-Time Instructors: *A Challenge*.

A second survey, in 1987, now provides additional information; most of the content of *Challenge* is included here, in *Responses to the Challenge*. A grant by the U.S. Department of Education through The Fund for the Improvement of Postsecondary Education (FIPSE) has provided funds to assist with compiling, editing and distributing both publications. In addition, a workshop at the Phoenix Joint Mathematics Meetings, January 1989, is partially supported by the FIPSE grant.

The Committee, now established as a joint AMS-MAA effort, will continue to monitor teaching by other than regular faculty. Attention will be given to the effects of legal and tax structures, of political and media attention, and to the need for more young Americans to study basic sciences. We hope our efforts will lead in the short term to more effective teaching. Taking the longer

view, it makes sense that graduate students for whom teaching was a positive experience will be more likely to choose academic careers, as will their own well-taught students. During the early critical years when pressures to publish are great, young mathematicians will better acquit themselves in teaching if they have, as graduate students, formed good habits and had excellent role models.

The committee sees further challenges ahead. As this report goes to press, it appears that a much closer look needs to be taken at the teaching by part-time and other temporary teachers who are not mathematics graduate students. Although our survey population was restricted by costs to graduate departments, our data and anecdotal information from two- and four-year colleges indicate a trend toward an increase in this staffing source. Legislation is pending in Congress which would affect the benefit status of part-time employees. Court cases establishing property rights for certain part-time teachers may influence departmental staffing patterns. We must be ready with our response to changing challenges.

Acknowledgments from *A Challenge*, 1987:

Each of the Committee members owes much to discussions with and support from colleagues. It would be difficult to single out all of the individuals whose influence is visible in my work on this project; I appreciate the contributions of colleagues in MAA and AMS and support from my department, college and university.

Lynn Steen and Al Schoenfeld facilitated the establishment of the Committee and shared valuable information and experience. Helpful early discussions took place in MAA committees and activities and in the Data Subcommittee of CEEP (AMS); I thank Don Bushaw, Jerry Goldstein, Billy Rhoades, Ken Ross, Don Rung, the Anaheim panelists, and others important in our early thinking.

The Committee and MAA thank the Department of Education for funding through FIPSE and Program Officers Diana Hayman and Helene Scher for advice and assistance. The work of Felix

Browder and Wendell Fleming on the committee during the first year was important. Recently there were fruitful discussions with representatives from departmental programs which will be described in the 1988 publication, and also with Dick Anderson and Ed Connors. Al Willcox, Dick Witter and the MAA staff at the Washington, D.C. office are always helpful. We thank Annette Blackwelder for data analysis and for working patiently with us on copywriting and editing; she held us to high standards. Josephine Curto and Moana Karsteter read the manuscript and helped us clarify and proof.

I personally thank the Committee members for their work and know that readers who find this material useful will thank Tom Banchoff, Phil Huneke and David Kraines. I am grateful for the efficient staff assistance from my department and in particular thank Karen Ball, Sheila Bernstein, Beth Hagler and Debbie Perry.

Additional Acknowledgements:

Most of those who are mentioned in the efforts before 1987 have continued helpful involvement. Each of the members of the committee has continued to give more intensive time over a much longer period than committees usually demand. Again, Jo Curto, Peter Renz and Al Willcox were there for us as we came down to the wire on deadlines. Two other time contributions have made it possible to meet publication dates: Annette Blackwelder, our Research Associate involved in data analysis, has collaborated on every aspect of the Committee's projects; my Department Chair, Ralph McWilliams, has generously taken on some of my usual responsibilities and arranged some released time for me in the last two months of manuscript preparation. Sheila Bernstein, again, attractively formatted a complicated manuscript; Deborah Russ has efficiently dealt with committee and Workshop publicity and correspondence; Kris Michaels has kept our accounts. I am grateful for all these efforts.

Contributions from many other colleagues are also represented here, including the faculty who are involved in orientation and support of TAs in my department. Each of the examples, models, and descriptions represents interviews, writings, and attention, generally by several individuals. This is a four year project that represents hundreds of

conversations and cabinets of correspondence; we appreciate the efforts of those whose names are missed, and apologize. But I must mention: John Addison, Dick Anderson, Henry Alder, Helmer Aslaksen, William Bade, Steve Bauman, William Beyer, Sarah Briggs, Lynn Cleary, Allen Cochran, Rob Cohen, Ed Connors, Janet Cooper, Jeff Cooper, Bill Derrick, Vivian Dennis, Steve Doblin, Peter Duren, Vincent Ferlini, Joe Fiedler, Jack Graver, Frederick Hartmann, Peter Hinman, Moe Hirsch, Guydo Lehner, Don Lewis, Dino Lorenzini, Dave Lutzer, Art Mattuck, Michael May, Marilyn McCollum, David McMichael, Mike Mears, Richard Millman, David Minda, Lang Moore, Shelba Morman, Maurice Nott, Al Ogus, Brad Osgood, Tom Read, Mike Reed, Tom Rischel, Steve Rodi, Ernie Ross, Doris Smith, Tommy Thompson, Constantine Tsinakis, Mimi Valek, Ray Wood, Terry Zachmanoglou.

Bettye Anne Case
November, 1988

Department of Mathematics
College of Arts and Sciences
The Florida State University
Tallahassee, FL 32306

Chapter 1

INTRODUCTION

This report holds few surprises for anyone familiar with current practices in college and university mathematics teaching in the United States today. Some two- and four-year college and graduate departments appoint on a part-time or temporary basis teachers without other employment who may teach part-time at several colleges and those who are sometimes called "gypsy scholars"; they also appoint moonlighters who are employed in government, industry, high schools or other colleges. In addition, graduate departments may appoint graduate students seeking degrees in other disciplines and may increase the involvement of undergraduates in mathematics instruction. More graduate teaching assistants are from other countries with different linguistic and cultural backgrounds.

The attention of the project, its publications and presentations, is to the broad spectrum of concerns associated with teaching by other than regular faculty. It is a sad reality that intense negative attention from politicians, parents, students and the media to "the foreign TA situation" may divert attention from constructive actions. The Committee feels a responsibility to try to prevent international graduate students from being made scapegoats for problems in U.S. education.

GENERAL REMARKS FROM THE COMMITTEE

In response to the challenge posed by these circumstances, mathematics departments have devised strategies for selecting potential teachers, for providing orientation and training as they begin teaching, for making appropriate assignments to them, for supervising and evaluating their work, and for establishing ways to support them, financially and personally, as they carry out their tasks. This report presents data analysis about the practices in graduate departments. But the main contribution is the presentation of a range of models of programs used by departments with very different characteristics. Components from these models will help two- and four-year departments as well as graduate departments. There is specialized information about the challenges of teaching by international graduate

students and the complex questions related to part-time instructors.

In most mathematics departments, there is an intent on the part of the regular faculty to assure competent teaching by graduate students and part-time and temporary instructors. But the number of faculty members involved in this effort is relatively small. The mathematician newly assuming such responsibility may be bewildered to find little agreement among colleagues about these matters, and little helpful information available..

The Committee on Teaching Assistants and Part-Time Instructors has assembled the data and analysis, descriptions and materials of this volume with the hope that it may have a multitude of applications. We have found a number of senior mathematicians who have cared about good teaching by new instructors and who have devoted time and influence to this effort for many years. They have generously given time to us to share their convictions, carefully honed practicality, anecdotes and writing. The first step in our learning process was to note the many similarities in apparently diverse departmental situations, and hence the potential value of sharing this informaiton with any interested department.

Some of the data may help you compare your situation to others—and perhaps to build a case to your administration for funds to pay stipends during orientation or to reduce loads or class sizes. The models provide descriptions of what works in some other departments; there is some technical information about language training and tests; there are sample evaluation forms and the results of their administrations. In the Appendix there are some materials that can be copied directly for new teachers, and there is also a syllabus for the professor teaching a course for teaching assistants.

In preparing this volume, we have attempted to assure accuracy and completeness. However, some sacrifice of polish has been made to bring the volume out in a timely manner.

Terminology

The Committee does not intend to show preference among the many synonyms in current use for the people and practices which are described. In each chapter of this volume, there is an attempt to describe the terms which are used and to give frequently used synonyms.

In particular, we use Teaching Assistant (TA) to mean a graduate student who has some classroom responsibility and awards grades (probably within carefully prescribed guidelines). Other studies have been careful to append *Graduate*; this is because undergraduates in some institutions who do not have classroom responsibility are called teaching assistants. In either case, the term "assistant" may be a poor label for one who is already beset with the insecurities of inexperience. Perhaps one of the other terms used now, or some new semantic invention which reflects all the right ideas, will become widely used.

Among the synonyms for TA which we have observed are: Graduate Instructor, Graduate Teaching Assistant, Graduate Student Instructor, Teaching Fellow (sometimes a true synonym; sometimes these awards are a partial fellowship), Graduate Assistant, Assistant Instructor. With restriction of duties to the lecture/recitation format, the terms Discussion Leader, Recitation Professor, and Recitation Instructor are used. Advanced TAs often help less experienced TAs learn their duties; they also assist faculty with orientation and training programs, classroom observations etc... Terms reported for TAs with these kinds of special duties are: TA Mentor, Assistant Mentor, Head Graduate Student Instructor, Master Teacher, Insurance TA, Big Buddy, Departmental Instructional Assistant.

As regards part-time or temporary instructors, many terms are again noted. In this volume, we use *part-time* generically to mean a person whose appointment is not considered full-time by the institution. We have found cases where people were classified part-time to indicate non-permanent (temporary) status even though their work load was full-time. Here, *temporary* is used when there is not the intent to reappoint the person indefinitely. A part-time instructor may not be temporary; some part-time instructors and their departments expect that the appointments will continue indefinitely. More discussion of titles for people in these categories is found in Chapter 5.

Concerning All Non-Regular Faculty

In identifying and selecting potential teachers, departments should use testing and screening procedures to assess teaching potential and communicative competence as well as academic potential. Orientation and training programs should familiarize participants with the expectations of the department and the needs of the students. It is desirable for programs to provide opportunities for acquisition and improvement of teaching skills by new teachers over an extended period of time (a semester or a year).

Supervision and evaluation procedures should provide monitoring of progress with respect to performance of teaching duties. In general, new personnel should be assigned in courses or formats with a high degree of structure and supervision (e.g., lecture/recitation model or multisection courses with fixed syllabi and common or monitored examinations).

There should be persons available to assist non-regular faculty in resolving problems and in meeting responsibilities, including regular faculty and, if possible, experienced teaching assistants or part-time instructors. Some undergraduates can be helpful, in many ways, from grading homework to (with close supervision) assisting in courses and recitation sections. This practice may have valuable long-range effects by introducing potential new members to the teaching profession. It also carries some risks: screening must reject those not sufficiently mature, mathematically or personally; teaching as an undergraduate must not hinder academic progress or excellence.

Some questions of academic integrity and institutional ethics in connection with part-time teachers are discussed in Chapter 5, and some references are cited there.

More about Teaching Assistants (TAs)

The dilemma of the beginning professor in our publish-fast-or-perish academic world is whether to devote time almost entirely to research or to put effort into teaching. Graduate teaching assistants, even more, must walk a thin line to acquit their teaching duties effectively and responsibly and enjoy teaching, while efficiently pursuing studies and research.

Coming of age as a teaching assistant in a U.S. college or university at the end of the Twentieth Century is a challenge which we hope will attract some of the best mathematical minds, both from the U.S. and other nations. But the prospect of teaching for the first time while simultaneously taking a collection of demanding and difficult graduate courses and adjusting to a new institutional environment can appear over-whelming. Graduate students deserve the full support of the departments that count on their teaching and of the faculty who educate them as future mathematicians, while working with them as they prepare to enter the teaching profession.

The challenges can be particularly acute for students facing a different linguistic or cultural environment for the first time. Teaching is communication, and acquiring communication skills necessary for effective teaching is itself a demanding task. As large percentages of our graduate classes continue to come from international backgrounds, it is especially important for us to pay attention to the pressure such students will face, and to provide support for them as they work to become effective teachers in our institutions.

Predictions of a need for many more Ph.D.'s in mathematics over the next decades make it imperative that mathematics departments recruit new candidates for the profession from among the most talented undergraduates. More scientists are now speaking out about the importance of these matters, and there is resulting press coverage. The interrelated concerns, a shortage of U.S. mathematics students and burgeoning numbers of international TAs, are receiving attention. (See [3], [4], [5], [6], and the section *Media Attention* in the References, this chapter.) The 1987 Colloquium, "Calculus for a New Century", at the National Academy of Sciences, included as a discussion session topic the Role of Teaching Assistants. [1] Conferences have been held about TAs which are not discipline-specific.[2]   It is important to continue to find new ways to encourage potential mathematics professors to view the TA process as their apprenticeship into the teaching profession.

Difficult Questions About
Teaching Ethics and Professionalism

Many of the materials described in Chapters 3, 4 and 5 and copied in the Appendix, make references to the obligations, ethics and professionalism related to an individual's teaching responsibilities. Since these matters deserve consideration by all teachers, there is some summary here; in the Appendix there is a collection of references to statements about these concerns. Situations arise where even an experienced teacher has doubts and a decision must be made which will have serious effects. The inexperienced will probably deal with these matters with more confidence if they have thought some about these things.

With highly uniform courses and examinations, there are many facets to academic honesty questions on the part of students and on the part of instructors privy to information which needs to be kept secure. Laws such as the "privacy act" have ramifications; sexual or other harrassment of students or actions reasonably perceived that way are not acceptable. Personal relationships with current students are always suspect. Matters of both form and substance fall under the topic "Professionalism"; it is important to follow local custom in any case. Professionalism includes the obvious—preparing for classes and being on time, meeting office hours, grading papers promptly—and also customs such as form of address and suitable classroom attire. (In the Appendix, see *Teaching Ethics and Professionalism*.)

TEACHING ASSISTANT AND ADJUNCT INSTRUCTOR COMPETENCY

Abridged Transcript of Panel Discussion
January 13, 1985
Anaheim Joint Mathematics Meetings

From *A Challenge*, MAA Notes, 1987

Bettye Anne Case, *Moderator*
Florida State University

The demographic characteristics of the graduate student population are rapidly changing. Increased percentages of the applicants for appointment as a TA are foreign nationals. Both language and cultural difficulties may impede the teaching process when these TAs are assigned as instructors without adequate screening and training. The American mathematics graduate student is less mathematically mature than twenty or even ten years ago; more careful training and supervision than in the past seem to be needed to assure effective teaching. The number of American TA applicants is much lower than the number of assistantships available at doctoral institutions. The necessity for hiring part-time and temporary instructional faculty to staff classes is a recent problem affecting many Ph.D. granting departments. A decade ago these departments had essentially no instruction by such teaching personnel. Two- and four-year college departments have also experienced a sharp rise in the percentage of instruction by non regular faculty.

When the Committee on the Teaching of Undergraduate Mathematics (MAA) last looked at the matter of training teaching assistants, the population was much less divergent than now and one "model plan" was suitable as a recommendation to all schools. As I now talk with colleagues working with TAs at various places, it is clear that such similarity no longer exists. Aware of this problem, CTUM is establishing a subcommittee to investigate; several of today's panelists will be on the committee, and I will chair.

At Florida State University, I direct the training and supervise the work of the teaching assistants and adjunct instructors; if there is time later this hour, I will describe our program. I'm eager to hear the others, and so we will begin with...

Wendell Fleming, *Brown University*

Some departments have well organized orientation and training programs and incentives for TAs to improve. It seems to me, as far as foreign TAs go, that they can be viewed as an asset to U.S. mathematics. Over the years we've owed a great deal to foreign mathematicians. In my own department, the Division of Applied Math at Brown University, nearly half of my colleagues come from other countries. We are competing for students with other science areas; stipends are set relatively high, and it is somewhat like recruiting star athletes. I think we're no longer in a situation where we have an adequate supply of bright, highly motivated U.S. students who are going on to graduate school; therefore, either we find some way to encourage them to do that or we're going to have to continue to find foreign students to help us carry on.

One of the things I did, being a department chairman, was to talk with each of our TAs individually just to get an idea of what was going on; about half our TAs are foreign students. Undergraduate students use the excuse that they are not doing well because of the TA's accent; however, from looking at the individual cases, I'm convinced that this is just an excuse. Okay, there are some really bad cases occasionally, but this is often not the meat of the problem. The cultural aspect is very important. Some foreign students misunderstand what our educational system is about and how weak the standard really is to let a student get by. Things like that I believe are as important as the accent. Some of our most successful TAs, in fact, have some kind of an accent.

Felix Browder
formerly: University of Chicago
now: Rutgers University

My remarks deal with the situation of the past few years at the University of Chicago with respect to the use of graduate students in teaching basic undergraduate courses.... One of its principal features is that no graduate student is allowed the responsibility of teaching a section on his or her own without going through a year of apprenticeship as a College Fellow during which time that student serves as an active junior collaborator of a regular faculty member in teaching an undergraduate course and is appointed Lecturer only if that faculty member will write an explicit formal letter of recommendation stating that the student has displayed the capacity for teaching with sufficient skill and understanding of the task. In particular, this process of apprenticeship, which occurs during the second year of the student's graduate work, makes it possible to have a clear and unambiguous perception of any personality difficulties or failures of linguistic skill in English, which may make the graduate student's teaching difficult, and to make sure that such difficulties or failures are corrected before the graduate student is given responsibility for a section. It is a very expensive system for the University, but it has had a very strong effect in getting rid of some of the most difficult problems which are often generated at the elementary level in university mathematics instruction....

Our ability to carry through... was undoubtedly influenced by the experience some members of the faculty had had with the Danforth Fellowship Program, which was an experiment in this direction. The importance of an apprenticeship or an acclimatization process in introducing people to teaching is evident from the experience of many. I can well remember the shock experienced when I began teaching for the first time at MIT, not having taught as a graduate student at Princeton.

When I consider the moral one can draw from the case history of the Chicago department, I think

the most important one from my point of view is the importance of an ethic of quality. Very few academic administrators are as idealistic in a practical sense on this point as [the Chicago President] Edward Levi was...Most administrators seem to appreciate only the bottom line... The endless public delcarations in favor of good teaching and better teaching deserve to be taken at this face value and answered with a very simple fact. Good teaching cannot be had cheaply. What you can have cheaply is the pretense of an effort at good teaching without basic alteration of the situation, the so-called good faith effort. The cynicism which this latter practice enforces on both mathematics departments and on university administrators is destructive in itself, quite apart from the constantly deteriorating situation in the classroom.

David Kraines, *Duke University*

Teaching duties for our graduate students consist of 4 hours per week teaching or equivalent service from first and second year students and 6 hours per week from more advanced students who take few courses. Until two years ago, first year students graded for advanced courses or conducted help sessions for first year calculus, and second year students taught a 4 hour section of pre-calculus or calculus of 25 to 35 students with minimal faculty supervision. Advanced graduate students typically taught two sections of the same three hour per week course. (added May 1987: Data from this report helped to persuade our administration to pay for a reduction of advanced graduate student teaching to one section per term.)

Although some of these graduate students began as excellent instructors, most of them had some trouble adjusting to the new pressures. The foreign-born graduate students seemed to have the most problems, although there have been a few notable exceptions. Many undergraduates complained that, in light of the high tuition that they (i.e. their parents) pay, they deserve to have more qualified instructors.

Lecture-Recitation Experiment. In the Fall of 1983, 180 first-semester calculus students were assigned at random to a lecture-recitation format. A faculty member, who had been a very popular teacher, lectured twice a week. First and second-year graduate students met sections of about 20 students twice a week. The lecturer met with the

TAs once a week to go over the material for the next week and also made up the hour exams and the homework assignments. This was considered to be a two-course load. The TAs did the bulk of the grading of tests, collected the homework and assigned course grades to their students under the supervision of the lecturer.

Note that this experiment was not done to save manpower. If the lecturer had taught two regular sections and his TAs had each taught one, then the 180 students would be divided into our usual sections of 30 students.

Advantages and Disadvantages. The main advantage of this experiemnt has been that the apprenticeship of the new graduate students has made the transition to their independent teaching far smoother. The better TAs were given their own classes the next semester and were replaced by the other first year graduate students who had been assigned to grading or other tasks. The weaker TAs might be kept as TAs for 2 or 3 semesters, or, in one extreme case, dropped from the graduate program. The undergraduates have had more uniformity in instruction. They have contact with a faculty member and have had personal attention in a smaller problem session.

On the negative side, the lecturer left the examples and the teaching of the computational techniques to the TAs and tended to emphasize more theory than most of the students could handle in one dose. It was difficult for him to coordinate the lessons with the TA's assignments. Some students complained that they felt uncomfortable asking questions in the large lectures. Other students felt that they were treated less fairly than their fellow students who were in classes of 30 taught by an experienced faculty member, instructor or graduate student.

Postscript. (Added May 1986.) After three years of experimentation with the large lecture recitation sections, we concluded that the drawbacks outweighed the benefits. Many students switched out of these sections into individual sections of 35 or more students even if the section were taught by a graduate student. The lecture-recitation sections actually became more expensive in faculty time than the regular classes. To ease the graduate students into teaching, the new students and those not proficient in English are now assigned as assistants in regular four hour per week calculus sections. They help to make up and grade tests and give occasional lectures under the supervision of the regular instructor. We are quite convinced that direct graduate student-undergraduate contact is very important before allowing the graduate student to teach his or her own section. We are still searching for better ways to increase this contact.

Leon Henkin, *University of California, Berkeley*

We have a very large operation.... Regular faculty take turns giving the large lecture courses which are split up into small sections met twice a week by a teaching assistant. Our teaching assistants must be graduate students; almost all are in a Ph.D. program; a few are seeking an M.A..

What happens in the TA-led sections is crucial to the learning of our students. I feel that the large lecture is one of the worst possbile ways of carrying out mathematics instruction. The only way that anyone really learns mathematics is by doing mathematics, not by sitting and listening to someone talk about it, especially if sitting too far away to see the board, or hear well, or feel that the instructor is talking directly to him or her. So really, the small [recitation] section is an opportunity for learning and I have always been anxious to have that opportunity realized to its fullest potential.

For this reason, for many years I've visited the small sections when I have had one of these large-lecture courses; I generally visit each TA-section twice during the term. By and large, I have found that the TA tends to do what every teacher at every grade-level from kindergarten through graduate school tends to do, which is to imitate the kind of teaching that s(he) has seen before. And, in particular, many TAs try to imitate the lecturer. This is a disaster. The students don't need another lecturer; they need to do mathematics. What really needs to happen is to have the TA activate the students.

After some years of observing the wide range of performance levels taking place naturally and spontaneously, I began to devote my efforts to bringing all of my TAs as far toward the highest level of teaching as I could. My first step was to see how well I could do as a TA myself. I made arrangements with my own TAs to let me take a turn at their work. We actually split the hour: I would meet them in advance, and they would tell me what they were planning to accomplish; sometimes I'd see if I could improve their ideas a little on that subject. At any rate, when the final lesson plan was decided, I'd ask if I could have a piece of the action.

Well, I found that it is very difficult to be a TA - certainly much more difficult than being a lecturer. To be a lecturer, you just see what it is you're supposed to say and you say it. And there's no back talk. To be a good TA, you have to crawl into the minds of every one of the students and see what's going on there, bring them into discussions, have them express their ideas and interact with the other students, get them as excited in talking about their mathematical ideas as research mathematicians are when they talk to each other....

[For TA training, we have] what we call a workshop. It is passed around among experienced faculty who conduct it as part of their teaching load. Unfortunately it varies in quality. When I conduct it, as I've done several times, I have a great many different kinds of activities for novice TAs. When I distribute evaluation forms at the end, I ask them to say which of those activities seem to work and which do not. Invariably the individual classroom visits I made with the members of the workshop, patterned after the visits I had been making to may own TAs, were judged to be the most valuable component of the workshop.

After the term just finished a few weeks ago, it occurred to me that, if these visits I'd developed in my own courses first and then tried out in the workshops seemed to be of some help, maybe we could institutionalize them. I had no way of persuading my colleagues to make such a commitment of time and effort, for very few of them are interested in lower-division courses to that extent. Some of them make occasional visits; some of them make no visits; none of them make these kinds of shared visits that I've been telling you about. Last term I went to a campus-wide committee interested in improving teaching and got some special funds to permit one specially-selected TA be given the title of Departmental Instructional Assistant and to devote his or her full teaching time to making visits of the kind that I had been doing as a faculty member ....

I concluded that this stirring up of interest in the teaching process had a salutary effect on both the TAs - future teaching - and their students. The bottom line is that activation of our students must lie at the heart of the teaching process, and we must inspire and guide our TAs to undertake this.

Professor Case

I'd like to add one thing which Professor Henkin told me earlier in the fall concerning dealing with their non-native English speakers. They require the Test of Spoken English which ETS has now developed and is being widely used....

Professor Henkin

Yes, Bettye Anne reminds me that in the fall of '83 our department faculty, after sustaining an avalanche of complaints for many years, voted as a departmental policy not to offer a TA position to a student coming from abroad for the first time unless evidence was presented that he or she was able to communicate effectively with spoken English at a level adequate for serving as a TA. Our Vice-Chair for Graduate Admissions was given the assignment of making this abstract policy into a practicum. He discovered that ETS has a test called Test of Spoken English; it is administered in countries all over the world. He got in touch with persons familiar with the test to determine the kind of score on this exam needed to be an effective TA.

What he did was to identify half a dozen of our foreign TAs about whom there were an excessive number of complaints. With funds to conduct an experiment, he went to our University Extension and asked them to prepare a special course which would improve the spoken English of these TAs. We required them, as a condition to continue in their jobs as TAs, to take that course. We got their scores on this ETS exam before the course and then we got their scores afterward. At the same time, we observed them in their work as teaching assistants, both before and after the course.

Based on this experience, we judged that the course our university extension had developed made a significant improvement in the ability of these people to serve as TAs. Secondly, we got scores that seemed reasonable to set as a requirement . First-time TAs who came in the fall of '84 had to produce those scores or else were not given TA assignments. We don't have, as the University of Chicago does, other ways of supporting graduate students; thus it meant that we did lose some of the foreign students whom we had chosen because we thought that they would do very well in our Ph.D. program.

John Philip Huneke, *Ohio State University*

Ohio State University is very large. The Department of Mathematics issues over 16,000 grades each Autumn Quarter.

With the size come some advantages and some disadvantages when compared with smaller departments. The Department needed to make 350 TA-type teaching assignments at OSU last quarter and would have had to double its number of graduate students to cover these assignments with Math graduate students. Departments of all sizes have the dilemma of hiring (and training) adjunct faculty to assist with teaching. We have found at OSU that, with careful hiring practices, the quality of teaching can be maintained and even improved while saving money.

At OSU we hire for particular jobs. Most of our individual sections of remedial courses are taught by former high school teachers earning an M.A. degree; lower level precalculus recitations are taught by undergraduate (predominately domestic) students selected through a competition; most other precalculus recitations are taught by graduate students from other departments (who are predominately international) hired competitively. Mathematics graduate students teach calculus recitations; international graduate students needing to strengthen their English grade papers and serve in an open tutor room; more experiences GTAs teach individual evening courses, etc. With different stipends for each category, the cost of instruction is minimized without loss of instructional quality.

Joseph Fiedler will describe the specifics of some of these staffing practices at the Ohio State University.

Joseph Fiedler, *Ohio State University*

Fiedler's comments are included in Chapter 3 as a significant portion of the *Ohio State University* section in *DOCTORAL PROGRAM MODELS*.

VIEWS FROM COMMITTEE MEMBERS
From *A Challenge,* MAA Notes, 1987

The Committee is only midway in its collection and dissemination of information which began with the initial Spring 1985 survey and will shift into continuing, but less intense effort, after a January 1989 workshop. We will then make recommendations to the Board of Governors for their action. The following are the early impressions, ideas and opinions of the individual committee members.

The comments at this point are closely related to the nature of the programs in the member's doctoral-granting departments. Later recommendations will be broadened to include more about PTIs and non-doctoral departments.

From Tom Banchoff
Brown University

(1) Faculty responsibility. A sort of community experience is a significant part of being a graduate TA. That concept has a lot to do with the attitude

of a department about teaching. If the TA gets the impression that the professors themselves, especially those in charge of the courses TAs teach, care about teaching effectively, then they are likely to develop well themselves and pass on their experience to those coming after them. On the other hand, if the faculty is perceived as uninterested in teaching or in working with students, it is this attitude that the students will take to their own TA jobs and pass on to the new graduate students.

(2) Peer assistance. Another sort of recommendation deals with the ongoing moral support of students teaching for the first time in an American college or university. A very important source of help comes from older graduate students or instructors who are teaching the same course or who have taught it previously. An atmosphere where graduate students discuss what they are doing in their teaching would be quite beneficial both to the new TA and to the more experienced

one. Such an arrangement can be formalized by having TAs work in teams, a natural situation if there are several graduate students handling recitation sections for a large course or teaching in courses with many parallel sections.

(3) <u>Evaluation of effectiveness</u>. Every program should include feedback mechanisms for all instructors involved in teaching. The TA should be constantly aware about his or her effectiveness in the classroom or the recitation section. Grading student papers or quizzes is a good way of seeing whether the material is getting across. In situations in which a TA is provided with an undergraduate grader, it is important that the TA look over the papers before returning them in order to get an idea of the difficulties the students are experiencing. First year graduate students preparing for teaching a course should spend some time grading. There should be feedback in any training session and after classroom observations by faculty.

(4) <u>TAs not ready for the classroom</u>. It might be better to have them grade and assist in the sorts of lower level courses that they will be teaching later, rather than assisting in a higher level course.

(5) <u>Undergraduates as TAs?</u> I know that in some schools the use of undergraduates in ways graduate TAs have normally functioned is totally forbidden, but it is possible that the times have changed sufficiently that such policies should be rethought. We will be facing an extreme teaching shortage in mathematics during the next decade. Anything we can do to encourage students to consider careers in teaching might work to alleviate this problem. A well-structured program where advanced undergraduates work with faculty, possibly along with graduate students, can provide a good experience of what teaching is about. I have worked with undergraduate assistants in every course I have taught with a computer laboratory component, precisely because it was these students who developed the software for my courses. Right now I have 96 students in such a course with three seniors and a junior as leaders of the weekly sections. They come to my lectures, and we have considerable discussion about the progress of the course. All four of them are showing strong leanings toward college or secondary school teaching, more so now than when they began.

(6) <u>Tax law changes</u>. Although stipends as such are not part of our committee concern, the changing

tax laws may affect whether or not teaching experience is to be considered a required part of graduate programs in mathematics and may have a definite effect on the morale of students.

From Phil Huneke
Ohio State University

(1) <u>Lecture-recitation format</u>. Lecture/recitation format can be a desirable mode of instruction, especially with lecture size of approximately 60 students (taught by an experienced faculty member lecturing two, resp. three, times each week in a four, resp. five credit course) which is split into a pair of recitations (taught by TAs as problem sessions each meeting twice a week.) The Lecturer need not teach daily, and the TA can concentrate on communicating with students instead of preparing lecture materials. The size of the Lecture can be doubled or tripled for financial savings, but only with very experienced lecturers and with the more routine performance-oriented courses. Graduate students develop well as classroom teachers if given a year or two experience teaching recitation sections before being offered individual classes.

(2) <u>TA loads</u>. Students in teaching roles must be carefully monitoried for work load so that the tendency for them to enjoy the teaching or the sheer burden of the work does not overshadow the importance of their own studies.

(3) <u>Class size</u>.. Individual classes should be limited to 30 students, in hopes of a steady state audience of not more than 25. Math at all levels is a participation sport in which instruction is productive only when the instructor has an opportunity to understand and to realize the confusion of individual students.

(4) <u>Training for teaching</u>. No one—TA, part-time faculty or new faculty—should be assigned a course without careful preparation describing the level of the student, content and expectations of the course, and other relevant pedantic information. Sample course syllabi and exams from previous offerings are crucial for lecturers and teachers of individual classes. TAs and other novices to the classroom should be given training involving multiple opportunities to present material in a mock teaching format.

(5) <u>Undergraduates as TAs?</u> The employment of undergraduate students in tutor rooms and assisting

regular faculty in recitation sections or problem sessions can have a very positive effect on the undergraduate instructor and can offer very inexpensive, high quality instruction, especially if jobs are competitively awarded to very select applicants.

Undergraduate students as instructors should not be asked to cover individual classes where significant preparation of lectures would compete for their time and they would have responsibility for determining grades for other undergraduates.

It is probably best if undergraduate TAs are not assigned to courses with students known to be at special risk, such as those in remedial courses, or to scientific calculus recitations. They should not be allowed to teach more than 4 to 6 contact hours per week, and then only in supportive roles.

Undergraduate students as instructors should be paid a reasonable professional stipend, thereby allowing for a presumed two hours of preparation for each contact hour. This is a very cheap mode of instruction, provided the jobs are highly desirable and highly competitive. Enthusiastic undergraduate senior math majors, if not overworked, can be the most effective precalculus recitation instructors available.

(6) PTI. In hiring graduate students from other departments as TAs and part-time instructors, advertising to draw an excess of applicants is desirable to create the atmosphere that to be able to teach is a privilege and an award. Selection should be made by two criteria: strength of resume and performance in several mock-teaching opportunities which can double as tryouts and teacher training sessions.

From David Kraines
Duke University

(1) Teaching evaluation and awards. The mathematics department should reward good teaching and punish bad teaching. Our supervisor of freshman instruction sends letters to all teaching assistants at the end of each term in which he thanks them for their work and points out—very tactfully—areas in which improvements can be made. A teaching award for graduate students, even if it means just a name on a plaque and a picture in the student newspaper, will spur on many graduate students. On the other hand, the

dismissal of one bad teaching assistant will have a sobering effect on the others.

(2) Training of TAs. Graduate students should have access to a wide variety of "How to Teach" booklets. Exam files are also very useful for new TAs. Encourage your faculty to contribute to them. Encourage new graduate students to attend lectures of professors who are excellent teachers. Videotape graduate students as they teach and let them see their weak points for themselves.

Provide opportunities for new graduate students to lecture. First, they can lecture to fellow students in a graduate class or in a special summer seminar in which new graduate students, especially the foreign speaking ones, "bone-up" on their mathematical vocabulary as well as their English proficiency. Then they might "guest lecture" in a recitation section or course at the level they will teach.

(3) First year assignments. First year graduate students rarely make good teachers. Try to find a way to avoid the use of first year students as instructors of individual classes. They should have some direct contact with undergraduates, perhaps as course assistants or recitation instructors. It is especially important to encourage foreign born students to speak English as much as possible.

(4) Teaching loads. The higher ranked universities tend to require their graduate students to teach only one course per term. Show this data to the dean and try to get more funding for instruction.

From Bettye Anne Case
Florida State University

I have had the advantage of reading my colleagues comments and hence will try to touch some additional points.

(1) Foreign TAs. All noncitizen TAs face cultural differences; additional problems may arise for those for whom English was not the language of the childhood home. These problems are sometimes exaggerated in student complaints and may be exacerbated by insensitive media reports or politically motivated comments. As the chair of this committee, questions I often hear from the media are "Why are there so many foreign TAs?" or even "Why are there too many...?" I answer that there are not too many, that we are fortunate that so many gifted young scholars want to study

with us. I point out that the current numbers of Americans with adequate mathematical talent in graduate mathematics programs indicate that our research and teaching institutions can not be properly staffed in the nineties, unless some of our visiting scholars stay here. We should welcome them graciously and help them adapt.

(2) Coping. Three main coping devices which departments use to meet staffing needs while productively employing foreign TAs have been around a long time and often were implemented for other reasons. Each of these devices requires careful monitoring:

(a) Departmental exams and highly departmentalized course syllabi and methods. (We do not want teaching so regimented that it is not enjoyable.)
(b) The lecture/recitation teaching mode. (The tutorial sections must provide good student-teacher interaction. More clock hours per week, adding lecture and recitation hours, may be needed than the credit hours of the course.)
(c) Employment of undergraduates and graduate students form other departments as TAs and of adjunct instructors. ("Strong senior majors" must not translate to "a pretty good engineering junior who really wants the work.")

(3) Language screening. No standardized language tests are currently sufficient for determining competency for teaching. There should be local screening, and it should follow several weeks of intensive training by the department in the company of new American TAs. The screening is most effective when done jointly by mathematics faculty and English Language specialists. Alternature work assignments and English courses are needed for TAs whose test scores seemed sufficient in advance but who fail at the local screening.

(4) PTI training and evaluation. Too often it is rationalized, because of problems in scheduling and paying for training time, that PTIs do not need training because they are already experienced teachers, usually at another level school, or because they were once TAs. Although fewer hours of training may be needed, the same careful explanation of policies, procedures and expectations is important.

(5) Supervision and evaluation. Evaluation of TAs and PTIs is often less careful than for regular faculty. Colleagues from a number of departments with established training programs have asked what they can do to help TAs continue to improve teaching effectiveness - what they can do to help the ones just barely good enough to get by. We do not have a lot of answers at this point, but regular classroom observation followed by discussion is essential as is some form of monitoring by regular faculty in courses which are not already highly departmentalized. Student evaluations should be carried out each semester and TA/PTI encouraged to salt lightly, then consider seriously.

REFERENCES AND RESOURCES

Many of the references cited in the other chapters are applicable to the overviews in this introductory chapter.

[1] Case, Bettye A. and Allan C. Cochran, "Role of Teaching Assistants." *Calculus for a New Century: A Pump, not a Filter.* MAA Notes No. 8 (1987) 76-77.

[2] Chism, Nancy Van Note, ed. *Institutional Responsibilities and Responses in the Employment and Education of Teaching Assistants.* Columbus, Ohio: Center for Teaching Excellence, The Ohio State University, 1986. (Readings from a national conference about teaching assistants held in November 1986. Tapes of the sessions are available at $15.00 each.)

[3] Finkbeiner, A.K. "Demographics or Market Forces?" *Mosaic* 18:1 (1987) 10-17.

[4] Neal, H.A.. *Report of the NSB Task Committee on Undergraduate Science, Mathematics and Engineering Education.* National Science Board, 1986.

[5] "Report of the Committee on American Graduate Mathematics Enrollments" (B. Simon, Chair; summary). *Notices of Amer. Math. Soc.* 34:5 (1987) 748-750.

[6] White, Robert M., "Calculus of Reality." *Calculus for a New Century: A Pump, not a Filter* MAA Notes No. 8 (1987) 6-9.

Recent Media Attention

The articles below deal with the declining and insufficient number of young Americans choosing careers in mathematics or the increasing numbers of international teaching assistants. Most recognize the relationship of these two concerns. Campus newspapers have also carried articles or series of articles dealing with perceptions about English language problems of international TAs. The typical campus spectrum of attention ranges from the editorial decrying the controversy as outright racism to the "my-daddy-pays-taxes-and-I-want-an-American-teacher" syndrome.

"A Losing Numbers game," *Buffalo News*, December 13, 1987.

"Americans Scarce in Math Grad Schools," (Kolata), *Science*, November 15, 1985.

"AMS Celebrates and Worries," (Cipra), *Science*, October 7, 1988.

"Beijing battles the brain drain," *U.S. News and World Report*, March 21, 1988.

"Beyond the David Report," (Steen), *Focus*, November-December, 1986.

"Call me if your college instructor can't speak English, Senator says," (Cotterell), *Tallahassee Democrat*, February 6, 1987.

"Contest winner sends an SOS to Congress," (Byrne), *The Scientist*, February 22, 1988.

"Demographics or Market Forces," (Finkbeiner), *Mosaic*, Spring, 1987.

"Document Confirms Beijing Shift in its Policy on Studying Overseas," *The NY Times,* April 4, 1988.

"Fewer and Fewer Americans Take Graduate Work in Mathematics," (Heller), *The Chronicle of Higher Education*, July 15, 1987.

"Fewer doctorates awarded in math: New book aims at reducing anxiety among college students," (Hechinger), *Union News*, January 15, 1988.

"Foreign Students Proliferate in Graduate Science Programs," (Vobejda), *Washington Post*, September 2, 1987.

"The 'foreign TA problem'-an update," (Constantinides), *NAFSA Newsletter*, March, 1987.

"Help Wanted: Math Majors," (Shearer), *Parade Magazine*, October 23, 1988.

"In Math, the Language of Science, Americans Grow Even Weaker," (Brown), *The New York Times*, October 1988.

"Foreign TAs, U.S. students fight culture shock" (Schwartz, *et al.*) *Newsweek On Campus*, December, 1985.

"The Rage to Study Abroad," (Yuchao), *China Reconstructs*, North America Edition, December 1986.

"Students get refunds because of teachers' accents," *AP dispatch*, February 2, 1985.

"Teaching Assistants Get Increased Training; Problems Arise in Foreign-Student Programs," (Heller and McMillen), *The Chronicle of Higher Education*, October 29, 1986. Letter to the Editor in Response, (Case), December 3, 1986.

"U.S. faces critical shortage of mathematicians in 1990s," (Freeman), *Springfield Union News*, December 28, 1987.

"U.S. mathematicians add few to their numbers," *The San Juan Star*, December 6, 1987. (AP dispatch by Tynan which was run under many headlines in different newspapers. Example: "Math profs becoming rare species," *Chicago Tribune*, December 6, 1987.

"Wanted: Students of science," (Washington Post), *Tallahassee Democrat,* February 11, 1988.

"We can't understand you, students tell foreign-born teachers," (Clifford), *Tallahassee Democrat*, August 2, 1987.

"We Should Train Our Undergraduates to Deal With International TAs," (Welsh), *The Chronicle of Higher Education*, November 26, 1986.

"When Teachers Can't Speak Clear English," (Fiske), *New York Times*, June 4, 1985.

Chapter 2

SURVEY RESPONSES

In 1985, the MAA conducted a general survey of graduate departments concerning graduate teaching assistants and other nonprofessorial teaching personnel. In 1987 a survey gathered more complete information about staffing and levels of courses taught. The response rates for both surveys were high and lead to confidence in the data implications. There is support for earlier anecdotal information about the significant portion of undergraduate mathematics taught by other than regular faculty. The later chapters in this monograph present some responses to the resulting challenges.

A summary of the data from the two surveys, with contrasts where interesting is in this chapter. The first of three introductory sections describe the survey population and response rates; there follow some descriptive statements and generalizations and a contents listing for the tables and figures. As much data analysis is included as possible; each reader will find the basis for summary statements about special concerns. Data are included on: the diversity of sourses of teachers; class formats; distribution of sections among teaching personnel; international TAs; and the orientation and training, selection, supervision and evaluation of nonprofessorial faculty.

DEMOGRAPHIC INFORMATION

The survey population was the departments with graduate programs in mathematics in the U.S., Canada, Costa Rica, Jamaica, Mexico and Trinidad. The survey included 464 departments for the 1985 survey, 472 departments for 1987.

Doctoral Department Responses

The U.S. doctoral programs are divided into three groups in keeping with the 1982 assessment of graduate programs conducted by the Conference Board of Associated Research Councils. [4], [5] The departmental rankings are based on the criterion, "Scholarly Quality of Program Faculty." The two cases of separate departments of mathematics and applied mathematics (Brown and Maryland) are combined so that Group 1 **(GP1)** consists of the 39 mathematics departments with scores of at least 3.0; Group 2 **(GP2)** consists of the 43 mathematics departments with scores 2.0 through 2.9; Group 3 **(GP3)** consists of the

remaining departments with a doctoral program in mathematics; 31 of these appear on the ranked list.

Several doctoral-granting departments in Canada and in Trinidad responded: 15 departments for 1985, 12 for 1987. To keep all the U.S. graduate programs together in the tables, we placed these other doctoral departments in the last column and denote it **C/T**.

The U.S. response rates yield almost census data: all but two of the 39 Group 1 departments and all but four of the 43 Group 2 departments responded to at least one of the surveys.

U.S. Doctoral Department Responses

	1985		1987	
Group 1 (GP1)	29 of 39	75%	32 of 39	82%
Group 2 (GP2)	35 of 43	81%	31 of 43	72%
Group 3 (GP3)	50 to 71	70%	48 of 81	59%

Master's Department Responses

About 300 of the surveyed mathematics departments offer a master's, but not the doctoral degree; 153 responded in 1985; 108 in 1987. Among master's departments, three from Canada, one from Costa Rica, and one from the Virgin Islands responded. Their input was appreciated. However, we have no comparative data from 1985 for these departments; the small number of responses is insufficient for analysis. The U.S. departments are divided into two groups based on whether or not they have TAs. The numbers of responses are:

1985	1987	
103	55	from departments which have TAs and may have other non-professorial faculty (**MAT**);
50	53	responses from departments which do not have TAs and may have other nonprofessorial faculty (**MAP**).

SOME OBSERVATIONS

This section is an attempt to pull out a few, perhaps superficial, conclusions about the data. What is *important* to readers depends on their particular points of view, needs and concerns. Conclusions will be most useful after a study of all the tables and figures with closely related information. Some descriptive analysis is presented in the context of each section. Terms used below:

GP1—top ranked 39 doctoral departments
GP2—next 43 departments
GP3—rest of doctoral departments
TA—mathematics graduate student who teaches

• The very high survey response rates from GP1 and GP2 indicates lively interest. This may be because just 40% of the instructors are professors. (This percentage rises to 50% for the remaining graduate departments.)

• There is a good chance university students will get their first glimpse of college mathematics from a TA. In each group of doctoral departments, TAs teach almost one section in three of total individual classes and one section in two of lower level individual classes. And TAs—or others recruited to fill the shortage in their ranks—teach almost all the recitation sections.

• International students, undergraduates, and part-time and temporary instructors are making up for the shortfall in U.S. mathematics students. *Undergraduates* teach 14% of the recitation sections in GP2 departments, though they teach few individual classes. *International TAs* teach more classes, although to the credit of many departments they are often given nonclassroom duties at first and/or recitation sections before individual classes. *Part- or full-time nonprofessorial instructors* are often used to staff individual classes including those of size around 100. GP2 schools report that 42% of their nonprofessorial teachers are not mathematics graduate students.

• For GP1, about half the total undergraduate sections and two-thirds of calculus are taught in the lecture/recitation format. The dependence on this format decreases but remains significant in the other doctoral departments. (Graduate departments may begin teaching more students in lecture and recitations if first year TAs are prohibited from teaching by accreditation regulations. In Chapter 5, see Regional Accreditation...)

• The GP2 and GP3 use of large individual classes (average size 100, without recitation sections) warrants watching.

• Although there is some reason for unhappiness with large class sizes in the U.S., in the reporting doctoral departments from Canada and Trinidad, average "regular" class size is about 50. More surprising, two students in five are taught in classes of over 100 (without recitations).

• There are active orientation and training programs for new TAs in most departments; advanced TAs are often used to work with new TAs in doctoral schools. It is unusual today if a graduate student is provided only with a text and syllabus and sent into the classroom, common practice a decade ago.

• Most departments show evidence of trying to help their new TAs. Often less time is devoted than needed however, to utilize the materials and innovations of the more active programs. Student evaluations are the norm; classroom observation of TAs by experienced TAs and faculty could be used more effectively in many departments.

LIST OF TABLES AND FIGURES

Teaching Personnel
1   Percentages of teaching personnel, 1987.
2   Percentage of departments with non-professorial instructors who are not mathematics TAs.
3   Percentage who are not mathematics TAs among nonprofessorial instructors.
4   Distribution of departments by group of percentages who are not mathematics TAs among NPI. (Also see Figure 5.)
6   Sources of part-time faculty as percentages of total part-time faculty, 1987. (Also see Figure 7.)

Class Formats
8   Numbers of sections reported at all levels by format, 1987.
9   Numbers of reported lectures and recitations by level, 1987.
10   Numbers of reported individual classes by level, 1987.
11   Percentage of sections taught in various formats by level of courses.
12   Percentage of departments using lecture/recitation format.
13   Percentage of large lectures and recitations at all levels taught by indicated personnel, 1987.
14   Percentage of lower level large lectures taught by indicated personnel, 1987.
15   Percentage of calculus large lectures and recitation taught by indicated personnel, 1987.
16   Percentage of upper level large lectures and recitations taught by indicated personnel, 1987.
17   Percentage of individual classes at all levels taught by indicated personnel, 1987.
18   Percentage of lower level individual classes taught by indicated personnel, 1987.
19   Percentage of calculus individual classes taught by indicated personnel, 1987.

20   Percentage of upper level individual classes taught by indicated personnel, 1987.
21   Percentage of lectures and recitations taught by nonprofessorial instructors. (Also see Figure 22.)
23   Percentage of individual classes taught by nonprofessorial instructors. (Also see Figure 24.)

Class Size
25   Average lower level class size, 1987.
26   Average calculus class size, 1987.
27   Average upper level class size, 1987.
28   Percentages of individual section class size, 1985.
29   Percentages of recitation section class size, 1985.

TA Work Load
30   Average contact hour work loads, 1987.
31   Departments' 1987 clock hour loads for math TAs.
32   Reported loads actually assigned to TAs, 1985.

International Graduate Students
33   Support of U.S. and international graduate students, 1987.
34   Support of U.S. and international graduate students, 1987; first year and advanced graduate students.
35   Numbers of departments with ITAs and percentages of ITAs among all TAs.
36   Departmental percentages of ITAs among all TAs. (Also see Figure 37.)
38   ITA qualifications, screening, and instruction.

Orientation, Supervision and Evaluation
39   Orientation/training program features.
40   Supervision and evaluation of nonprofessorial instructors, 1987.
41   Supervision and evaluation, 1985.

Prohibitions
42   Number of departments reporting particular undergraduate duty assignments.

TEACHING PERSONNEL

Descriptions of positions and responsibilities of teaching personnel vary in different institutions. To assure some uniformity, the following definitions were adopted for the 1987 survey:

Mathematics Instruction (MI) refers to teaching individual classes or holding recitation (tutorial) sections from large lectures with grading responsibility. It does not include those who work exclusively in a "mathematics laboratory" setting or those who grade without classroom or tutorial section teaching responsibility.

Professor refers to all faculty in professorial ranks, whether or not tenured, including visiting, research and partially-retired professors.

Full-time doctoral refers to nonprofessorial personnel who hold a doctorate in the mathematical sciences and are appointed on a full-time basis. They usually hold the titles lecturer or instructor.

Full-time nondoctoral refers to nonprofessorial MI personnel who do not hold a doctorate in the mathematical sciences.

Math TA refers to mathematics graduate students involved in MI. (Mathematics education graduate students are included here only when their program is a mathematics department degree program.)

Other Graduate TA refers to graduate students, seeking degrees in departments other than mathematics, involved in MI. (Graduate students in mathematics education are presumed reported in this category when their degree is to be in a College of Education.)

Undergraduate TA refers to any student without a bachelor's degree involved in MI. Undergraduates who work exclusively in a "mathematics laboratory" setting or grade under the direction of regular faculty or TAs are excluded.

Part-Time Instructor refers to all other persons involved in MI.

The percentage of teaching personnel in each of the above categories among responding departments is given. (For GP1, GP2, GP3 see [4], [5]; MAT refers to master's only departments which have TAs and may have other non-professorial faculty and MAP refers to master's only departments which do not have TAs but may have other nonprofessorial faculty.)

Table 1

1987 Percentages of Teaching Personnel

	GP1	GP2	GP3	MAT	MAP	C/T
Professor	39%	40%	46%	50%	59%	55%
Full-time doctoral	3%	1%	<1%	6%	5%	4%
Full-time non-doctoral	1%	6%	4%	7%	8%	2%
Math TA	47%	36%	35%	16%	0%	22%
Other graduate TA	5%	6%	3%	1%	<1%	1%
Undergraduate TA	3%	6%	3%	1%	0%	11%
Part-time instructor	2%	5%	9%	18%	28%	5%
Useable Responses:	27	29	44	52	47	12

How Many Of The Teaching Personnel Are Neither TAs Nor Professors?

On the 1985 survey, departments were asked to report only the categories Math TA and PTI. Information about teaching by professorial faculty was not collected. Also, the category PTI in 1985 included most of the personnel in five 1987 categories. In addition to the 1987 category "Part-Time Instructor", PTI in 1985 included the 1987 categories: Full-time Doctoral, Full-time Nondoctoral, Other Graduate TA and Undergraduate TA. The numbers in these five categories are added to the numbers of graduate mathematics TAs to form the 1987 super-category Non Professorial Instructors (NPI).

In Chapter 7, the terminology Nontraditional Teaching Personnel is used. It represents *almost* "NPI less Math TAs". To explain the *almost*: In addition to teaching by professorial faculty, in graduate departments there has long been some teaching by not only mathematics graduate teaching assistants but also by post-doctoral fellows who teach small loads in favor of research time. A difference between the directions on the survey instruments complicates comparison: In 1985, post docs were specifically excluded from the category PTI. In 1987 some respondents, especially GP1 departments, reported traditional post doctoral fellows as Full-Time Doctoral. Others may have considered post-docs as visiting faculty and counted them as professors. This anomaly will not significantly affect the overall percentages except possibly as discussed below in GP1 departments.

Table 2 shows that three-quarters of the GP1 and virtually all other departments have some instructors other than professorial faculty and mathematics graduate students, and that this is significantly up from 1985, a comparison to discern possible trends would be helpful.

A small amount of the increase shown in *Table 3* from 1985 to 1987 in GP1 may be considered due to the above described anomaly. The large increase in GP2, with some increase in GP1, indicates that these doctoral departments have increasing difficulty staffing classes with traditional personnel; the increases are consistent with the increase in C/T doctoral departments. In GP3 and MAT these percentages of teaching personnel were already high; there is consistency in the above

percentages with anecdotal information that these departments are being successful in recruiting more graduate students, often among their own undergraduates.

Table 2

Percentage of Departments with NPIs Who Are Not Mathematics TAs

	GP1	GP2	GP3	MAT	MAP	C/T
1987	75%	97%	98%	96%	96%	100%
1985	71%	83%	90%	89%	74%	93%

Table 3

Percentage Who Are Not Mathematics TAs Among NPI

	GP1	GP2	GP3	MAT	MAP	C/T
1987	27%	42%	37%	63%	100%	51%
1985	20%	26%	36%	67%	100%	45%

In order to illustrate the variation among departments in each group, this percentage was computed for each department. The distribution over intervals is given as *Table 4* and *Figure 5*. A larger percentage of departments in 1987 than in 1985 report that more than 20% of their NPIs are not mathematics TAs. This is true in all groups, even MAT, which shows a decrease in the overall percentage of non-TAs.

On the other end of the scale, departments with over half of the NPIs not mathematics TAs increased significantly and, for the GP2 and C/T departments there is much heavier dependence on these nontraditional instructors.

Table 6 and *Figure 7* give some information about other employment of part-time faculty members. It does not reflect the part-time faculty who teach part-time at several colleges and universities, a phenomenon rumored to be increasing. It does given lie to the often heard misconception that most part-time instructors are otherwise employed as secondary school teachers or in government or industry. This is supported in the data of [1] where on page 141 a similar figure is given for sources of part-time two-year college faculty.

Table 4 and Figure 5

Percentage of Departments by Group

			GP1	GP2	GP3	MAT	MAP	C/T
		0%-20%	46%	26%	26%	11%	0%	17%
	1987	21%-50%	39%	48%	50%	35%	0%	33%
Percentage Who Are		51%-100%	15%	26%	24%	54%	100%	50%
Not Mathematics TAs								
Among NPI		0%-20%	67%	52%	39%	17%	0%	21%
	1985	21%-50%	29%	39%	40%	21%	0%	46%
		51%-100%	4%	9%	21%	62%	100%	33%

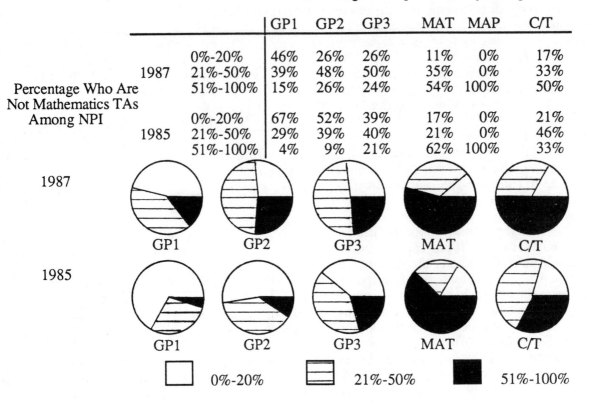

1987

GP1   GP2   GP3   MAT   C/T

1985

GP1   GP2   GP3   MAT   C/T

☐ 0%-20%   ▥ 21%-50%   ■ 51%-100%

Table 6 and Figure 7

1987 Sources of Part-Time Faculty as Percentages of Total Part-Time Faculty

	GP1	GP2	GP3	MAT	MAP	C/T
Secondary school teacher						
with MA or higher	10%	5%	29%	33%	34%	22%
Other degree	0%	5%	1%	1%	<1%	7%
Government or industry based						
with MA or higher	0%	8%	15%	15%	12%	0%
Other degree	0%	3%	3%	2%	<1%	0%
Others						
with MA or higher	67%	71%	44%	41%	51%	71%
Other degree	23%	8%	8%	8%	3%	0%

GP1   GP2   GP3   MAT   MAP   C/T

■ Secondary School Teacher   ▤ Government/Industry   ☐ Other

CLASS FORMATS

The level and content of courses with the same name may vary from institution to institution. The following classifications were used on the 1987 survey. (Note: Since some mathematics departments are joint departments with computer science, it should be pointed out that the data does not include computer science courses or their teachers which were excluded on the survey instrument.)

Lower Level refers to non-calculus based courses with titles such as college algebra, trigonometry, analytic geometry, finite math, liberal arts math, remedial math, elementary statistics; excluding computer science.

Calculus refers to any course which is primarily elementary calculus of functions of one or more variables.

Upper Level refers to all other undergraduate mathematics courses (linear algebra, differential equations, discrete mathematics, advanced calculus, statistics using calculus). Computer science is excluded.

Additionally, the mathematics courses taught at each level are described by teaching format: lectures with associated recitation sections, individual classes of size less than or equal to 65 (called *regular size individual classes*), and individual classes of size greater than 65 (called *large individual classes*). Classes referred to as individual in this chapter may be multisection courses but are taught independently as distinguished from those in lecture/recitation format. The term lecture/discussion is sometimes used and a distinction between lecture/discussion and lecture/recitation is drawn. No distinction was made on the survey instrument. These are combined and are referred to as lecture/recitation.

Numbers of sections reported in responding departments are listed, first combined and then separated by these levels and formats. Although the numbers reflect the reported data and are not extrapolated to the population, these total numbers are meaningful in combination with the useable responses information.

Table 8 shows that in GP1 almost as many undergraduate sections are taught in lecture/recitation format as in individual classes. More discussion about lecture/recitation format sections is given in relation to *Table 12*.

There is anecdotal information that the incidence is increasing of large individual classes which do not have associated recitation sections. The 1985 survey design does not permit comparative data which might show this trend. One use of these large classes has been to meet the demand for calculus for business students.

Table 10 indicates that the total number of students taught in these large individual classes, assumed to be of average size 100-150, is not very large in comparison to the number taught in lecture/recitation or regular size individual classes. But the data (see *Table 10*) does confirm their existence in all groups of departments.

Some other indications from *Tables 10, 17, 18, 19* and *20* in the data about use of the large individual class format:

• The C/T doctoral departments, proportionately, use this staffing exigency more than the U.S. graduate departments.

• In U.S. schools the format is proportionately more prevalent in GP3 departments, and there, largely for lower level classes.

• There is significant use of this mode in calculus in GP2 departments with an estimated average of 250 students per department taught during a fall term in these large individual classes.

• At the calculus level, only 60 percent of these large individual classes in GP2 are staffed by professors. This is possibly confirmation of the anecdotal information: A lot of departments are putting business calculus into large individual class format and staffing with part-time and temporary instructors (see *Table 19*.)

• GP1 and GP2 departments staff less than half of these large individual classes at the lower level with professorial faculty.

Table 8

1987 Numbers of Sections Reported at All Levels by Format

	GP1	GP2	GP3	MAT	MAP	C/T
Individual Classes (all sizes)	2679	3044	3790	3680	2250	618
Large Lectures	500	230	178	28	1	139
Associated Recitations	2409	1378	960	127	2	323
Useable Responses	30	29	45	52	40	10

Table 9

1987 Numbers of Reported Lectures and Recitations by Level

	Large Lectures						Associated Recitations					
	GP1	GP2	GP3	MAT	MAP	C/T	GP1	GP2	GP3	MAT	MAP	C/T
Lower Level	109	104	73	24	1	9	590	674	487	115	2	31
Calculus	340	124	96	4	0	79	1604	697	442	12	0	198
Upper Level	51	2	9	0	0	51	215	7	31	0	0	94
All Levels	500	230	178	28	1	139	2409	1378	960	127	2	323
Useable Responses	30	29	45	52	40	10	30	29	45	52	40	10

Table 10

1987 Numbers of Reported Individual Classes by Level

	Individual Classes (Size ≤ 65)						Individual Classes (Size > 65)					
	GP1	GP2	GP3	MAT	MAP	C/T	GP1	GP2	GP3	MAT	MAP	C/T
Lower Level	868	1075	1574	1826	1302	37	37	31	158	78	30	20
Calculus	725	1081	984	928	453	130	13	69	39	28	9	67
Upper Level	1021	775	1022	819	456	306	15	13	13	1	0	58
All Levels	2614	2931	3580	3573	2211	473	65	113	210	107	39	145
Useable Responses	30	29	45	52	40	10	30	29	45	52	40	10

The percentage of students taught in the various teaching formats at each level is given in *Table 11*. The percentage of total sections taught at each level is counted as the sum of recitation sections plus individual classes of both sizes. The last two entries in *Table 11*, a comparison between 1987 and 1985 data, are for *All Levels* combined. (The reported numbers of sections are given in *Tables 8, 9,* and *10*).

GP1 departments use the lecture/recitation format more than in any of the other categories, with well over 50 percent of the calculus sections taught in that mode. The ratio of lecture/recitation to individual sections drops a little at GP2 but not significantly until GP3, where only 1 in 5 undergraduate is taught in this mode.

The percentage of departments using the lecture/recitation teaching format is given in *Table 12*. There is significant use of this format in all the doctoral categories, with a modest upward trend in GP2 and a startling increase in C/T departments. Lecture/recitation format is clearly a good starting point for the assignment of international TAs. MAP schools indicate use of this format in 1987 but not in 1985.

The use of lecture/recitation is not as prevalent in master's departments; also, many master's departments do not accept international graduate students unless their English skills are sufficient to teach individual classes. The Southern Association of Colleges and Schools now requires that graduate teaching assistants have completed 18 hours of graduate mathematics before being assigned individual classes. More GP3 and master's departments may be pushed toward the lecture/recitation format if such regulations are widely adopted.

Table 11

Percentage of Sections Taught in Various Formats by Level of Courses

	GP1	GP2	GP3	MAT	MAP	C/T
1987 Lower Level						
Lecture/Recitation	40%	38%	22%	6%	<1%	35%
Individual Classes (Size ≤ 65)	58%	60%	71%	90%	98%	42%
Individual Classes (Size > 65)	2%	2%	7%	4%	2%	23%
1987 Calculus						
Lecture/Recitation	68%	38%	30%	1%	0%	50%
Individual Classes (Size ≤ 65)	31%	58%	67%	96%	98%	33%
Individual Classes (Size > 65)	1%	4%	3%	3%	2%	17%
1987 Upper Level						
Lecture/Recitation	17%	1%	3%	0%	0%	20%
Individual Classes (Size ≤ 65)	82%	97%	96%	100%	100%	67%
Individual Classes (Size > 65)	1%	2%	1%	<1%	0%	13%
1987 *All Levels*						
Lecture/Recitation	47%	31%	20%	3%	<1%	34%
Individual Classes (all sizes)	53%	69%	80%	97%	100%	66%
1985 *All Levels*						
Lecture/Recitation	50%	22%	20%	3%	0%	14%
Individual Classes (all sizes)	50%	78%	80%	97%	100%	86%

Table 12

Percentage of Departments Using Lecture/Recitation Format

	GP1	GP2	GP3	MAT	MAP	C/T
1987 Lower Level	69%	70%	24%	4%	7%	33%
1987 Calculus	91%	64%	31%	5%	0%	40%
1987 Upper Level	33%	0%	3%	0%	0%	50%
1987 *All Levels*	91%	70%	31%	5%	7%	50%
1985 *All Levels*	90%	64%	46%	14%	0%	20%

Distribution of 1987 Lectures and Recitations Among Teaching Personnel

The percentages of lecture and recitation sections taught by the different categories of teaching personnel are given in *Tables 13, 14 , 15* and *16*. *Table 13* gives the data combined for all levels. Tables *14, 15,* and *16* present the data for the three levels: lower, calculus and upper, respectively. Where there are no entries in a column for a group of departments, no department in that group reported any sections. (For example, there were no large lectures of calculus reported by MAP departments.)

Virtually all of the calculus and upper level large lectures are taught by faculty in professorial ranks. The only significant percentages of lectures taught by TAs are lower level GP1 sections which are numerically significant although lower than calculus section numbers in those departments. Of the lower level lectures, a significant proportion in doctoral departments are taught by Full-time Nondoctoral and Part-Time Instructors. Mathematics TAs teach over 60% of the recitation sections at all levels. There is significant use of nonmathematics graduate students and undergraduates in recitation sections. In the lower level recitations for GP1 and GP2, they teach nearly a third of the recitation sections. Not many calculus recitations are staffed by Other Graduate TAs, although GP2 and GP3 each show 13% of the calculus recitations staffed by undergraduates. There is little incidence of teaching by these categories of teaching personnel at the upper level.

Table 13

	Percentage of Large Lectures at *All Levels* Taught by Indicated Personnel						Percentage of Recitations at *All Levels* Taught by Indicated Personnel					
	GP1	GP2	GP3	MAT	MAP	C/T	GP1	GP2	GP3	MAT	MAP	C/T
Professor	68%	74%	90%	89%	100%	85%	1%	0%	<1%	12%	100%	0%
Full-time doctoral	16%	0%	0%	11%	0%	7%	1%	0%	0%	0%	0%	0%
Full-time non-doctoral	6%	20%	6%	0%	0%	1%	<1%	<1%	0%	6%	0%	7%
Math TA	6%	2%	1%	0%	0%	2%	84%	70%	85%	73%	0%	61%
Other Graduate TA	0%	0%	0%	0%	0%	1%	9%	14%	6%	0%	0%	0%
Undergraduate TA	0%	0%	0%	0%	0%	0%	4%	14%	9%	0%	0%	32%
Part-Time Instructor	4%	4%	3%	0%	0%	4%	1%	2%	0%	9%	0%	0%

Table 14

	Percentage of Lower Level Large Lectures Taught by Indicated Personnel						Percentage of Lower Level Recitations Taught by Indicated Personnel					
	GP1	GP2	GP3	MAT	MAP	C/T	GP1	GP2	GP3	MAT	MAP	C/T
Professor	22%	61%	80%	80%	100%	75%	<1%	0%	0%	0%	100%	0%
Full-time doctoral	20%	0%	0%	20%	0%	0%	<1%	0%	0%	0%	0%	0%
Full-time non-doctoral	22%	29%	12%	0%	0%	0%	<1%	<1%	0%	9%	0%	0%
Math TA	26%	2%	2%	0%	0%	0%	63%	60%	87%	77%	0%	25%
Other Graduate TA	0%	0%	0%	0%	0%	0%	24%	21%	8%	0%	0%	0%
Undergraduate TA	0%	0%	0%	0%	0%	0%	11%	16%	5%	0%	0%	75%
Part-Time Instructor	10%	8%	6%	0%	0%	25%	2%	3%	0%	14%	0%	0%

Table 15

	Percentage of Calculus Large Lectures Taught by Indicated Personnel						Percentage of Calculus Recitations Taught by Indicated Personnel					
	GP1	GP2	GP3	MAT	MAP	C/T	GP1	GP2	GP3	MAT	MAP	C/T
Professor	80%	87%	99%	100%		82%	1%	0%	<1%	33%		0%
Full-time doctoral	15%	0%	0%	0%		8%	2%	0%	0%	0%		0%
Full-time non-doctoral	2%	12%	1%	0%		2%	<1%	0%	0%	0%		6%
Math TA	<1%	1%	0%	0%		3%	91%	81%	83%	67%		61%
Other Graduate TA	0%	0%	0%	0%		2%	4%	6%	4%	0%		0%
Undergraduate TA	0%	0%	0%	0%		0%	1%	13%	13%	0%		33%
Part-Time Instructor	3%	0%	0%	0%		3%	1%	0%	0%	0%		0%

Table 16

	Percentage of Upper Level Large Lectures Taught by Indicated Personnel						Percentage of Upper Level Recitations Taught by Indicated Personnel					
	GP1	GP2	GP3	MAT	MAP	C/T	GP1	GP2	GP3	MAT	MAP	C/T
Professor	100%	*	83%			90%	8%	*	0%			0%
Full-time doctoral	0%	*	0%			6%	6%	*	0%			0%
Full-time non-doctoral	0%	*	0%			0%	0%	*	0%			11%
Math TA	0%	*	17%			0%	82%	*	100%			66%
Other Graduate TA	0%	*	0%			0%	2%	*	0%			0%
Undergraduate TA	0%	*	0%			0%	0%	*	0%			23%
Part-Time Instructor	0%	*	0%			4%	2%	*	0%			0%

*There was one report of incidence but a breakdown by teaching personnel was not given.

Distribution of 1987 Individual Classes Among Teaching Personnel

The percentages of individual classes taught by the different categories of teaching personnel are given in *Tables 17, 18, 19,* and *20. Table 17* gives the data combined for all levels. *Tables 18, 19* and *20* present the data for the three levels: lower, calculus, and upper, respectively. Where there are no entries in a column for a group of departments, no department in that group reported any sections.

Over 50% of the regular size lower level classes, in the responding doctoral departments are taught by mathematics TAs. It must be noted again that there are a number of lower level sections taught in the doctoral departments. Nonprofessorial instructors are teaching approximately 50% of the regular size calculus individual classes in doctoral departments. Part-time instructors are involved in teaching individual classes of both sizes at the lower and calculus level in all the groups of departments. There is a more complete analysis of the incidence and staffing of large individual classes which precedes *Tables 8, 9* and *10.*

Table 17

	Percentage of Individual Classes (Size ≤ 65) at *All Levels* Taught by Indicated Personnel						Percentage of Individual Classes (Size > 65) at *All Levels* Taught by Indicated Personnel					
	GP1	GP2	GP3	MAT	MAP	C/T	GP1	GP2	GP3	MAT	MAP	C/T
Professor	55%	46%	47%	61%	61%	81%	65%	60%	62%	72%	67%	86%
Full-time doctoral	4%	1%	1%	3%	4%	7%	4%	1%	1%	3%	5%	6%
Full-time non-doctoral	1%	10%	8%	11%	15%	4%	0%	17%	18%	18%	11%	2%
Math TA	31%	32%	31%	11%	0%	7%	13%	10%	4%	3%	0%	0%
Other Graduate TA	3%	4%	4%	1%	0%	<1%	0%	0%	0%	0%	0%	0%
Undergraduate TA	0%	<1%	<1%	0%	0%	0%	0%	0%	0%	0%	0%	0%
Part-Time Instructor	6%	7%	9%	13%	20%	1%	18%	12%	15%	4%	17%	6%

Table 18

	Percentage of Lower Level Individual Classes (Size ≤ 65) Taught by Indicated Personnel						Percentage of Lower Level Individual Classes (Size > 65) Taught by Indicated Personnel					
	GP1	GP2	GP3	MAT	MAP	C/T	GP1	GP2	GP3	MAT	MAP	C/T
Professor	16%	9%	14%	33%	47%	65%	40%	43%	62%	66%	63%	75%
Full-time doctoral	3%	<1%	1%	6%	4%	3%	7%	0%	1%	4%	7%	5%
Full-time non-doctoral	6%	19%	12%	17%	20%	24%	0%	27%	16%	22%	15%	0%
Math TA	55%	51%	52%	23%	0%	5%	23%	23%	3%	4%	0%	0%
Other Graduate TA	2%	9%	8%	1%	0%	0%	0%	0%	0%	0%	0%	0%
Undergraduate TA	0%	<1%	<1%	0%	0%	0%	0%	0%	0%	0%	0%	0%
Part-Time Instructor	18%	11%	13%	20%	29%	3%	30%	7%	18%	4%	15%	20%

Table 19

	Percentage of Calculus Individual Classes (Size ≤ 65) Taught by Indicated Personnel						Percentage of Calculus Individual Classes (Size > 65) Taught By Indicated Personnel					
	GP1	GP2	GP3	MAT	MAP	C/T	GP1	GP2	GP3	MAT	MAP	C/T
Professor	38%	53%	58%	77%	75%	70%	100%	60%	69%	89%	78%	84%
Full-time doctoral	4%	2%	3%	1%	3%	2%	0%	2%	3%	0%	0%	9%
Full-time non-doctoral	1%	8%	6%	8%	11%	4%	0%	16%	9%	7%	0%	3%
Math TA	53%	30%	25%	3%	0%	22%	0%	5%	5%	0%	0%	0%
Other Graduate TA	1%	1%	1%	<1%	0%	0%	0%	0%	0%	0%	0%	0%
Undergraduate TA	0%	<1%	0%	0%	0%	0%	0%	0%	0%	0%	0%	0%
Part-Time Instructor	3%	6%	7%	11%	11%	2%	0%	17%	14%	4%	22%	4%

Table 20

	Percentage of Upper Level Individual Classes (Size ≤ 65) Taught By Indicated Personnel						Percentage of Upper Level Individual Classes (Size > 65) Taught By Indicated Personnel					
	GP1	GP2	GP3	MAT	MAP	C/T	GP1	GP2	GP3	MAT	MAP	C/T
Professor	93%	94%	93%	94%	85%	88%	100%	100%	100%	100%		93%
Full-time doctoral	4%	<1%	1%	1%	5%	10%	0%	0%	0%	0%		2%
Full-time non-doctoral	<1%	<1%	2%	2%	6%	1%	0%	0%	0%	0%		3%
Math TA	2%	4%	2%	<1%	0%	1%	0%	0%	0%	0%		0%
Other Graduate TA	0%	0%	0%	<1%	0%	<1%	0%	0%	0%	0%		0%
Undergraduate TA	0%	0%	0%	0%	0%	0%	0%	0%	0%	0%		0%
Part-Time Instructor	1%	2%	2%	3%	4%	<1%	0%	0%	0%	0%		2%

In *Tables 21* and *23*, *Figures 22* and *24*, percentages of total sections taught by other than professorial faculty are given, by teaching format, for lectures, recitations and individual classes. There is comparison data for 1985 and 1987. There was a significant increase from 1985 to 1987 in teaching large lectures in both GP1 and GP2 departments by nonprofessorial instructors who are not mathematics TAs. There was also an increase from 1985 to 1987 in the involvement of these people in teaching individual classes in <u>all</u> groups of departments and in the involvement of mathematics TAs in all the U.S. departments. As would be expected, the data show that virtually all recitation sections are taught by NPIs. From 1985 to 1987 the percentage of individual classes taught by mathematics TAs increased.

Table 21 and Figure 22

Percentage of Lectures and Recitations Taught by Nonprofessorial Instructors

	Percentage of Large Lectures Taught by						Percentage of Associated Recitations Taught by					
	GP1	GP2	GP3	MAT	MAP	C/T	GP1	GP2	GP3	MAT	MAP	C/T
1987												
Math TA	6%	2%	1%	0%		2%	82%	70%	85%	74%		61%
Other NPI	26%	24%	8%	11%	0%	13%	15%	30%	15%	15%	0%	39%
Total	32%	26%	9%	11%	0%	15%	97%	100%	100%	89%	0%	100%
1985												
Math TA	3%	4%	2%	0%		0%	76%	86%	67%	37%		51%
Other NPI	3%	14%	14%	21%		0%	21%	12%	24%	54%		18%
Total	6%	18%	16%	21%		0%	97%	98%	91%	91%		69%

Lectures 1987: GP1, GP2, GP3, MAT, C/T

1985: GP1, GP2, GP3, MAT, C/T

Recitations 1987: GP1, GP2, GP3, MAT, C/T

1985: GP1, GP2, GP3, MAT, C/T

■ Math TA   ▤ Other NPI   □ Professor

Table 23 and Figure 24

Percentage of Individual Classes Taught by Nonprofessorial Instructors

	GP1	GP2	GP3	MAT	MAP	C/T
1987						
Math TA	30%	31%	29%	11%		5%
Other NPI	15%	22%	22%	28%	39%	13%
Total	45%	53%	51%	39%	39%	18%
1985						
Math TA	30%	21%	21%	9%		10%
Other NPI	11%	15%	16%	24%	29%	10%
Total	41%	36%	37%	33%	29%	20%

1987

GP1   GP2   GP3   MAT   MAP   C/T

1985

GP1   GP2   GP3   MAT   MAP   C/T

■ Math TA   ▤ Other NPI   □ Professor

CLASS SIZE

The differences in the 1985 and 1987 survey questions about class size make direct comparison of the resulting data impossible. *Tables 25, 26* and *27* give, for 1987 data, average sizes for large lectures, for the associated recitation sections and for regular and large individual classes. On the 1985 survey, departments were asked to give the distribution of sections (individual classes and recitations) taught by TAs and other non professorial instructors over class size intervals. No distinction was made between levels or between sizes of individual classes. The 1985 data are given in *Tables 28* and *29*.

Table 25

1987 Average Lower Level Class Size

	GP1	GP2	GP3	MAT	MAP	C/T
Large Lecture	169	200	178	163	35	80
Associated Recitation	30	29	26	41	17	28
Individual Classes (Size $\leq$ 65)	34	35	37	35	34	60
Individual Classes (Size > 65)	113	120	121	115	156	92
Useable responses	21	24	37	46	38	9

Table 26

1987 Average Calculus Class Size

	GP1	GP2	GP3	MAT	MAP	C/T
Large Lecture	113	157	121	100		86
Associated Recitation	28	26	26	33		30
Individual Classes (Size $\leq$ 65)	29	32	36	32	31	47
Individual Classes (Size > 65)	106	86	105	74	124	113
Useable Responses	23	27	36	44	39	9

Table 27

1987 Average Upper Level Class Size

	GP1	GP2	GP3	MAT	MAP	C/T
Large Lecture	96	89	107			89
Associated Recitation	26	25	29			30
Individual Classes (Size $\leq$ 65)	30	25	26	20	19	31
Individual Classes (Size > 65)	85	95	115			91
Useable Responses	20	24	37	43	39	7

Table 28

1985 Percentages of Individual Section Class Size

		GP1	GP2	GP3	MAT	MAP	C/T
	0-20	7%	3%	2%	7%	15%	9%
Class	21-30	28%	24%	15%	28%	36%	34%
Size	31-40	57%	63%	54%	44%	26%	22%
	41-65	8%	10%	28%	19%	18%	29%
	65-	0%	0%	1%	2%	5%	6%

Table 29

1985 Percentages of Recitation Section Class Size

		GP1	GP2	GP3	MAT	C/T
	0-20	8%	11%	7%	24%	28%
Class	21-30	50%	30%	56%	44%	45%
Size	31-40	41%	54%	22%	21%	3%
	41-65	1%	5%	4%	9%	5%
	65-	0%	0%	11%	2%	19%

TA WORK LOAD

As the 1987 survey was framed, it was apparent that the duty assignments of TAs were far more complex than the previous survey items could reveal. Even if the correct equation were determined relating the total time expenditure involved in recitation section teaching to the total time required to teach individual classes, TAs with mixed-duty assignments complicate reporting. One TA may be assigned a mixture of teaching individual classes and/or recitations, staffing a tutorial center, or grading for a professor. Also, in some departments TAs have undergraduates or new international TAs assigned as readers for homework or graders so contact hours of teaching do not give the whole picture.

The newly-designed 1987 items were difficult and time-consuming to complete and required a more careful analysis of each TA's efforts than some departments could complete. Also, when departments equated individual class hours to clock hours there resulted more diversity than the reality of time required for the duties can account for.

(This could be an example of reporting influenced by what *someone* wants to hear. Perhaps the someone is a state regulartory agency requiring certain hours of work for fee waivers; perhaps prospective graduate students who like to hear that work loads will be low.)

Departments were asked to report the total number of hours work required per week for a "full assistantship." A distribution over intervals is given in *Table 31*. More GP1 departments report loads in the lowest hour category and fewest in the higher 16-20 hours per week category. Thus, overall teaching loads in GP1 appear from *Table 31* to be significantly lower loads than in GP2. The average individual class contact hours reported by departments are very similar for GP1 and GP2 as shown in *Table 30*.

Recitation contact hours are even more elusive than individual class contact hours to use for comparing loads. In addition to some differences in division of responsibility between lecturers and recitation instructors, some lecture/recitation formats have one recitation hour per week, and some have two. (Is it more work to hold four contact hours per week but grade for just two sections, or to hold three contact hours per week while grading for three sections?) *Table 30* gives the average contact hours for recitation duties.

Table 30

1987 Average Contact Hour Work Loads

	GP1	GP2	GP3	MAT
Individual classes	4.6	4.5	5.8	5.3
Recitation sections	3.6	4.4	4.7	5.8

1987 Survey responses indicate a definite attempt to set first year work loads lower than those of experienced TAs. This is more common in GP2 and GP3 (30% and 25% of reporting departments, respectively). The smaller percentage of these reports from GP1 is probably attributable to their overall lower work load.

Table 32 reflects the data from the only item on the 1985 survey about the amount of work required of mathematics TAs. Departments were asked to give a distribution of their TAs over intervals of hours worked per week; the suggested conversion from individual class contact hours to total clock hours of effort was 3 class contact hours equals 10 clock hours of total effort.

Caveat: A superficial look at *Tables 31* and *32* might cause a jump to the conclusion that work loads over 20 hours per week no longer exist. The 1987 numbers reflect the standard departmental full assignment loads. The 1985 numbers reflect the actual average hours of assignment to an individual TA. Some TAs take on, even request, extra assignments to be able to make more money and some departments permit this. It is to be hoped that the work loads have gone down, and there is a lot of anecdotal information to support this. It is a fact that three of the university models described in Chapter 3 have reduced TA work loads during the time of this project.

Table 31

Departments' 1987 Clock Hour Loads for Math TAs

Percentage of Departments Reporting Indicated

		GP1	GP2	GP3	MAT	C/T
Clock Hour Load	1-10 hrs	44%	35%	21%	23%	44%
of	11-15 hrs	30%	15%	26%	40%	44%
Math TAs	16-20 hrs	26%	50%	53%	37%	12%
Per Week	21- hrs	0%	0%	0%	0%	0%

Table 32

Reported Loads Actually Assigned to TAs, 1985

Percentage of Total Number of TAs

		GP1	GP2	GP3	MAT	C/T
Load	1-10 hrs	16%	23%	11%	31%	62%
of	11-15 hrs	18%	22%	21%	21%	20%
TA	16-20 hrs	56%	47%	66%	40%	10%
Per Week	21- hrs	10%	8%	2%	8%	8%

INTERNATIONAL GRADUATE STUDENTS

Several items on each survey related to international graduate students. Comparisons are made in the data analysis to indicate trends. The 1985 survey asked for information about "international teaching assistants." At the time that survey was framed, it was not apparent that a more complete picture about IGSs was needed. In particular, it is important to know the numbers of those who were supported but did not actually teach, and who were enrolled but not receiving any support. The 1987 survey was framed to reflect this larger picture.

To make table headings in this section more convenient, some abbreviations are used. (**TA** is always used to mean a mathematics graduate student involved in actual teaching.)
IGS—International graduate student. Any international student enrolled as a mathematics graduate student.
ITA—International teaching assistant. An IGS who is actually involved in teaching recitation or individual class sections.

Table 33 gives the total reported numbers of graduate students, and of those the numbers who are IGSs. Then, this group and subgroup are each classified as one of: TAs and ITAs—those who teach; those who are supported (fellowships, research, grading, tutorials) but do not teach recitations or individual classes; those with no support from that institution. The data are given by category of department for U.S. and C/T departments, and then combined for U.S. doctoral departments.

The raw numbers are given in *Table 33* as reported, along with the number of useable responses for each category. The percentages given are computed from the reported data. They follow closely anecdotal information and related but not directly comparable data from other sources. (See [2], [6], [7].) Although the reporting rate for these surveys is very high, especially in GP1 and GP2, some self-selection is still present; there are no extrapolations because questions of the validity of extrapolation require the attention of an expert in statistical inference.

As expected, the data in *Table 33* show that about half of the mathematics graduate students in the leading graduate departments are international students. For GP1 departments, half of the mathematics graduate students who are teaching are international students; in GP2 departments this falls a little, to 44 percent. It is relevant to remember here two kinds of information from earlier sections. First, there are international students who are seeking degrees in other disciplines and are reported in *Table 1* as Other Graduate TAs. This means that the *numbers* of international students who are teaching are actually higher than shown. But the *percentages* are not higher; in fact, there are likely more U.S. than international students among the graduate students from other academic disciplines teaching mathematics. Secondly, as regards the higher percentage of teaching in GP1 done by ITAs, it is interesting to review the data which show the percentages of GP1 sections which are in lecture/recitation format. (See *Tables 11* and *12* and the discussion which preceeds these tables.)

It should be noted that there is probably underreporting in the category "No Support"; departmental staff who complete such surveys often do not known the graduate students who do not receive departmental or institutional support, and fail to report them. There is not an obvious reason why the underreporting would be biased toward either U.S. or international students, though, and the *percentages* are likely representative.

Departments were also asked to break the information in *Table 33* separately for graduate students in their first year of study and those beyond the first year. Some departments reflected in *Table 33* did not provide useable responses on this breakdown. The data received are reported in *Table 34* and the desirable conclusion, that international first year students are often assigned duties other than teaching, with progression to teaching in later years, may be made. In GP1 and GP2, the small computed differences in the percentages of international first year graduate students and international advanced graduate students may not be significant; better trend data is provided by comparison of the 1985 and 1987 surveys in *Tables 35* and *36 and Figure 37*. It is interesting that both the slight decrease in new international graduate students in GP1 and the slight increase in GP2 have been reported anecdotally as a trend by several departments from each category.

Table 33

Support of U.S. and International Graduate Students, 1987

	GP1				GP2			
	TA	Non-TA	No Support	Total	TA	Non-TA	No Support	Total
Number of graduate students	1505	430	151	2086	1126	185	66	1377
Number of IGS	748	277	51	1076	500	109	24	633
Percentage of IGS	50%	64%	34%	52%	44%	59%	36%	46%

22 Useable Responses   28 Useable Responses
of 39 Departments   of 43 Departments

	GP3				U.S. Doctoral Departments			
	TA	Non-TA	No Support	Total	TA	Non-TA	No Support	Total
Number of graduate students	868	199	90	1157	3499	814	307	4620
Number of IGS	318	146	28	492	1566	532	103	2201
Percentage of IGS	37%	73%	31%	43%	45%	65%	34%	48%

37 Useable Responses
of 81 Departments

	C/T				MAT			
	TA	Non-TA	No Support	Total	TA	Non-TA	No Support	Total
Number of graduate students	97	49	7	153	285	76	9	370
Number of IGS	59	32	1	92	58	46	9	113
Percentage of IGS	61%	65%	14%	60%	20%	61%	100%	30%

6 Useable Responses   44 Useable Responses

Table 34

Support of U.S. and International Graduate Students, 1987
First Year and Advanced Graduate Students

GP1

	Students in First Year			Students Beyond First Year		
	TA	Non-TA	No Support	TA	Non-TA	No Support
Number of graduate students	407	140	52	843	245	94
Number of IGS	179	98	19	439	142	27
Percentage of IGS	44%	70%	37%	52%	58%	29%

19 Useable Responses of 39 Departments

GP2

	Students in First Year			Students Beyond First Year		
	TA	Non-TA	No Support	TA	Non-TA	No Support
Number of graduate students	300	95	30	583	69	36
Number of IGS	126	46	11	225	46	13
Percentage of IGS	42%	48%	37%	39%	67%	36%

23 Useable Responses of 43 Departments

GP3

	Students in First Year			Students Beyond First Year		
	TA	Non-TA	No Support	TA	Non-TA	No Support
Number of graduate students	266	82	39	520	116	51
Number of IGS	82	61	11	217	82	26
Percentage of IGS	31%	74%	28%	42%	71%	51%

32 Useable Responses of 81 Departments

MAT

	Students in First Year			Students Beyond First Year		
	TA	Non-TA	No Support	TA	Non-TA	No Support
Number of graduate students	78	13	0	91	13	0
Number of IGS	18	11	0	16	7	0
Percentage of IGS	23%	85%	-	18%	54%	-

22 Useable Responses

More About IGSs Who Teach: ITAs

In this section the focus is on information about graduate students who teach, and the term TA and ITA are reserved for that meaning. The first table below lists the numbers of departments reporting TAs, reporting ITAs, and the numbers of those ITAs. For the 1987 data in each category of department, these are the same numbers of ITAs reflected in the first column of *Table 33*. *Table 35* also gives comparison data from the 1985 survey. Again, raw numbers cannot be compared due to differences in report rates. But the percentages are not suspect and reflect more teaching by international graduate students than a few years ago.

In order to illustrate the variation among departments in the percentage of teaching by ITAs, that percentage was computed for each department and a frequency distribution prepared based on the data from each survey (see *Table 36 and Figure 37*). In 1985, 45% of all graduate departments reported the lowest class, 0-30% of ITAs among all TAs. But in 1987, the percentage of departments in this class is much lower and includes only 9% of the GP1 departments. On the other end of the scale, the overall numbers of departments reporting 71-100% ITAs stayed about the same, but more were from GP1 in 1987, with reduced percentages of GP2 and GP3 departments indicating such a large percentage of ITAs.

Departments were also asked to respond to several other questions about international graduate students. A few of these items refer to admissions criteria, but the screening/testing and English instruction items are applicable as qualifications for teaching; hence the inclusion of *Table 38* in this section about international graduate students who teach. Comparative data is given for the two surveys when the framing of the items permits. The TOEFL remains the language test most often required before arrival for international graduate students seeking to teach, but language testing or "screening" after arrival on campus is usually required. Because of the intense media and political attention to "the foreign TA situation", the committee followed a suggestion and framed a question on the 1987 survey "Have you had complaints from students about 'foreign accents'? (Yes/No)".

All reporting GP1 and GP2 departments answered "yes," and 91%, 70% and 67%, respectively, of the GP3, MAT and C/T departments. Comments from respondents indicate the feeling that many complaints mask intolerance or are a convenient scapegoat for lack of effort or talent in the course; one reply said "one angry complaint last year was about a young woman from Scotland". (And there is the story now folklore of complaints at a midwestern university about a TA who turned out, upon investigation, to be from Brooklyn! Sadder, because of its racist overtones, was the complaint at a western university about a TA of oriental extraction who spoke no language but English and had grown up in, and graduated from a public high school in California.)

In an open question, departments were asked to report duties of non-native speakers of English who are not involved in teaching. The most common assignments involve grading and/or working in a mathematics tutorial center and/or with students registered for self-paced courses.

Table 35

Numbers of Departments With ITAs and
Percentages of ITAs Among All TAs

	GP1	GP2	GP3	MAT	C/T
1987					
Number of departments reporting TAs	22	28	37	38	6
Number of departments with ITAs	22	27	36	25	5
Number of ITAs reported	748	500	318	58	59
Total number of TAs reported	1505	1126	868	285	97
Percentage of TAs who are ITAs	50%	44%	37%	20%	61%
1985					
Number of departments reporting TAs	28	34	48	103	15
Number of departments with ITAs	23	32	33	44	11
Number of ITAs reported	549	475	278	103	98
Total Number of TAs reported	1446	1264	777	346	210
Percentage of TAs who are ITAs	38%	38%	36%	30%	47%

Table 36 and Figure 37

Departmental Percentages of ITAs Among All TAs

		GP1	GP2	GP3	MAT	C/T
	1987					
	0%-30%	9%	39%	38%	68%	50%
	31%-50%	41%	21%	35%	26%	0%
Percentage of	51%-70%	41%	36%	20%	3%	33%
ITAs of ALL	71%-100%	9%	4%	6%	3%	17%
TAs in	**1985**					
Department	0%-30%	46%	44%	46%	46%	9%
	31%-50%	46%	38%	30%	36%	64%
	51%-70%	4%	9%	12%	7%	27%
	71%-100%	4%	9%	12%	11%	0%

1987

GP1   GP2   GP3   MAT   C/T

1985

GP1   GP2   GP3   MAT   C/T

☐ 0%-30%   ▤ 31%-50%   ▥ 51%-70%   ■ 71%-100%

Table 38

ITA Qualifications, Screening, and Instruction

	Percentage of Departments by Group				
	GP1	GP2	GP3	MAT	C/T
Exam required for nonnative speakers of English					
1987	89%	88%	91%	63%	89%
1985	82%	88%	80%	56%	27%
TOEFL arithmetic mean					
1987	556	545	544	543	559
1985	575	565	545	539	-
TOEFL reported range					
1987	500-600	500-600	500-580	500-600	520-650
1985	530-600	525-650	500-600	500-550	-
On-campus English screening/testing required, 1987	76%	85%	91%	46%	22%
On-campus screening/testing involves:					
Individual interviewing	56%	61%	77%	50%	100%
Testing by language expert	69%	61%	33%	39%	50%
Testing by language expert and mathematics faculty together	19%	13%	13%	0%	0%
Testing by mathematics department faculty	0%	17%	15%	17%	0%
Other	6%	35%	5%	0%	0%
Require English instruction if needed					
1987	72%	70%	71%	57%	33%
1985	60%	71%	68%	25%	33%

ORIENTATION, SUPERVISION AND EVALUATION

Extensive descriptions of programs for the orientation of new teachers in graduate departments, and for their later supervision and evaluation, are described in Chapter 3. Chapter 5 includes similar ideas and practices related to part-time and temporary teachers in any department.

Both of the surveys of graduate departments asked for information about policies and practices. The first table below, *Table 39* shows that most departments require some kind of orientation or training, and generally run a program. There is an increase between the 1985 and 1987 surveys in the percentage of departments reporting orientation or training programs. The table reflects data for each survey, showing comparisons when appropriate. Most programs involve new teachers in some activities before the first day of teaching.

Simulated classes and videotaping are commonly used. Experienced TAs are often utilized to help new TAs learn to ropes.

The information in *Table 40* is based on the 1987 survey items about supervision, evaluation and observation, and retention of TAs and other nonprofessorial faculty. The data in response to the somewhat different items on the 1985 survey follow in *Table 41*. The top section of *Table 41* gives the data for mathematics graduate students who have teaching assignments. In the lower section, the answers to the same questions refer to all other nonprofessorial teachers.

Table 39

Orientation/Training Program Features

Percentage of Departments by Group

	GP1	GP2	GP3	MAT
Departments with orientation/training program				
1987	79%	86%	76%	51%
1985	66%	88%	64%	44%
Percentage of departments requiring the above, 1987	83%	92%	91%	89%
Departments with indicated training feature:				
Conducted before selection				
1987	13%	12%	6%	4%
1985	5%	0%	0%	3%
Conducted before first teaching				
1987	74%	88%	85%	89%
1985	74%	86%	83%	88%
Conducted concurrently with first teaching				
1987	39%	64%	47%	56%
1985	68%	43%	57%	59%
Involves experienced TAs				
1987	52%	56%	53%	4%
1985	68%	46%	60%	24%
Uses simulated classes				
1987	61%	64%	47%	22%
1985	63%	43%	40%	14%
Uses videotape				
1987	57%	60%	35%	0%
1985	42%	32%	67%	2%
Uses MAA's "How to Teach..."				
1987	49%	40%	35%	48%
1985	58%	43%	47%	27%
Uses other publications				
1987	35%	28%	24%	15%
1985	47%	43%	25%	19%
Involves group discussions with faculty and advanced graduate students, 1987	70%	64%	53%	67%
Personnel involved with TA training, 1987				
Professorial faculty	83%	88%	76%	96%
Non-professorial faculty	35%	20%	41%	26%
Professionals from other departments	26%	12%	29%	7%
Non-faculty professionals	22%	12%	24%	4%

Table 40

Supervision and Evaluation of Nonprofessorial Instructors, 1987

	Percentage of Departments					
	GP1	GP2	GP3	MAT	MAP	C/T
Supervision is solely the responsibility of one faculty member.	10%	20%	13%	27%	59%	10%
General coordination of duties and supervision is responsibility of one faculty member, but others are involved.	66%	77%	72%	63%	56%	60%
Others involved are:						
Professorial faculty	68%	83%	85%	97%	87%	83%
Non professorial faculty	26%	17%	15%	12%	20%	50%
Professionals in other departments	11%	4%	9%	6%	0%	0%
Supervision in small groups, each by one faculty member.	24%	20%	26%	13%	0%	10%
Faculty members work with at most one NPI at a time.	17%	13%	11%	13%	4%	20%
Classroom observations by faculty member, then discussed with NPI.	62%	67%	67%	67%	30%	10%
Observations are completed:						
First term teaching	39%	45%	16%	26%	25%	100%
Each term	11%	45%	61%	63%	75%	100%
Student evaluations required:	79%	90%	80%	90%	59%	60%
Each term	74%	85%	76%	79%	75%	83%
Student evaluations reviewed by faculty and discussed with NPI.	61%	85%	84%	68%	67%	50%
NPI selected by graduate coordinator or department chair.	52%	57%	67%	63%	74%	50%
NPI selected by committee.	31%	47%	41%	31%	15%	10%
Departmental course outlines provided.	86%	93%	91%	92%	70%	70%
Suggested departmental problem list provided.	41%	43%	26%	23%	15%	10%
Uniform final exams used.	76%	77%	59%	42%	19%	80%
Uniform final exams group graded.	79%	60%	43%	23%	4%	50%
Uniform scale for grades of all sections used.	52%	57%	35%	15%	0%	80%
NPI may be removed from classroom or not renewed as a result of evaluation procedures.	80%	83%	88%	77%	100%	86%

Table 41

Supervision and Evaluation, 1985

Mathematics Graduate Student TAs	Percentage of Departments				
	GP1	GP2	GP3	MAT	C/T
Supervision is the responsibility of one faculty member.	21%	15%	19%	44%	36%
General coordination of duties and supervision is responsibility of one faculty member, but others are involved.	61%	62%	52%	36%	50%
Supervision of small groups, each by one faculty member.	25%	29%	40%	15%	14%
Faculty members work with at most one TA at a time.	18%	6%	23%	10%	0%
Classroom observations are made by regular faculty member, and discussed with TA.	46%	62%	52%	37%	14%
Student evaluations administered, reviewed by faculty, and discussed with TA.	39%	74%	67%	50%	36%
TA may be removed from classroom or not renewed as a result of evaluation procedures.	46%	62%	56%	46%	36%

Nonprofessorial Instructors Who Are Not Mathematics TAs	Percentage of Departments					
	GP1	GP2	GP3	MAT	MAP	C/T
Supervision is the responsibility of one faculty member.	25%	17%	16%	35%	40%	38%
General coordination of duties and supervision is responsibility of one faculty member, but others are involved	60%	59%	44%	35%	24%	46%
Supervision of small groups, each by one faculty member.	15%	14%	33%	14%	2%	8%
Faculty members work with at most one instructor at a time.	20%	7%	13%	4%	2%	0%
Classroom observations are made by regular faculty member, and discussed with instructor.	55%	48%	43%	32%	18%	8%
Student evaluations administered, reviewed by faculty, and discussed with instructor.	40%	66%	58%	52%	46%	23%
Instructor may be removed from classroom or not renewed as a result of evaluation procedures.	55%	55%	53%	49%	38%	46%

PROHIBITIONS

Institutional or department policies, or state or accreditation regulations, sometimes impose restrictions on the assignments. Sometimes, despite these restrictions, such assignments are made. When that is the case, it is hoped that the anonymity assured for the data in the preceding sections has prompted reporting which reflects the actual situation. On each of the surveys, some questions were framed to determine whether there were certain prohibitions, whatever the actual reality of the situation.

In 1985, departments were asked whether TAs were prohibited from either teaching individual classes or lecturing in large lectures. The departments answering "yes" in Groups 1, 2 and 3 were, respectively, 26%, 12%, 8%. It seemed clear from anecdotes that the prohibition on TAs being assigned to large lectures was actually much greater than those reports. The 1987 survey asked the question separately for individual classes and large lectures and the individual class prohibition percentages were low. However, on this second survey, TAs were listed as prohibited from lecturing in large lectures in 60% of the doctoral departments.

Many institutions and/or departments have a policy against using undergraduates for classroom instruction; about 60% of the doctoral departments report this prohibition and even more of the master's departments. There were strong anecdotal indications of increasing assignments to

undergraduates of classroom duties of the type traditionally given to graduate teaching assistants. In 1987 a question was included which asked for all duty assignments to undergraduates which the department may make without regard to whether any individuals were actually appointed in that category at the time. *Table 42* reflects those responses. It is interesting that a number of the graduate departments indicating that they may assign undergraduates to classroom duties actually report on earlier pages of the survey no such assignments. Of the 10, 12, 18 and 6 Group 1, 2, 3 and MAT departments listing these duties, 4, 4, 8 and 4 report no such actual assignments.

Table 42

Number of Departments Reporting Particular
Undergraduate Duty Assignments

	GP1	GP2	GP3	MAT	MAP	C/T
Instruction:						
Conduct recitations or tutorial sections with grading responsibility.	9	11	14	5	1	6
Teach and grade individual sections.	3	3	8	2	0	1
Other:						
Grade for regular faculty.	29	25	38	45	29	9
Grade for TAs.	19	9	13	11	1	4
Conduct tutorials which are not associated with particular sections.	9	13	25	37	22	5
Miscellaneous.	1	5	8	10	9	1

REFERENCES AND RESOURCES

[1] Albers, D.J., R.D. Anderson, and D.O. Loftsgaarden; *Undergraduate Programs in the Mathematical and Computer Sciences: The 1985-1986 Survey.* MAA Notes No. 7, Mathematical Association of America, 1987.

[2] Finkbeiner, A.K. "Demographics or Market Forces?" *Mosaic* 18:1 (1987) 10-17.

[3] Neal, H.A., chair. *"Report of the NSB Task Committee on Undergraduate Science, Mathematics and Engineering Education.* National Science Board, 1986.

[4] Rung, D.C. "CEEP Data Reports: New

Classification of Graduate Departments." *Notices of Amer. Math. Soc.* 30:4 (1983) 392-393.

[5] Rung, D.C. "Newest Ratings of Graduate Programs in Mathematics." *Notices of Amer. Math. Soc.* 30:3 (1983) 257-259.

[6] Simon, B., chair. "Report of the Committee on American Graduate Mathematics Enrollments" (a summary). *Notices of Amer. Math. Soc.* 34:5 (1987) 748-750.

[7] Simon, B., chair. *Report of the Committee on American Graduate Mathematics*

Chapter 3

MODELS OF PROGRAMS

This chapter describes some methods developed by mathematicians in graduate departments to help as their TAs or part-time instructors begin teaching. There are representative samples of materials that mathematics departments or institutions use in their programs. Most of these materials were developed for graduate teaching assistants. The chapter concludes with two sections containing descriptions of some programs which respond to the challenge in those departments. Components from these well-established programs may provide new ideas for other established programs or a basis for setting up a balanced plan for orientation, training, supervision and evaluation. Grant support for the project made possible a number of site visits. The collegiality, along with openness and full cooperation, made these visits enjoyable and informative. University departments visited were:

Cornell University	Stanford University
Brown University	University of California, Berkeley
Duke University	University of Akron
Florida State University	University of Cincinnati
University of Maryland	University of Michigan
Ohio State University	Vanderbilt University

HISTORICAL PERSPECTIVE

Until recent years, teaching assistants were considered primarily as apprentices, novices preparing to enter the profession. They were willing to go through a training process in teaching and in research in order to become, themselves, faculty members. As graders they were literally assistants; as teachers of traditional individual classes, they followed the strong role models of regular faculty. They were learning the trade so that they would be considered for employment on the basis of their already developed skills. Several factors have brought about a change in that pattern.

For one thing, in many institutions there are now more graduate students whose primary interest is not an academic career. Some students in master's programs or in applied mathematics and statistics are preparing for a career in business or industry where it is not apparent that teaching experience is important. Some intend to return to

another country and do not want to divert the necessary energy from mathematics study to achieve spoken English levels needed for effective teaching. Such students may view the teaching obligation as a burden, "just a job", a means of earning support or avoiding payment of tuition. There is little motivation for them to treat an apprenticeship seriously. Some are graduate students in other departments, further complicating matters.

In recent years, there has been a marked increase in the number of international students pursuing graduate degrees in the United States; in contrast with some previous times, many of them want or may be required to return to their home countries. Their backgrounds are different from those of most American students, and they will go back to a different kind of educational system. They are at a comparative disadvantage when it comes to dealing with undergraduates, and there is little incentive for them to improve the situation beyond their need of financial assistance.

The "traditional" university mathematics teaching model involves several facets. Class size is about thirty. Instructors are tenure track faculty or graduate students. There are usually a prescribed text and, in sequential courses, a list of topics or sections which are to be covered. Most course policies are determined by the individual instructors, within institution guidelines. Tests are made and graded by the instructor. Grades are awarded without reference to performance of students in other sections. It is not clear that this ideal model termed traditional was either universal or completely successful!

Many variants from this traditional pattern are found in the teaching offered by departments today. Most of these "coping devices" have arisen due to pressures on departments and changing student needs.

Some of these devices are not new, as the part- and full-time teaching described in the biographies of Julia Robinson and Mary Ellen Rudin attest. In professional society publications and committee reports, the pros and cons of some of these variants have been discussed. Each affects the climate in which non-professorial teachers function.

Some variations from the traditional mathematics class model are described below. It is not the purpose here to pass judgment on the relative or absolute merits of any of these coping devices.

Coordinated Multisection Courses

Frequently courses with multiple sections are run under the supervision of a faculty member who has primary responsibility for determining the syllabus, policies of the course and sometimes the homework assignments. Usually there will be common final examinations and sometimes other common tests, often graded in common. Sections may be taught by several different kinds of teaching personnel. Sometimes a particular coordinated course may be offered in two or more of the class modes described below: Independent Sections, Large Classes, and Lecture with Recitation.

Independent Sections

The traditional mathematics teaching model is one type of individual, or independent, section model. More frequently today these sections, when assigned to other than professorial faculty, seem to be for coordinated multisection courses. Usually those with some previous classroom experience teach courses in which they have full classroom responsibility.

Large Classes

This variation of the independent class involves a section with 50 or more students taught by an instructor who is usually supported with graders and/or test proctors. An advantage of the Large Class format, besides its financial efficiency, is the opportunity of productive work for graders and proctors who might not be employable as teachers; in particular, work is provided for the international teaching assistant whose English skills do not yet permit classroom assignments or for the undergraduate student. Such employment offers beneficial apprenticeships for those assisting the lecturer.

Large Lectures with Recitation Sections

In classes of size ranging from 60 to 700, a faculty member delivers lectures two to four hours per week, and students meet in a recitation section for an additional one or two hours, optimally with fewer than twenty-five students. The total number of weekly clock hours which a student spends in a lecture course with recitation often exceeds the credit value of the course. In most cases the recitation instructors are graduate TAs, but sections might be led by other faculty, part-time instructors or, in a few instances, undergraduate teaching assistants. In some universities, the only TA assignments are such recitation sections. Recitations are tied closely to lectures by a faculty member. Such assignments are an appropriate first task for a TA, before consideration for the assignment of an independent section.

Full-Time Faculty not on Tenure Track

In addition to post-doctoral appointments, most graduate departments employ one or more instructors or lecturers with master's or doctoral degrees who have full-time teaching positions which are not tenure track.

Part-Time Instructors

Part-time instructors often have other full-time employment or teaching. Whether part-timers teach for the extra money or because they especially enjoy it, they rarely have a long term commitment to the profession under these circumstances. Even when they are experienced teachers, they may be unfamiliar with the institutional and departmental standards and goals. The ethical problems associated with the employment of part-time and temporary instructors are too rarely addressed. They may continue to teach for many years, often at more than one

institution in the same time period, without faculty benefits. (See further discussion in Chapter 5.)

Undergraduates and Graduate Students from other Departments

A good case can be made for involvement of undergraduates in teaching, but it is often an inadequate supply of mathematics graduate students with sufficient English skills which leads to teaching appointments for undergraduates. Graduate students from other departments are sometimes found to be surprisingly weak in basic mathematical skills; even when they are excellent teachers and grateful for the support, their academic allegiance is to another department. (See Chapter 5 for one regional accreditation association's related requirements.)

CHARACTERISTICS OF THE MODELS

Various pressures demand more care in the selection, training, evaluation and supervision of graduate students and part-time instructors. With increases in the cost of education, students and their parents are more vocal in their objections when a teacher is unprepared or unable to handle a college or university class. Political pressure on public institutions is strong. National reports demand more quality control in education. All these things call for a closer examination of the training models which have been used successfully in various types of departments.

Due to the diversity of characteristics of teaching personnel, sizes of departments and varying financial resources, there is no single model which could be recommended for all current situations. Therefore, several models of programs will be described; a mix of components and ideas taken from several models, with modifications as situations change, provides a flexible base program.

From the results of the survey described in the previous chapter, materials returned with the surveys and from site visits made by members of the committee, characteristics of many different programs were collected and compiled. Described are things like when the orientation and training are held, how long it lasts, whether or not it is required, who conducts it, and what techniques are

used. The department models exhibit various choices within these characteristic features.

Time Frame

Generally, programs for new TAs are organized one of two ways. There is the mode in which all activities takes place before teaching begins. This is usually during a summer workshop. The other mode includes a fraction of the training before the term begins (several days, or only a few hours) and additional activities concurrent with the first term duties. Both modes may include some type of help and maintenance supervision when additional responsibility is assumed during successive terms of appointment.

Joint or Departmental Activities?

Some colleges and universities have well-established programs for acquainting new teachers with matters of interest in all teaching areas—general institutional information, grading regulations, payroll procedures, etc. These institution-wide programs are also helpful in fostering institutional loyalties and in creating enthusiasm for good teaching. Since the number of mathematics TAs is relatively large, it is efficient to give some sessions which are discipline-specific, even when there is a good institution-wide program.

Most mathematics departments prefer that a large proportion of the time spent with new teachers take place in programs specifically for mathematics or for mathematics and related disciplines. There is wide variation in the degree of support for university-wide training programs. Some mathematics departments tell their TAs about university-wide programs but make no attempt to encourage or monitor participation. However, there are other cases where departments enthusiastically cooperate in university-wide programs and require participation on the part of their TAs. (In Appendix, See Syracuse: *University-Wide Orientation.* with a relatively short departmental segment. Also see Florida State: *TA Orientation Schedule for Faculty/Mentors/Staff* which describes a departmental program covering some of the things included in the extensive university-wide program at Syracuse; at FSU the university-wide orientation is a short segment.)

A number of institutions have developed supplementary training programs to prepare international students as TAs, often with the assistance of professionals in language and communication skills. There is strong feeling that such programs should not isolate the international students from American graduate students preparing to be TAs in mathematics, but rather that departments should foster cooperation and collegiality among all their students. (See Chapter 4 which specifically treats international TAs.)

Who Works with New Teachers?

Selection, orientation and training, assignment, supervision and evaluation of teaching personnel not on tenure tracks: Who does what? Departments have evolved many different models for dealing with TAs. Patterns which may appear dissimilar may work well.

Selection of TAs from among available candidates is usually carried out by a faculty member or, especially in cases where there are many more applicants than positions, a faculty committee. For international TAs often there is some screening based on language ability. Competition is keen for support at top-ranked departments; other departments recruit vigorously. Selection of part-time and temporary instructors, based primarily on teaching needs of the department, is usually the responsibility of a departmental administrator.

Orientation and training of new TAs is usually carried out by one or more faculty members, often in conjunction with experienced TAs. The faculty members involved may be professorial faculty, others for whom this is a major component of their duties, or a combination. Especially when the lecture and recitation format is used, all training after a few orientation sessions may be in the context of duties.

Assignment, supervision and evaluation of TAs are generally the responsibilities of faculty members involved in the teaching of related courses. A faculty member responsible for a course will monitor progress in the syllabus and handle complaints. Frequently faculty or experienced TAs visit classes taught by TAs and record their observations, often discussing them with the TA. It is common nowadays to seek evaluations by students.

Major differences appear in the supervision and evaluation modes, often as a function of the duties expected of TAs. When most TAs continue to be assigned only recitation sections during the entire tenure, or the same course, there is less need to continually monitor effectiveness than when the assignments require increasing teaching skills. Some degree of ongoing supervision is automatic in coordinated multisection courses.

Generally, part-time instructors tend to be more experienced teachers who receive little by way of orientation, training or supervision, unless there are complaints.

An unsuccessful TA can sometimes be moved to a job that does not require student contact or can be given more assistance to overcome limitations. An unsuccessful part-time or temporary instructor is usually not renewed.

Types of Activities

1. *Orientation.* There are usually sessions as soon as new TAs arrive in which college and departmental policies and procedures are described. Sometimes the schedule is sent before TAs arrive on campus. Often the activities are tightly packed and are an intensive cram course to prepare new TAs for their first classroom days.

2. *Social and mixer activities*. In addition to the expected inclusion in various schedules of TA activities such as "reception" or "refreshments", several offer livelier activities: Pub Crawl, Wine and Cheese Party, Movies, Math Grad Student Beer Party, Pot Luck Supper. Faculty working with TAs at Maryland developed an introductions game which works similar to the Virginia Reel. The TAs seem to enjoy it; with over 20 in a room it sounds like the Tower of Babel! There was a report of TAs playing the commercial game *Pictionary;* although it was not the intent, the international TAs learned many new words.

3. *Simulated lessons*. For this activity, the TA is assigned a problem or lesson to present as though teaching a class. The observers may watch silently and critique the performance afterwards, or they may pretend to be students and ask the types of questions that students often ask. The observers may be faculty and advanced TAs in addition to the other new TAs.

4. *"What-not-to-do" sessions*. Most training sessions include discussion of things a TA should not do in the classroom. North Carolina State has prepared a clever video tape using faculty as actors. Staged mock sessions are reported.

5. *Video taping*. A simulated lesson or the TA in the actual classroom is video taped. The tapes are viewed by the individual for self evaluation, or they may be viewed by the TA and faculty leader for discussion, or they may be viewed in the group with group discussion.

Other innovative uses of this technology are a video tape of students asking international TAs questions during language screening and the video tape mentioned in 4. above.

6. *Interviews*. An interview is conducted by one or more faculty members with a new TA to check that TA's understanding of policies, to learn more about the TA's academic background, and, for international students, to assess spoken English skills.

7. *Group discussions*. The new teachers meet as a group in a classroom setting and a faculty member describes or explains a given topic (such as professional ethics) or encourages discussion of recent classroom experiences.

8. *Specific course information and practice*. Some institutions assign all new TAs to teach the same course. In this case, their training may be very specifically aimed at helping with that course; lessons are prepared and practiced beginning with the first day. In other cases, help is given concerning course content when there are courses of a type not typically taken in their undergraduate days by the TAs—finite mathematics topics for business majors, for example.

9. *Interning with an advanced TA*. New TAs attend classes of advanced TAs, learn testing policies, grading skills, and even practice teaching lessons. The advanced TA may later serve as a mentor during the new TA's first teaching term.

10. *Classroom observation and discussion*. When the TAs teach during their first term of appointment, they are often observed by an experienced TA or faculty member during the first or second hour of teaching; improvements are suggested and the observation is repeated.

11. *Reading assignments*. Most institutions assign various readings for the new TAs. A list of the popular materials and copies of some materials appear in the appendix.

12. *Written assignments*. In several cases, the TAs are asked to react to and assess their orientation and training activities.

MATERIALS

Many interesting materials were sent with survey responses and collected at site visits. A sampling is included in the Appendix. Most items are for TAs; many items for international graduate students were received, but only a few materials about part-time instructors. Although most of the Appendix materials were written for local use, they give ideas to others.

1. *Getting started*. Two interesting ideas from Maryland for early use with new TAs: A packet of miscellaneous materials is described which is sent a

couple of months before arrival. There is a game developed for the first orientation session which facilitates introductions and also gives an indication of spoken English skills to those conducting the session.

2. Materials about *orientation and training sessions*. These schedules will reveal the number of hours or days and types of activities at several different institutions. It is usually not clear how many sessions other than at the beginning of a term, are held with new TAs. There are several schedules in the Appendix which describe the early activities in detail, and one which describes a summer TA workshop. A "mathematical exposition seminar" held in three summer weeks is also described. One innovation is a Workshop evaluation form to gather input from the TAs for improving future workshops. Schedule-related materials are copied from Duke, Florida State, North Carolina State, Oregon State, Syracuse, Villanova and Wisconsin—Madison.

3. *TA workshop syllabus*. These instructions are written as guidelines for the professor conducting the TA workshop at Berkeley. In the past, orientation and training programs were often set up and made effective through the efforts of one or two professors. When a sabbatical or other respite for the innovator occurred, the wheel was reinvented. A few departments now have these "how to" directions—some acknowledging that the idea to write things down comes from these Berkeley notes.

4. *Departmental or locally prepared information*. A wide variety of materials are prepared for TAs by departments and there are sometimes helpful university-wide publications. Some of the materials are applicable for other part-time or temporary instructors. In the workshop syllabus described in 3. above, there is a handout of "Do's and Don'ts" which supplements the department's information guidebook. A videotape which was produced at North Carolina State is an innovative way of presenting classroom management information. Many departments have a general guidebook with teaching suggestions and a description of departmental expectations. Two guides which have served their departments well for a number of years are copied: *Helpful Hints* from Wisconsin, and *Torch or Firehose* from MIT;

also for consideration are the topics on the Appendix pages titled *Teaching Ethics and Professionalism*.

When the lecture/recitation format is used, a careful description of the division of responsibilities is usually specified. This may be in the general guide—see *Helpful Hints* and *Torch*—or it may be discussed in a TA Workshop (see 3. above); more specifically, note the *Responsibilities of*...lists from Florida State. Sometimes the general information is given in each of a number of course guides for nonprofessorial instructors. Course guides also include information about text and material to be covered, and are often very specific and include a daily pacing schedule, homework lists, sample tests and section-by-section hints about the text. Examples are given of course guides from North Carolina State and Ohio State.

5. *Professional Society or other "slick" booklets*. Instructions, tips and advice for being a good teacher are given. See Reznick: *Chalking It Up*; also, available from the MAA, *College Mathematics*: *Suggestions on How To Teach It*. For international TAs the guide by Althen and one from Berkeley are helpful.

6. *Evaluation and Classroom Observation forms*. These are the forms that an observer (faculty, advanced TA) completes upon observing a teacher. These may consist of broad open questions, or they may indicate a lot of suggested responses or include a checklist. The last two pages of MIT's *Torch or Firehose* is the latter type; other examples given are from Clemson and Ohio State.

7. *Student evaluation forms*. These are the forms that students fill out to evaluate their instructor. This practice is required at some institutions, and the generic form is used by all departments. In other cases, it is required or encouraged in the mathematics department, and the form is more specific. Two examples are reprinted: The form from Clarkson is used there throughout the sciences, and the Ohio State example is departmental.

DOCTORAL PROGRAM MODELS

The programs in many different types of departments offering the Ph.D. in mathematics were studied. (Some programs in departments offering a master's, but not the doctoral degree, are described in the next section.) The materials and descriptions sent with the surveys were interesting and informative. Several university programs have been selected to illustrate a variety of the characteristics listed above. A brief description of these departments and some details about the outstanding features of each program are given. The institutions are identified with permission. Representative materials collected during site visits or sent with surveys are described above and copied in the Appendix. These materials provide the best picture of what is being done and the best ideas for components which may help departments setting up or modifying programs. The data given for each model are numbers representative of a Fall term, and are updated versions of the survey responses. (Copies of the 1985 and the 1987 Committee surveys are shown in the Appendix.) Descriptions follow for:

> University of California...Berkeley, Calfornia
> Cornell University...Ithaca, New York
> Duke University...Durham, North Carolina
> Florida State University...Tallahassee, Florida
> University of Michigan...Ann Arbor, Michigan
> University of Montana...Missoula, Montana
> Ohio State University...Columbus, Ohio
> Syracuse University...Syracuse, New York
> Vanderbilt University...Nashville, Tennessee
> University of Wisconsin...Madison, Wisconsin

University of California, Berkeley

Mathematics at Berkeley is a large department with more graduate students than it supports with teaching or research. There is keen competition for support. (Some mathematics graduate students teach physics or computer science.) In terms of numbers, there are:

79	in professorial ranks
13	full-time, with doctorates, not in professorial ranks
115	mathematics graduate student teaching assistants
22	mathematics graduate student research assistants
31	mathematics graduate students supported and assigned other duties
16	mathematics graduate students with non-duty fellowships
5846	undergraduates taught, all levels, Fall term

Some characteristics of the Berkeley program are:

- TA load is two recitation sections of two contact hours each per week.
- Average recitation section size is 23; undergraduates grade homework for TAs.
- Strong department training: TA Workshop, begun in 1975, is required for all TAs.
- Legislative action in 1987 obliges the university to ensure oral English proficiency.

Almost all of the teaching involving TAs is in the lecture-recitation format; the TAs are called Graduate Student Instructors. There are no part-time instructors, but there are post-docs. A few experienced TAs have special duties as "Head GSIs" helping lecturers deal with administrative details. (One or two exceptional TAs may lecture in some of the few noncalculus—precalculus, math for teachers, linear algebra—sections.) Undergraduates are not appointed as recitation instructors but do serve in self-paced classes where their duties are largely the same as those of TAs. They also work as graders for the graduate student TAs.

Since the 1987 legislation about oral English proficiency, the Dean of the Graduate Division requires that TSE scores be provided in advance where possible. (See Chapter 4 for a description of the TSE.) Departments are also required to set up a screening mechanism in a microteaching format involving a department faculty member and an ESL expert. The university with the Graduate Assembly provides a day-long orientation for new TAs. There is also a program of topics seminars related to teaching during the term. Excellent programs are available for international TAs; mathematics TAs are encouraged but not required to participate in these activities. The publications are made available to all the TAs. (Parts of *Teaching at Berkeley* are in the Appendix.)

For new TAs, participation in the department's Orientation and TA Workshop (MATH 300), and attendance at every session, are strictly required. The TA Workshop Syllabus (copied as an Appendix) helps a well-intentioned faculty member accepting this teaching assignment to profit from previous experience and not have to reinvent the wheel. It helps the professor undertstand that, although the usefulness of the Workshop is well accepted, participation by some of the TAs is less than enthusiastic, whatever the directing professor's efforts. It counsels the Workshop professor to help TAs cope with situations in which the course professor to whom they are assigned does not fulfill the responsibilities that the TAs have been told are those of the course professor! (One of the readings for the new TAs is "Helpful Hints to Instructors of Large Lecture Courses;" the reasoning is that it helps the recitation instructors to know what the professor has been told.) This Workshop Syllabus reflects a slice of the real world and is full of useful information; it is a reminder that our situations from one university to another are more alike than they are different.

A detailed description of procedures and resources is provided in the Information Manual for GSIs.

Faculty lecturers supervise the TAs assigned as their recitation instructors. Lecturers observe classes and these faculty evaluations are included in TA files. Student evaluations are also made available to the TA and to the Graduate Appointments Committee.

In the Appendix:

Course Guideline for the TA Workshop (MATH 300): *Manual for the Instructor.*

Several articles from *Teaching at Berkeley: A Guide for Foreign Teaching Assistants.*

Cornell University

Cornell University, one of the "Ivy" institutions, is a moderately large university with some public colleges but is mostly a private university; students and faculty are strong. There are:

60	in professorial ranks
1	full-time, with doctorates, not in professorial ranks
65	mathematics graduate student teaching assistants
10	mathematics graduate student research assistants
5	mathematics graduate students with non-duty fellowships
22	graduate teaching assistants seeking degrees in other departments
4	part-time instructors who are not students
3598	undergraduates taught, all levels, Fall term

Some characteristics at Cornell:

- Graduate students from many other departments teach mathematics.
- Videotaping was tried but not found helpful, overall.
- A foundation grant has been received for training foreign TAs.

All of the 20-30 new TAs are required to attend a week long orientation session prior to the Fall semester. The faculty coordinator and some of the advanced graduate students demonstrate some of the problems that can arise in teaching. This is done in a lighthearted manner designed to help relieve the anxiety of the new TAs. Although each new TA is asked to do some practice teaching, the object is not so much to train the graduate students to teach as to give them a taste of what to expect in the classroom. Criticism is always presented in the most positive and constructive way possible. Videotaping was tried but stopped because some of the new TAs became too self conscious about minor aspects such as their arm movements.

First- and second-year mathematics graduate students who are reasonably proficient in English are assigned two recitation sections of calculus (4 contact hours); they grade the homework and the tests, and hold office hours. More advanced students teach one section of calculus with about 20-25 students for 4 contact hours per week. They have an undergraduate paper grader, but make up and grade the hour tests. A syllabus and departmental final are provided by the course coordinator. Those first year students who are not ready to teach are given jobs grading papers and holding office hours for advanced undergraduate courses. It is estimated that a TA works a total of 10-15 hours per week.

Although the University runs a program for foreign born TAs, this program is more appropriate for the discussion format in social sciences and humanities than for the typical math format. A grant has been received to better train foreign TAs in mathematics; the program began in Fall, 1988.

Duke University

Only about three percent of mathematics teaching at Duke is at the precalculus level and more than half of Duke students have had a year of calculus in high school. In this small department, the number and the nationality of the TAs have varied considerably from year to year. There are:

28	in professorial ranks
3	full-time, without doctorates, not in professorial ranks
21	mathematics graduate student teaching assistants
4	mathematics graduate student research assistants
6	mathematics graduate students with non-duty fellowships
2	part-time instructors who are not students
1967	undergraduates taught, all levels, Fall term

Some characteristics of the Duke program are:

- A number of the elementary math courses are taught by full time teaching instructors.
- TAs teach at most one section per term and have undergraduates to grade homework.
- Some non-teaching support is available for first-year international TAs.

During their first year of graduate study, most TAs hold a problem session once a week for a 4 hour per week calculus course or they run an evening help session for calculus and precalculus students. Beginning in their second year, TAs teach one section per term with about 30 students, typically a calculus class meeting 3 hours per week. These courses have a standard syllabus and a block final exam, but all the lectures are given by the TA; the lecture recitation mode is not used.

A teacher training program coordinated by a full-time instructor who works in conjunction with the chair, director of graduate studies and the supervisor of freshman instruction has recently been instituted. First year graduate students are expected to spend an average of an hour per week on teacher training in addition to the duties mentioned above. Their training activities include:

(1) observation and discussion of lessons taught by experienced teachers;

(2) readings from some of the "how to teach" booklets. (see Appendix);

(3) group discussions on grading policies, making up exams, planning lectures, and departmental and university policies;

(4) presentation of two lectures, one for practice and one as a guest lecturer in a regular class. The students in this class fill out a short evaluation form for the TA;

Most graduate students begin teaching on their own in their second year. They are observed by regular instructor early in the term. If the quality of teaching is satisfactory, then no more observations are made. One of the classes taught by the TA is videotaped. The videotape is viewed by the TA and the coordinator of teacher training who will discuss the TA's strengths and weaknesses.

Each of the graduate students teaching for the first time will be assigned a mentor from among the experienced teachers in the department. The mentor will be available for advice and to approve tests designed by the TA. At the end of each term, the TA will be given his or her Teacher-Course evaluation forms and will be asked to write a self evaluation describing his or her strengths and weaknesses and discussing ways to improve the teaching.

As elsewhere, Duke has a problem with the spoken English of international graduate students, although several of them have become exemplary teachers. A first year student without sufficient command of English will do no practice teaching until the summer before the second year. At that time there is in an intensive training session with similar TAs in which they will practice lecturing and will learn about the peculiarities of American students. Some TAs must take ESL courses and the department delays the time when they enter the classroom to their fourth semester of graduate work. A successful Mathematical Exposition Seminar, in which the international students lectured on mathematics of interest to them and not just on calculus, was held during the summer of 1987 (see Appendix).

In the Appendix:

Report on Mathematical Exposition Seminar.

Florida State University

The administration has been supportive in regard to spoken English and other orientation efforts, but the department's service teaching obligation is exacerbated by effects of at least seven state government acts or regulations in recent years. There are:

35	in professorial ranks
3	full-time, with doctorates, not in professorial ranks
3	full-time, without doctorates, not in professorial ranks
53	mathematics graduate student teaching assistants
6	mathematics graduate student research assistants
6	mathematics graduate students supported and assigned other duties
1	mathematics graduate students with non-duty fellowships
19	graduate teaching assistants seeking degrees in other departments
1	undergraduate students teaching mathematics
9	part-time instructors who are not students
6452	undergraduates taught, all levels, Fall term

The program at FSU has benefited greatly from ideas collected in the early stages of this TA/PTI project. Some characteristics:

- Support is available for a 6-week TA orientation during the summer preceeding first teaching.
- There is close cooperation with spoken English instructor.
- Mathematics education TAs are classified as seeking degrees from another department.
- Two instructors who work with TAs were themselves TAs in the program

Mathematics in Florida's state colleges and universities has been affected by several regulations since 1980. Students face a sophomore test; the test "competencies" influence some courses content. A six hour course requirement has ballooned student numbers. The legislated freshman and sophomore maximum average class size is 27. The formats of course offerings and staffing are also impacted by a fluency requirement in spoken English (interpreted as a 220 TSE score) and a Southern Association of Colleges and Schools requirement that TAs have completed 18 graduate mathematics hours before independent section assignment.

The resulting program for working with those who teach lower level courses builds on earlier departmental and institutional foundations, but there is increased emphasis on working with all new TAs, and a much improved spoken English program. The part-timers (adjunct instructors) reported are mostly faculty at the local community college who teach lower level lectures; if new part-time instructors were appointed, they would be asked to participate in relevant parts of the activities for TAs.

Orientation sessions prior to the first teaching are required for all new TAs. The six week summer workshop, called the "TA class", is recommended for all new TAs; it is required if the home language was not English. For international TAs, about half the time is in the spoken English program; American TAs use this time to take a mathematics course from the regular summer offerings. (See Appendix, *Class Notes*—Florida State.)

In a Fall or Spring semester, the orientation days include, along with academic advisement, activities from the TA Class necessary before classes begin: discussion of practices and policies, grading and quiz making techniques, and simulated lessons with video taping. During these three to six days, the faculty and TA mentors are assessing the capabilities of the new TAs so appropriate assignments can be made. (See Appendix, *TA Orientation Schedule for Faculty/Mentors/Staff.*) The new TAs all register for the TA Class, which meets nine hours during the term and covers the remaining activities indicated on the summer Class Notes, plus discussions of classroom situations encountered.

A question often asked by colleagues is "How do you get your administration to pay a 6-week stipend?" Fortunately, there was local precedent: The Department of English had supported new TAs in the summer for many years on the rationale that they needed to teach them to grade essays and proctor tests at orientations for new freshmen. Factors which seemed to loosen the purse strings for mathematics were an obvious need to work with international and inexperienced TAs and lots of freshmen testing to proctor. A mechanical boost is provided by the July 1 beginning of the fiscal year and a "new" (not yet exhausted) budget. The Dean of Arts and Sciences and the central administration are supportive.

There is a written assignment for the TA class; ideas and enthusiasm are encouraged on this assignment—not polished writing. It is described in the *Class Notes*, Appendix.

New TAs are typically assigned to recitation sections of college algebra or business precalculus/finite mathematics. These courses are closely structured with a mix of test item types and formats: departmental multiple-choice, test bank free response, instructor-written items. The intent is that TAs learn to judge length and learn weighting skills in a controlled setting before assuming more responsibility in other courses. (See Appendix, *Lecturer and Recitation Duties*.)

For their first "solo" teaching, TAs often are assigned one recitation section (to see the experienced lecturer in action and serve as a procedural check) and one solo (full responsibility) section. A few new TAs, who have been TAs in a master's program, are assigned solo sections in the first term.

Experienced TAs observe classes and serve as mentors for new TAs. If the need is anticipated in advance or if there are student complaints, an experienced TA may be paired with a recitation instructor to give necessary support and encouragement; this may include "rehearsing" for the actual recitation, and multiple observations with feed-back.

As TAs gain experience, assignment made is to a course with a Coordinator who provides detailed "Instructor Notes" and pacing and a ditto master form for the Student Syllabus. Tests are written by the TA but checked before administration. Advanced graduate student TAs teaching calculus follow the Course Outline which faculty use, and are assigned a faculty mentor with whom they can discuss problems.

New TAs are observed by faculty and TA mentors several times; each TA and adjunct is observed at least once during each academic year, or whenever assigned to a new course, or if student input warrants. Student evaluations are administered each semester; results are discussed if needed. Both the classroom observation form and the student evaluation summary are kept in the departmental file.

The proportion of international TAs is fairly high. The relatively new spoken English program was developed largely in response to the needs of mathematics TAs. A portion of Chapter 4 of this volume is written by the director of that program.

In the Appendix:

Class Notes, MAT 5941, *Internship in the Teaching of College Mathematics* .

TA Orientation Schedule for Faculty/Mentors/Staff.

Lecturer and Recitation Responsibilties.

University of Michigan

Michigan has strong undergraduate students; considerable success is reported in recent years in attracting strong American graduate students, and a number of international graduate students come with their own support. There are:

68	in professorial ranks
2	full-time, without doctorates, not in professorial ranks
91	mathematics graduate student teaching assistants
6	mathematics graduate student research assistants
3	mathematics graduate students supported and assigned other duties
7	mathematics graduate students with non-duty fellowships
1	graduate teaching assistants seeking degrees in other departments
3	part-time instructors who are not students
6025	undergraduates taught, all levels, Fall term

Some features of the program:

- Single-variable calculus is not taught in Lecture-Recitation format.
- Self-paced mode is offered in several courses.
- Admitted students are normally assumed to be ready to enter the calculus sequence.

The English Language Institute (ELI) at Michigan is well known, and the Michigan Test, which has a larger component of spoken English than the TOEFL, is administered. The screening of international graduate students who wish to be appointed as TAs is administered at ELI and the panels include a mathematics faculty representative. An interesting set of video tapes has been developed for this screening procedure. Students recruited through the Student Government serve as actors; they ask questions of the type which might be asked in class, before or after a class, or at office hours. The screening panel observes as the prospective TA tries to respond to the film. The mathematics faculty participant deals with the appropriateness of the questions for a mathematics TA. The prospective TA also gives a short mathematical presentation at the blackboard while the examiners imitate student responses.

Because of the department's commitment to the traditional class mode for single variable calculus, first year graduate students who are expected to have difficulty with spoken English are rarely awarded teaching assistantships. (Grading or work with self-paced classes is available for at most one support.) International TAs may be accepted if mathematically strong enough and are encouraged to come with their own support. The language personnel interviewed during the Michigan site visits stress the time necessary to bring proficiency to the point for adequate teaching communication—mentioning one to three years.

Departmental administrators determine policy matters concerned with TAs. The Associate chairman for Graduate Studies and a training staff consisting of a lecturer and advanced TAs work with new TAs. An initial one week orientation (three weeks for international TAs) is followed by close monitoring and individual work with faculty and TA mentors during the first semester's teaching. Very detailed instructor notes are provided for courses assigned. Uniform final examinations, group graded, are used for all courses taught by TAs. TAs are observed in the classroom during at least their first semester of teaching. Student evaluations are required for each term of teaching.

Both faculty mentors and TAs called assistant mentors serve an important role with TAs beyond the first semester; they monitor standards and provide advice. There is university-wide training for TAs which is of a general nature. All TAs are provided with information about these sessions; the participation rate is not very high.

University of Montana

Although the location may be considered rather isolated, the town is attractive and the regular faculty is supplemented by local instructors. There are:

18	in professorial ranks
21	mathematics graduate student teaching assistants
1	mathematics graduate student research assistants
1	mathematics graduate students supported and assigned other duties
4	part-time instructors who are not students
2741	undergraduates taught, all levels, Fall term

Some results of being a comparatively small state university are:

- The small university summer budget precludes summer training.
- Lecture-recitation format is not used except for statistics.
- A member of the regular faculty coordinates each course.

All six to eight new TAs are required to attend a two-day orientation session prior to the Fall quarter. The first day sessions includes: (a) hand-out materials—acquainting the new students with departmental office policies, such as location and availability of texts and supplies, explanations on how to use office equipment, what they need to do to get their paychecks, where to find the syllabus for courses they teach, etc.; (b) graduate requirements—handbooks explaining what needs to be done to get their degree; (c) academic advising—this is done in two parts: the first hour is a rap session between the new TAs and more advanced TAs. They discuss their experiences, background, rigor of the courses, enabling the new TAs to obtain an idea of what will be expected. The advanced TAs are paired off with the new TAs as "big buddies" providing the new TA with someone with whom to discuss their individual situation. (The "big buddy" continues to provide advice during the first year). This is followed by a coffee-donut break, and then faculty describe the graduate courses they plan to teach.

On the second day the focus returns to the teaching duties. A fauclty member and some of the advanced graduate students reminisce about some of the problems that may arise in teaching. They discuss such items as grading (which may be very different from the TAs previous experience), examinations (number per term, difficulty), office hours, what to tell their students about how the class will be graded on the first class day, and how much homework to require. These discussions are often lighthearted and anecdotal. The new TA orientation ends with a cook-out for all the TAs and faculty. Three other events are scheduled during the year for faculty/TA socializing: a Christmas party, a Spring softball game, and an end of the academic year party.

All new TAs are required to take a 1-credit graduate course in the teaching of mathematics during the Fall quarter. This course is taught in the department by an experienced and gifted faculty member. Other faculty are called upon to provide other aspects of instruction and information to the new TAs. There are also lectures and discussions with university personnel. For example, the EEO officer discusses such topics as harassment, ethics, and the professional standards required of a TA. The course also involves the presentation of "mock" lectures, with supportive criticism from other observers. International TAs with TOEFL scores less than 580 are required to take a course in English taught by the Linguistics department. During Winter quarter, all new TAs are required to take a course in research methods in mathematics which includes a written term project.

International first year graduate students are assigned only one 2-3 credit course (usually a recitation section of applied calculus) the first quarter. These students are also expected to do paper grading. All other TAs are assigned 5-6 credit hours of work: either a 5-credit course or two identical 3-credit courses. Class size is generally 35 (if larger a grader is assigned). These courses are usually intermediate algebra, college algebra, trigonometry, or finite math for business. All TAs have office hours, grade their own homework, and frequently make up their own tests and grade them.

The degree of independence in each course depends on the course coordinator, a faculty member that is in charge of the particular course. The course coordinator may decide to have uniform exams, may review each TAs tests prior to their being given, or may allow them to proceed on their own with minimal supervision. Course coordinators prepare a syllabus for the course, set its policies, and on occasion assign the homework. The course coordinator is also the first person to whom undergraduate students bring complaints. Every new TA is visited during Fall quarter by a faculty member who assesses the TA's performance. These written reports form part of the basis for continued renewal of the TA. The reports are given to the TA, and help is provided in overcoming problems. Student evaluations are completed every quarter; the student evaluation summary becomes part of every TA's file.

The proportion of international TAs is 50%. The selection process is very competitive: most international TAs are expected to have TOEFL scores over 580 and GRE scores over 700.

The part-time instructors work exclusively for us. They are a dedicated group of professionals who perform very specialized duties: one works 3/4 time in the remediation program where she is very talented in alleviating "math anxiety", the other two work 1/2 time and provide assistance with statistical computing or geometry. All instructors working 1/2 time or more at the university receive full benefits. They are selected by the chairman based on the needs of the department. They are given complete latitude in the teaching of their courses.

Ohio State University

Since Ohio State University maintains an "open admission" policy, the department must juggle priorities from remedial arithmetic for freshmen to completion of 6-10 Ph.D.'s each year. There are:

84	in professorial ranks
31	full-time, with doctorates, not in professorial ranks
1	full-time, without doctorates, not in professorial ranks
153	mathematics graduate student teaching assistants
2	mathematics graduate student research assistants
22	mathematics graduate students supported and assigned other duties
9	mathematics graduate students with non-duty fellowships
60	graduate teaching assistants seeking degrees in other departments
29	undergraduate students teaching mathematics
7	part-time instructors who are not students
15,700	undergraduates taught, all levels, Fall term

With many sections of most undergraduate courses, O.S.U. offers extensive support to its instructional staff:

- "*How to Teach Math xxx*" books assist instructors of each multi-section course.
- Non-classroom GTA grading and tutoring assignments are available.
- Summer Fellowship or GTA support is available for nearly all mathematics graduate students.
- The individual small class format is used for all remedial, evening, honors and post-calculus classes.

The staffing of classes taught by other than regular faculty was described at the 1985 Anaheim panel by Joseph Fiedler. (See Chapter 1.) His updated version of those remarks describes the six groups of non-faculty teachers which are distinct by background, selection, teaching assignments and compensation:

"The incoming M.S./Ph.D. students teach through both of our major calculus sequences. Since our students know what to expect for their first two years, they develop a better sense of the courses and their students than is usual. Our M.A. students provide the core of instructors for our extensive remedial program. In addition to graduate students from mathematics, we screen, train and hire some 70 graduate students from other departments and 35 undergraduate students to work as classroom teaching assistants in our precalculus and, to a degree, in our remedial courses.

"An enrollment support program furnishes Head Start Fellowships to encourage our new graduate students to begin Summer Quarter. We require our incoming international students to enroll for a "Head Start" summer quarter under this special fellowship, thereby giving them time to adjust to a new country and academic system. Additionally, we require that they participate in a 6-week teacher training program run by one faculty member and one senior TA, our Master Teacher. By structuring the program about mock recitation sections in the mainline calculus sequence, we are able to target the course and format very tightly in which they will first teach. Our domestic M.S./Ph.D. graduate students are strongly urged to come in the summer so that we may adequately assess their background and preparation; they too are given a six-week teacher training course, designed around the business calculus sequence which they are first assigned. Our M.A. program is designed for inservice secondary mathematics teachers; no explicit training program is organized for them although some participate in the teacher training program for domestic M.S./Ph.D. students.

"The pool of graduate students from outside of our department represents an experienced and valuable resource for us. Most of these are foreign (this year 37 nationalities are represented), older (usually 25-35), and experienced (many have had extensive teaching experience in their native countries). Our relationship with this group is quite different from that with our own graduate students. We are their employers and their continued employment depends on their classroom performance. The average tenure of this group is five years so that we need to replace 10-20 each year. These jobs are sought after on campus as one of very few sources of support not tied to a student's discipline. During the winter we review 100-150 applications and invite 30-60 trainees for a three-week training and tryout program in which the trainees, in groups of five, are evaluated in six mock recitation sessions each. These sessions are conducted by a concerned faculty member and constitute half his/her teaching assignment.

"We also hire a sizable number of professionals from the community as adjunct instructors for our evening program (typical enrollment is 1200). A list of some 50 names is maintained with replacements recruited via vitae and interviews as needed. An impressive group, these supply an invaluable cadre of skilled and reliable instructors. We require student evaluations of all of the instructors in the evening and weekend programs, and they are consistently rated excellent by their students.

"One other group of TAs for us are the undergraduates who do recitations in our precalculus. They are trained and selected by a process similar to that for outside graduate students. Usually we must replace 25 students each year and receive 250 applications from which we select 60 for training. It is not unusual for members of this group to be stronger in mathematics than are some of our new M.S. students. The training is conducted by our Winter/Spring Master Teacher, a senior graduate student past his or her qualifying exams and experienced in both our calculus and precalculus courses (often as a lecturer in one of the 200 size precalculus lectures). This group of TAs, a very select one, has enormous energy and enthusiasm. Being students themselves, they have an empathy with and understanding of their students that is not available from the rest of our staff. Our undergraduates need closer supervision (largely for their own protection so that they do not take on too much), but add a valuable leavening to our program."

Mathematics teaching at Ohio State University is organized efficiently. The faculty frequently enjoy recitation section or grader assistance, or departmental exams in large precalculus lectures. The TAs primarily assist and have well prepared course materials. The students are all in an individual or recitation class with at most 30 others and have lectures by a regular faculty member except in the two remedial courses. Incentives for effective TA teaching involve summer support for math graduate students and a well advertised intense competition for other TA-type positions. Student and faculty evaluations are required each quarter for all non-faculty teachers. Evaluation forms for this purpose evolved over several years; a copy of each is included in the Appendix.

The TAs can draw on a variety of departmental support including the Master Teacher, an experienced TA whose assignment for two consecutive quarters is to be peer support for other TAs, to be a firefighter for TA classroom dilemmas, and to direct teaching training programs. All instructors have the support of detailed "*How to Teach Math xyz*" books which include a history of grades, former exams for the course, a syllabus, suggested pacing of material and comments on the calibre and motivation of students in the class. A sample of these books is included in the Appendix.

In the Appendix:

How to Teach Math 151.

Student Evaluation.

Evaluation of Recitation Instructor.

Syracuse University

Syracuse University, founded in 1870, is an independent privately endowed university. There are approximately 16,000 students enrolled; 12,000 are undergraduates.

There are:

35	in professorial ranks
2	full-time, with doctorates, not in professorial ranks
1	full-time, without doctorates, not in professorial ranks
65	mathematics graduate student teaching assistants
2	mathematics graduate students supported and assigned other duties
3	mathematics graduate students with non-duty fellowships
4900	undergraduates taught, all levels, Fall term

The department was recently an enthusiastic participant in setting up a university-wide orientation program. Other features:

- Only non-science track "service" offerings are taught in lecture-recitation format.
- TAs are moved to courses with more responsibility as they gain experience.
- The department was an enthusiastic participant in helping set up a new orientation program for TAs.

About one-third of all undergraduates in mathematics courses are taught in the lecture-recitation format. The courses taught in this format include: elementary statistics, introduction to modern mathematics (a course to fulfill arts and sciences distributive requirements), finite mathematics for business students, and calculus for business students. All of the remedial algebra, precalculus sections and most of the scientific calculus sections are taught by graduate teaching assistants.

The orientation and training of new Fall TAs begin over two weeks before the start of classes with the requirement of participation in a university wide TA orientation program. (The schedule of the 1988 program is included in the Appendix.) Some important features of the program are:

•TAs are housed and fed by the university during the program;

•TAs attending the program get an early pay check; however, the total stipend is not increased;

•two days are set aside for departmental orientation programs;

•new TAs assist with testing and advisement of freshmen;

•the departments are expected to hold follow-up sessions to the orientation sessions during the first semester.

New TAs who have no previous teaching experience usually start as recitation section instructors. A recitation instructor meets with each of 3 or 4 recitation sections one hour a week. The total enrollment in these sections will run between 75 and 100. By their second year, almost all TAs are teaching sections of their own. Normally, a TA would teach two sections of the same course with an average enrollment of 25 per section. Each course is under the supervision of a faculty member (the course chair). The course chair produces the course syllabus, is available to help the TAs in planning lectures and designing tests and quizzes, and deals with students complaints in all sections of the course. The final exam is the same for all sections and along with assignment of final grade is under the supervision of the course chairman.

All TAs are evaluated each semester by their students and their faculty supervisor (the course chair, if they teach sections; the lecturer, if they teach recitation sections). These evaluations are discussed with the TA.

In the Appendix:

Fall TA Orientation, Syracuse.

Vanderbilt University

Vanderbilt has an exceptionally large enrollment in the undergraduate major which is drawn upon as a supplement to graduate students for service teaching. There are:

29	in professorial ranks
25	mathematics graduate student teaching assistants
1	mathematics graduate students with non-duty fellowships
28	undergraduate students teaching mathematics
11	part-time instructors who are not students (4 with the doctorate)
2702	undergraduates taught, all levels, Fall term

Some characteristics:

• Of over 400 undergraduate math majors, over 100 also have engineering majors.
• There is limited financial support for international TAs.
• All but two of the graduate students are Ph.D. candidates.

The new Fall TAs participate in an orientation involving a general workshop presented by faculty, simulated teaching and video taping. The undergraduates, a few juniors but mostly seniors, serve primarily as recitation instructors for the service statistics course. (The comment was made that overall their performance is better than that of graduate students who never took this type course.) The load of undergraduates is limited to one recitation section and grading. The big service courses are statistics, finite mathematics and business calculus. All of these are taught in the lecture-recitation mode. It is hoped that a side effect of this involvement with teaching may be that some decide to teach, and/or decide to pursue mathematics or statistics at the graduate level.

Graduate students teach three sections spread over their first two semesters, and two sections each semester in subsequent years. Only those past the master's degree level teach scientific calculus. During the first semester of teaching, there is a special mid-semester evaluation which involves a classroom observation and a session with discussion of classroom matters in which all new TAs participate. Student evaluations are required during each term of teaching.

University of Wisconsin

The strong Wisconsin department reports more mathematics graduate student TAs than any other department. Detailed written materials, training sessions and careful TA supervision are a long tradition. There are:

84	in professorial ranks
1	full-time, without doctorates, not in professorial ranks
160	mathematics graduate student teaching assistants
6	mathematics graduate student research assistants
6	mathematics graduate students supported and assigned other duties
4	mathematics graduate students with non-duty fellowships
2	graduate teaching assistants seeking degrees in other departments
10	part-time instructors who are not students
10,111	undergraduates taught, all levels, Fall term

Some characteristics of Wisconsin's program:

• Most TA teaching during the academic year is as recitation instructors.
• There is not much summer work for TAs.
• Advanced TAs are involved in the orientation for new TAs.

An eight week long summer program is strongly encouraged for newly appointed international TAs. One advanced TA and faculty members work with the new TAs. Groups of two or three TAs are attached to small calculus classes taught by faculty. They hold office hours, review sessions and help grade papers and exams. They may give one or two presentations during summer. The English as a Second Language responsibilities are handled within the Department of English. Each international TA has an assessment of English and most attend a course in ESL specifically designed for international TAs. If language warrants, TAs may be asked to take additional courses in English. Participation in this assessment and following the recommendations based on it are required of mathematics TAs.

The international TAs are joined by newly-appointed American TAs for a one-week orientation before the fall term. All new TAs are assigned recitations (which meet two hours per week in conjunction with three hours of lecture) for Calculus I. A TAs load is two of these recitation sections. Each lecture has 4-5 new TAs and one experienced TA as coordinator.

Only the college algebra and trigonometry course among the courses which TAs are assigned is taught in the traditional class mode. Student evaluations are required each semester, and there is classroom observation during the first term of teaching.

TAs are sometimes reminded that information about their teaching may go into their permanent departmental file if they request or if it is unsatisfactory. Records of TA evaluations are kept in the department. Departments with vacant positions often do inquire about teaching abilities. File this anecdote under "it's a small world": Early in the preparation of this volume, the editor's department was interviewing for a junior analyst; she ran across a recommendation from the TA supervisor at Wisconsin. With permission it is copied in the appendix.

In the Appendix:

Helpful Hints to Good Teaching.

Letter of Recommendation for...

MASTER'S PROGRAM MODELS

TAs and other nonprofessorial instructors in departments with master's, but not doctoral programs, often teach the same courses taught by such instructors in doctoral departments. Their orientation and training for their duties share many similar components. But the physical restriction of the length of their expected tenure make difficult the progressions upon which doctoral departments often depend. A non-teaching assignment in the first year is usually not economically feasible in a one to two and one-half year program, making it more difficult to support international TAs. Also, American TAs in these departments are less likely to have time to progress to teaching more mathematically advanced courses.

Another characteristic of most master's-only programs is a more direct involvement of the department chair than in doctoral departments. In some cases, the department chair carries out most of the functions involved with working with nonprofessorial teachers. In most cases, the chair has a direct involvement and exercises close supervision of practice as well as policy. There are usually not TAs with adequate experience to assist with the orientation of new TAs.

Three models follow which describe the situation at departments with master's programs which have both TAs and part-time instructors. In Chapter 2 these master's departments with TAs are grouped as MAT. Information in Chapter 5 is applicable for the MAP departments which have no TAs but depend on part-time instructors for some of their teaching.

Miami University . . . Oxford, Ohio
University of Southern Mississippi . . . Hattiesburg, Mississippi
Western Washington University . . . Bellingham, Washington

Miami University, Ohio

First term graduate students do not teach independent classes. During their first semester as GTAs, they enroll in a one-credit teaching seminar, assist a faculty member in a large lecture, and tutor in the departmental sessions for calculus and pre-calculus students.

The teaching seminar meets fifteen hours during the first semester, with two-thirds of the meetings scheduled during the week before classes begin. The initial week is a survival course for the classroom. The course is run by a faculty member, chosen by the department chair, based on a reputation for quality teaching (as reflected in student teaching evaluations). GTAs are introduced to such basics as where to find and how to use the duplicators, and to the department's policies on grading, missed classes and office hours. They spend some time learning how to make and use transparencies and other A-V aids. Use of classroom computer systems is discussed.

Teaching and testing techniques are discussed, and each new TA in the teaching seminar prepares a short presentation on material from the textbook being used in the course in which the student will be assisting. These presentations are videotaped and discussed by the members of the seminar. Members of the seminar also make up mock quizzes over various chapters in the precalculus course, and these quizzes are presented and discussed. Selected second year GTAs are asked to return to the seminar to discuss their own teaching experiences and selected faculty attend seminar meetings to discuss special teaching skills.

During the four or five seminar meetings during the regular semester, graduate students discuss their own experiences as assistants in classes taught by Miami faculty. Faculty members who are assisted by first term GTAs are asked to make sure that their GTAs get some teaching experience, e.g., through explaining homework sets, holding office hours and grading parts of examinations.

After the first semester in the teaching seminar and as large course assistants, most GTAs are assigned their own classes. (Some do continue as assistants, however, based on recommendations from the teaching seminar leader, from the faculty members for whom they worked as assistants, and from faculty members who taught graduate courses in which they were enrolled.) The normal teaching assignment is a 3-hour precalculus course (enrollment not over 32), plus several hours per week spent in departmental help sessions. GTAs normally teach precalculus sections. A few of the strongest may get to teach calculus, and a few second year statistics graduate students teach a service statistics course.

In each successive semester, the GTA is assigned a Faculty Supervisor who typically supervises two GTAs. The duties of the Faculty Supervisor are:

- to review and approve each of the GTAs' tests before they are given;
- to review and approve the final grades for each GTA's class before they are handed in;
- to visit each GTA's class at least twice in the semester.

Miami University, in accord with state law in Ohio, must certify that a foreign born student is appropriately fluent in English before assigning teaching duties. This is accomplished by a screening based on TOEFL scores before the student is admitted and by an on-campus language competency test after the student arrives. The university offers special language classes to help graduate students who have difficulty with the local examination, but the bottom line is that prospective GTAs must pass the local examination before they are allowed to teach.

Part-time instructors are assigned only for precalculus sections on the branch campuses. There is little turnover in these experienced teachers; in one case where negative student reaction at mid-course evaluation was confirmed by faculty peer evaluation, the part-time instructor was replaced in mid semester.

A guide to departmental procedures has in its Contents:

GUIDELINES FOR THE TEACHING OF COURSES
TAUGHT BY GRADUATE ASSISTANTS . . . . . .

SUPPLEMENTARY NOTES TO GRADUATE ASSISTANTS
WHO ARE TEACHING COURSES . . . .

SUPPLEMENTARY NOTES TO GRADUATE ASSISTANTS
WHO ARE ASSISTING IN LARGE LECTURES . . . . .

DROP AND ADD PROCEDURES . . . .

University of Southern Mississippi

Graduate teaching assistants are instructors of record, usually of sections of College Algebra. Second-year students sometimes teach Trigonometry, Finite Math, or Developmental Math. Enrollment in sections taught by GTA's is usually 40-50. Each GTA teaches one section the first semester, and one or two after the first term. Foreign nationals usually grade papers for faculty members, tutor in the Math Learning Center, and/or substitute for instructors who are sick or out of town. A graduate teaching assistantship is considered a half-time position, approximately 20 hours per week. Some part-time instructors have been employed, but an attempt is now made instead to hire visiting full-time faculty, who teach twelve hours of classes.

Some of the GTAs have numerous years of high school teaching experience, and present few training problems. The fact that GTAs are only in the master's program two years has proven a major obstacle in working with foreign TAs from countries where people typically have poor English backgrounds.

The GTA training program is coordinated by a full-time faculty member, a master's level mathematician who carries the rank of Instructor. She is in a non-tenure track position, has been on the faculty for approximately 20 years, and is an outstanding teacher. She has one course each semester of released-time to operate, refine, and expand the GTA training program.

The training program has the following components:

(1)   A orientation program for new GTAs is conducted during the registration period each semester.

(2)   A GTA notebook is provided which presently contains:

• locally-generated handbook with departmental information and suggestions on effective mathematics teaching;
• a typical course policy handout;
• a detailed (i.e., sections to cover, problems to assign) syllabus for the assigned course;
• a copy of *College Mathematics: Suggestions On How To Teach It.*;
• a detailed (i.e., explanation of topics, worked-out problems, examples suggested to illustrate various points) set of lesson plans for the assigned course.

(3)   A Resource Notebook which contains additional articles and booklets dealing with effective teaching is maintained.

(4)   A file of College Algebra tests is available from previous semesters is available for reference.

(5)   Class visits are made by the GTA Coordinator and followed up with one-on-one evaluations and critiques.

(6)   There is continuing assistance by the Coordinator throughout the semester. This includes individual and group help with lesson preparation, teaching methods, test construction, evaluation of test results, assignment of grades,

and possible classroom problems (.e.g., absences, cheating, disruptions, drop and add).

(7)   The administration of student evaluations is required. These are reviewed and critiqued by the Coordinator in one-on-one sessions and reviewed again by the Department Chair. Future course assignments are based, in part, on the results of these evaluations.

Western Washington University

Approximately half of the TAs were undergraduates at Western; the others come mostly from colleges and universities in the Pacific Northwest. One current TA is Canadian; the others are all American citizens. There is a policy not to give assistantships to foreign students in their first quarter unless they have spent a significant period in the United States and it is verified that their language skills are adequate.

Each TA is responsible for the classroom duties of one lecture section of a precalculus course (Intermediate Algebra, College Algebra, Trigonometry, Precalculus) or of Business Algebra or Business Calculus. One section of each of these courses is taught by a faculty member who is responsible for the examinations (common among all sections), for the pace at which the material is covered, and for determining the grading scale at the end of the quarter. The faculty instructor meets several times a week with the TAs to discuss both general teaching techniques and specific problems with current course material. For Intermediate Algebra this is formalized as a one credit course which each TA much take during his first or second quarter.

Preliminary orientation of new TAs takes place during two days of the registration period. In addition to information about the graduate program and the various detials of life as a graduate student, the Graduate Advisor discusses at some length the principles and practices of good teaching. A number of written materials are given to the TAs. On the second day, each student is given a small topic (e.g., the quadratic formula) about which to pepare a ten-minute lecture suitable for an algebra class. These lectures are delivered to the TAs and selected faculty members and then analyzed very thoroughly by the audience.

Once the quarter begins, each TA's class is visited at least once by the Graduate Advisor and, if possible, by the faculty supervisor. There are more visits if this seems advisable. Each TA is required to administer the standard university teaching evaluation to his or her class each quarter and to deliver a copy of the evaluation to the Graduate Advisor. Most of the TAs are regarded very favorably by their students.

The department attempts to restrict teaching to tenure-track or visiting faculty members and to graduate students. There is a small pool of individuals living in Bellingham who have an M.S. in mathematics and extensive teaching experience at the two-year or four-year college level. These individuals are hired on a quarterly (or occasionally annual) basis to fill last minute gaps. Since they are already experienced teachers no training program is necessary. In fact, these part-time instructors frequently do a better and more enthusiastic job in pre-calculus courses than some regular faculty. There is consideration of more regular status for some of them.

REFERENCES AND RESOURCES

The materials section of this chapter, and the Appendix, contain many references and resources. See also the references in Chapters 1, 4 and 5, as well as the survey analysis and references of Chapter 2.

[1] Case, Bettye A., et. al. *Teaching Assistants and Part-Time Instructors: Responses to the Challenge*. MAA Notes, 1988. (This volume.)

[2] *College Mathematics: Suggestions on How To Teach It*. Mathematical Association of America, CTUM, 1987. (The much used product of an MAA Committee. Addresses "real world" issues of the mathematics classroom and offers practical suggestions. $2.00.)

[3] Perry, William and Doug Jewett. *The TAs Handbook: A Guide to Teaching Calculus*. New York, NY: Saunders College Publishing, 1985. (Addresses topics unique to the calculus classroom: epsilons and deltas, related rates, techniques of integration, series. Offers practical suggestions; includes example problems and applications. $8.00.)

Chapter 4

INTERNATIONAL TEACHING ASSISTANTS

All noncitizen TAs face cultural differences; additional problems may arise for those for whom English was not the language of the childhood home. Most universities have some kind of specific training in English available for or required of international teaching assistants. This chapter contains some general information and specific facts about testing and training for spoken English competency.

Factors related to the relatively large numbers of international TAs are discussed in the context of other chapters: In Chapter 1, Wendell Fleming's presentation at the 1985 Joint Meeting is a department chair's view. In Chapter 2, the collected data concerning numbers and duties of international TAs are presented. Chapter 3 makes special mention of international TAs in the context of the orientation, training, supervision, and evaluation of all TAs. In Chapter 6, results of some perception surveys of TAs and of their students are analyzed, when appropriate to highlight the situation as regards to international TAs.

INTRODUCTION

Both spoken English classes and sessions for orientation in teaching mathematics appropriately address adjustment difficulties and culture shock. The Committee on Teaching Assistants and Part-time Instructors feels that actual "training to teach" best involves citizen and international TAs together to foster mutually rewarding collegiality. But the special role played by dedicated language specialists in facilitating adaptation is often critical. The Committee has collected information from language experts at a number of the university sites visited on the grant project.

Roger Ponder, the instructor of spoken English at Florida State University, has acted as a consultant to the committee, has visited a number of exemplary language programs for TAs, and has "translated" his findings for the Committee. That translation was helpful, and the Committee feels that, since information about language training and the possibilities for mathematics TAs is not available in our professional literature, some information is appropriate here. Such information is difficult for mathematics faculty or administrators to glean from the professional writings of the language disciplines.

One activity in Dr. Ponder's work with TAs at Florida State is the classroom observation of the recitation and independent sections of his present and past spoken English students. He has observed a number of U.S. TAs so that he now has a strong basis for comparison and realistic expectations. After these observations, he writes a report—a "Verbal Snapshot" of the class. These are useful to the TAs and to those supervising them. With identifying material removed, some samples of this useful photography technique close the chapter.

The Committee asked Dr. Ponder to put together several kinds of information geared for the use of mathematicians: technical information about terminology and practices within the academic area known as *English as a Second Language*; information about standardized language tests; additional and self-help activities; a resource list.

The Committee found his presentation of information interesting and useful, but for space considerations has abridged his report which comprises the remainder of this chapter.

ENGLISH AS A SECOND LANGUAGE (ESL)

ESL instructors are usually members of the professional organization TESOL (Teachers of English to Speakers of Other Languages); many are also affiliated with NAFSA (National Association for Foreign Student Affairs).

ESL specialists and mathematics departments are primarily interested in what individuals can *do* with language. Two concepts from second language pedagogy are helpful in developing language courses for international TAs in mathematics: "communicative competence" and English for Specific Purposes (ESP).

Communicative Competence

This phrase may be taken to represent the broadest goal of ESL instruction. The language learner must master not only the grammar and vocabulary of the second language, but must also acquire a knowledge of the appropriateness and functional purpose associated with situational discourse in the new language. *Communicative competence*, then, is very different from what is often labeled simply "language proficiency". Though the two concepts are certainly related, communicative competence encompasses a much wider range of skills than those normally associated with linguistic competence alone.

English for Specific Purposes

For mathematics TAs, communicative competence means being able to effectively use English in ways specifically suited to mathematics instruction. Language classes for foreign TAs in mathematics must emphasize language which is appropriate for this purpose—English for Special Purposes (ESP)—not "general purpose" English. The starting point for designing ESP instruction must be the concrete, real-life situations in which the mathematics TA is expected to function. The initial question is "What exactly does a mathematics TA do?", followed immediately by "What kind of language is appropriate for these actions in the environment in which the actions occur?"

SURVEYING THE GROUND

Mathematics in the Academic World

The duties and responsibilities of mathematics TAs vary widely, from personally interacting with individual students at an advanced level to managing large impersonal groups of students in basic courses. The language demands of these extremes are also quite different: conversational, interpersonal, and elicitation skills, as opposed to skills in addressing large assemblies, giving explicit instructions, and, in general, handling the demands of room-sized classes filled with anonymous undergraduates.

The Language of the Mathematics Classroom

The language of the mathematics classroom, especially under the lecture-recitation format, is often of that type which Hatch and Long [1] call "unequal power discourse," that is, exchanges between an authority figure and a subordinate. Unstructured "discussions" among equals — of the kind found in many liberal arts classrooms — rarely occur. Yet, as an important study of discourse in the mathematics classroom (See [2], [3].) demonstrates, skillful language use is as much a distinguishing characteristic of effective mathematics instruction as of other kinds of instruction. In her study of mathematics classroom language, Rounds distinguished between what had to be said (the mathematics) and what was added (elaboration). Classroom talk, she concluded, "must go beyond the mere mouthing of the mathematics symbols". [3] Some of the elements which she sees making up effective language include defining and describing processes, verbally distinguishing major points, and organizing topics in an explicit fashion.

Graduate School as a Language Learning Environment

From a language learning perspective, the prevailing environment of graduate studies in mathematics would seem to impose several handicaps on language learners. The discipline itself has the reputation of being inherently "language-poor". Indeed, the concepts of graduate-level mathematics can often be satisfactorily conveyed without much verbal elaboration. (Perhaps it is not surprising to find individuals who insist that mathematicians do not really need much English.) Electives which require intensive language use, thereby helping students to develop language skills, are generally precluded by departmental requirements. Finally, mathematics TAs are busy people, under pressure in their own graduate courses and extensively involved in preparing materials, administering tests, and teaching their assigned classes. The bottom line is that language learning requires time and concentration, two elements which graduate school does not often provide.

Five Points About International TAs

In some respects international teaching assistants are like American teaching assistants. That is, it is easy to recognize in them familiar personality types and universal human characteristics. It is dangerous, however, to overgeneralize. Following are five important points to keep in mind:

1. *Individuals from different language backgrounds have different kinds of language learning problems.*

The distinguishing characteristics of a given language (and language type) create unique learning problems when it becomes necessary to modify these characteristics in order to acquire a second language. Thus, many language-learning difficulties are language-specific. The linguistic problems Chinese-speakers face in mastering English are very different from those encountered by Spanish- or Turkish-speakers. A new set of difficulties altogether would emerge if speakers of these languages were attempting to learn another language besides English. In sum, problems of mastering English can often be attributed to language background; it is a grave error to suppose that the language learning process is the same for all non-native speakers.

2. *In moving from one culture and language to another, individuals often experience "culture shock."*

A common reaction is a retreat into the familiar, the known, the old ways. There may be a period during which anything American seems to become a source of annoyance. Unfortunately, a few individuals never fully enter the mainstream. By personal preference and attitude, they remain culturally estranged. Though they may perform satisfactorily in many ways, linguistically they generally lag behind their peers.

3. *Native (or near native) speakers of English can be just as incomprehensible to other native speakers as foreigners.*

Comprehensibility may be an issue for students from the Indian subcontinent, the Carribbean, Africa and for that matter certain regions of the United States and the British Isles. Unfamiliar varieties of English may elicit the same kind of negative comments as those provoked by a foreign accent. The problem for speakers of these varieties of English is not to learn a new language, but to become bidialectical. Pronunciation practice may help.

4. *An individual's score on a written language test may be a poor indicator of speaking ability.*

Mathematics TAs on the whole appear to have a very good language proficiency level as measured by mostly pencil-and-paper tests (e.g., TOEFL). The problem is not that previous language training is inadequate, but that it is incomplete. Most ESL instruction for academic preparation does not stress spoken English. And students arriving in the U.S. directly from overseas are even less likely to have had experience in speaking than those who have spent time in intensive (or other) courses somewhere in the U.S.

5. *Some individuals with mediocre scores on written proficiency tests may or may not be good communicators.*

The bottom line is how well the individual gets messages across: *communicative competence*. Personality, experience, mannerisms, and even appearance are all elements to consider. Individuals who are good mathematicians and good language learners may finally be judged lacking as TAs.

What an ESL Course Can Offer

1. *Language practice*. The course meetings themselves can offer practice. For some students ESL class may be the only time in a busy schedule when language receives their primary attention. For others, it may be the only place outside a real classroom to practice speaking in front of a group.

2. *Instruction*. ESL classes can offer valuable instruction about language use, etiquette, meanings, and pragmatics, as well as factual and cultural information. (The first verbal snapshot example contains examples of "gambits," a type of language phenomenon which ESL classes might emphasize.)

3. *Individual help*. Sessions can be scheduled on an individual basis or in small groups (formed according to language background, for example). A self-help format in the language laboratory can be useful, both as a volunteer activity and for specially assigned work.

4. *Reinforcement of teaching skills*. Delivery, preparation, logical presentation of ideas, thoughtful discussion--these skills build confidence necessary for effective teaching. The ESL teacher can tie these activities directly to pedagogy by classroom observations and critiques. (The sample observations illustrate the kinds of comments a visiting ESL teacher might make.)

DEPARTMENTAL AND INDIVIDUAL ACTIVITIES

ESL courses or other language training is not available at some institutions. Departments can compensate through various initiatives. Several of these are useful continuation projects after formal training.

Use English! Language learning should never stop. Students should be encouraged to read, watch television, observe and—most importantly—participate in the world around them.

Class Observation. Teachers-to-be can watch other teachers' classes. An experienced teaching assistant can be a good role model for a beginner.

Provide Language and Social Events. Both directly and indirectly, departments can create favorable conditions for language practice. Socialization is an important component in the task of language learning; unstructured events such as departmental mixers, graduate student get-togethers, and summer outings can provide opportunities for foreign students to meet Americans and practice English. In universities which have host family programs, foreign students should be encouraged to participate.

Literacy Programs. In many parts of the country with recent immigrants, ESL instruction is available through libraries or adult education. Many literacy programs emphasize the teaching of reading, but tutors are sometimes available for conversation practice. The local public library can provide further information.

Campus/Community Exchanges. Foreign students are an invaluable campus resource, because they represent other countries, cultures, and world views. The information they provide (through interviews, giving talks, etc.) can be used by academics, students, clubs, and other organizations.

Self-Help. A self-service tape library can provide help with pronunciation, vocabulary, and grammar. The campus language laboratory is the logical housing for this operation but, within limits, a helpful self-service tape library can be assembled within a department.

For listening only, a regular tape recorded is adequate. Recorders which utilize double tracks (i.e., an unerasable master track and a student track which can be recorded) are available from two manufacturers only (Sony, Tandberg) at a cost of about $500 to $600 per unit.

PLATO. This computer-assisted learning system contains many kinds of materials which may be helpful for some students. The English Department or the campus Computer Center can also provide details of suitable materials.

Other Resources. A short list of materials is given at the end of this chapter. Many other materials can be found by consulting publishers catalogs.

SOME COMMON LANGUAGE TESTS

TOEFL

TOEFL: Test of English as a Foreign Language. Since 1973, the TOEFL program has been operated jointly by Educational Testing Service (ETS), the College Board, and the Graduate Record Examinations Board. Today, it is administered on a regularly scheduled basis in most of the countries of the world.

The TOEFL is a multiple-choice test containing 150 items (see below). Raw scores are calculated for each section, then converted to standard scores using formulas furnished by ETS for each administration of the test. The final score is the average of the three standardized section scores (multiplied by 10).

Section I	Listening Comprehension		50 items (about) 25 minutes
	Content:	Part A (20 items): Single sentences Part B (15 items): Dialogues Part C (15 items): Short talks and conversations	
Section II	Structure and Written Expression		40 items 25 minutes
	Content:	Part A (15 items): Sentence competion Part B (25 items): Error identification	
Section III	Reading Comprehension and Vocabulary		60 items 45 minutes
	Content:	Part A (30 items): Vocabulary Part B (30 items): Reading comprehension	

TSE

TSE: Test of Spoken English. This test was developed under the auspices of the TOEFL program at ETS. Individuals taking the test hear a series of taped instructions and questions, to which they respond verbally. Trained raters score the recorded answers at a later time. Pronunciation, Fluency, and Grammar scores are reported using a scale from 0.0 to 3.0.

An Overall Comprehensibility score is reported using a scale from 000-300. Of note is the fact that, whereas the overall TOEFL score is calculated directly as the average of the three section scores, the Overall Comprehensibility score of the TSE is an independent measure not directly related to the other three scores reported for the test. The TSE contains seven sections from which four scores are derived; the sections are described in a table below.

An audio tape with graded performance samples and a TSE User's Manual are available on request from Educational Testing Service. (See the **Resources** list.)

SPEAK

SPEAK: Spoken English Assessment Kit. This product is commerically available from ETS. The tests for use with the kit are retired forms of the TSE. Thus, SPEAK is identical to TSE in form and general content. However, since the reliability of these tests depends upon inter-rater reliability, each institution administered the SPEAK is responsible for determining the statistical soundness of its own testing procedures. (As in the case of all the tests it manufactures, ETS assumes these responsibilities for the TSE.) The initial kit includes training tapes, instructional manuals, and other materials necessary for training raters to score TSE recordings.

Sections of the TSE and SPEAK

Scored for

(P=Pronunciation, G=Grammar,
F=Fluency, C=Comprehensibility)

		P	G	F	C
1	Informal Questions (a "warmup")		Not scored		
2	Reading Aloud After a short period of study the examinee reads aloud a printed passage of about 125 words.	P		F	C
3	Sentence Completion The examinee completes ten partial sentences.		G		C
4	Picture Sequence After a short period of study, the examinee must "tell the story" shown in the picture sequence.	P		F	C
5	Single Picture The exmainee is given a brief time to study a single picture, and is then asked four questions about it.	P	G	F	C
6	Free Response Questions The examinee is required to talk for about a minute each on three topics of general interest.	P		F	C
7	Schedule The examinee must orally transmit information to an imaginary class.	P		F	C

A VERBAL SNAPSHOT:
THE CLASSROOM OBSERVATION

A report on an observation represents the personal viewpoint of the observer, not necessarily an objective evaluation. In some cases, the judgements expressed may not be shared by the instructor's home department. Nonetheless, an outsider's personal, descriptive account of a lesson can help mathematics departments to see their TAs in a new light. At a minimum, a report should probably include the following elements:

Descriptive Information

These details clearly identify the class (name of instructor, course and section).

Narration/Chronology

This information provides an indication of the general content of the lesson and a record of the events which occurred during the observation (the *Lesson Plan* section of the sample observations).

Setting

The purpose of these remarks is to suggest the atmosphere in which the instructor worked: the physical surroundings, the behavior of the students, the appearance of the instructor, and so on. Here, the observer attempts to convey what it was like to be in the room during the lesson. (See the *Class Atmosphere* section of the sample observations.)

Teaching

These remarks describe actions (including language) the instructor engages in to convey information to the class. These include organizational features of discourse, communicative behaviors, personal idiosyncrasies, voice projection, and so on. It is helpful to convey some idea of how well the instructor was able to gauge the appropriate level of instruction for the class, meet the expectations of the students, and cope with the background and abilities of the group.

Language

Statements in this section identify major linguistic problems (e.g., pronunciation, grammatical accuracy), if possible, with reference to specific difficulties (consonant contrasts, subject-verb agreement, verb endings, etc.). Some comment on overall comprehensibility is helpful.

Suggestions

Some advice can be easily dispensed and almost as easily acted upon: the vocabulary of classroom management and a single pattern of mispronunciations, for example. Many problems, however, are not amenable to one-time suggestions. It may not be possible to make suggestions which can be realistically helpful without major commitments of time and effort.

A VERBAL SNAPSHOT:
EXAMPLES

The comments in italics at the beginning of each Example are explanatory and not part of the feedback to departments and TAs.

Example 1.

This individual lacked command of language suitable for classroom management. It was fairly easy to offer concrete suggestions.

TA Identification. It is early in the TAs' second graduate year and first "solo" section teaching; the TSE was passed some months before, and there was a successful year as a recitation instructor.

Class Description. Precalculus and Finite Mathematics for Business. September 3, 1987, 7-8:15 p.m. 26 students: 7 women, 19 men.

Lesson Plan. The lesson covered aspects of combinations and permutations. After passing out the last quiz (using students' names), the teacher went over homework and then presented his lesson. Essentially, it was a continuation of material introduced earlier. In his presentation, the teacher referred to notes which from a distance appeared to be fairly detailed. At any rate, the continuity of the presentation was logical and smooth.

Class Atmosphere. Almost all the students arrived in the last five or six minutes before the class. At that hour of the day, there was little joy in their gait or manner. Two or three carried on conversations, but most of the others took their places laconically and waited for the arrival of the teacher.

The teacher did not announce his presence verbally. He walked to the front of the room and, without discernible comment, began to pass out the last quiz results, calling each student's name as he came upon it in the pile. He left the papers of the latecomers in a stack on the front desk and they were free to walk up and claim them as they entered the classroom.

During the lesson, most students seemed to be paying attention. Others obviously had other matters to think about. Two young women in the back row spent the hour writing notes to each other and passing them back and forth. About twenty minutes before the end of the class, one of them quietly picked up her things, checked to see if the teacher was going to object (He didn't.), then quickly slipped out into the hallway. A couple of male students about halfway back from the front sporadically carried on a conversation which was quite audible all the way at the back of the room. Fortunately, they fell silent pretty regularly, and so allowed their neighbors to concentrate on the lesson. Throughout, the hall door remained opened. On the back row, the noise coming into the room was extremely distracting, but there was no way to judge the effect in other locations in the room.

The students, then, were a real mixed bag: some halfheartedly participating, others quiet and attentive. If this bothered the teacher, he gave no sign of it.

Teaching. The teacher was very well-prepared. He spoke from notes written on letter-sized paper which appeared to be filled with writing. Though he referred to these notes, he did not read them, and on occasion spoke very confidently without them.

The explanations seemed to be very complete. At one point, there was an elaboration of a formula in the form of a lengthy aside which seemed to have little to do with the formula itself. As it turned out, however, the apparently irrelevant material was actually another version of the formula. The teacher thus showed how two quite different looking equations were really related. Later in the lesson, a diagram was extremely useful in making sense of a problem dealing with sets.

It was easy to read the teacher's writing on the blackboard, even on the back row. His use of the writing space was clear and logical.

Comments. *Kudos*

1.   The preparation for the course was complete and well-planned. And I think the teacher presented it to the class exactly as he had imagined it at his work desk. Overall, the lesson probably went according to the teacher's plan (which was a good one.)

2.   The materials ventured away from the textbook presentation. Given the constraints of the course syllabus, it isn't possible to get too far off track. But bringing in a slightly different approach to problems shows a good instinct for high interest material which students seem to appreciate.

3.   The teacher's appearance and demeanor were mature and professional. I stress this point because I know that he is very conscious of his age (which is only slightly greater than the average age of his students). He wishes to present himself as an authority figure (rightly so, I think), and did so in the hour I observed him.

Hard Knocks

4.   In the effort to appear professorial, the teacher seems to have lost his amiability. In the entire hour, I don't think I spotted a single smile. Though he may have fancied that he was conveying a sense of power, the message transmitted to the class was that he was tense and

nervous, maybe even a little afraid of these large, informal American young people. (They are like circus lions: though they appear to be docile, they will attach if they sense fear.) To improve, the teacher will have to find a way to appear relaxed and friendly, but at the same time forceful. For now, my suggesions is: Smile more often!

5. Again, motivated by a desire to create an appropriate techer image, the instructor used language very sparsely. As the lesson progressed, I became aware that something was "unusual". At first I was not able to discover exactly what was going on. Then, I realized that something was not going on; that is, the teacher simply wasn't saying very much. The element I was missing was the banter usually associated with classroom business. I learned afterward that this silence was a technique which the teacher invoked deliberately. It is an avoidance strategy which keeps him from having to deal directly with students.

(These comments should not be too negatively interpreted. The teacher maintained good eye contact with his students and responded to questions and comments with sensitivity.)

Suggestions. Talk more. In your case, silence is not golden. Students and teachers the world over expect a certain jargon in the classroom. In a sense, such language is "meaningless," and you were able to accomplish a lot by using it very little. But it is familiar to us and puts us (and the speaker) at ease. It also enables us to shift our attention from one thing to another and to organize our thoughts more efficiently.

As concrete suggestions, I would recommend the deliberate use of several "gambits" (manuvering phrases), some "soliciting" and some "structuring."

Soliciting gambits:
"We want to do X. How do you think we might do this?"
"If X is true, is Y also true?"
"But if we use X, what will happen?"

Structuring gambits:
To shift attention:
"Now let's look at another problem like this."
"Now that we've seen a problem involving X, let's look at a situation where Y is unknown."

To review:
"There we have it. We do A,B, and C, and come up with X."
"Now this problem resembles the problem we did yesterday where we found A, B, and C."
To verify:
"Did everybody get that?"
"Is this clear to everybody?"

Language. The teacher's speech retains many characteristics of his natural language: the articulation of vowels and consonants is not altogether that of American English; vowels especially are sometimes more typical of staccato Chinese vowels than fluid, variable-length American vowels. Stree and intonation are fairly accurate imitations of typical American language. I found comprehensibility to be acceptable, though I hope to see much more improvement in the future as the teacher gains in experience and confidence.

In a sense the "silent teaching" technique avoids the language problem. The teacher knows better than I the linguistic frontiers he can venture into. As he becomes surer of his performance pedagogically, perhaps he will venture out farther into the language.

There is a lot of potential here. I hope that the speaker realizes that and will keep striving to get better.

Example 2.

This individual was a competent, sensitive teacher who wished very much to be a successful teacher. He represents a very common type of language user: a fluent speaker whose English is, in reality, quite poor. He is glib, but often incomprehensible.

TA Identification: An advanced graduate student who has successfully taught a number of courses in "solo" mode; before passing TSE, spent terms as a grader and teaching recitations.

Class Description: Calculus II. May 25, 1988, 11:00-12:15 p.m. 24 students: 16 men, 8 women.

Lesson Plan. The instructor began the hour by discussing the upcoming test. He said that several students had expressed concern about what they would be responsible for. Accordingly, he said that, at the next class, he would be giving some help. He then asked for questions from the class and wrote the numbers of the problems to be worked. After about an hour, he introduced the new material from the textbook. The rest of the period was spent explaining the concepts involved and working problems using the latest acquisitions.

Class Atmosphere. The teacher looked comfortable and confident, and his class reflected this approach. There were frequent questions, and a number of volunteers spoke up with answers to problems or sections of problems. Without calling any of his students by name, the teacher seemed to know several of them. He often looked at individuals as he spoke and seemed to welcome input from all those present.

Teaching. As the round of requests to do homework problems began, the instructor said that he wasn't sure he could work out every problem on the spur of the moment. (They were the types of problems where one has to "hunt" for a solution.) This element of uncertainty became the vehicle for his presentation. As each problem came up, he was able to show what steps one goes through in order to discover a solution, how to identify unworkable solutions, and how to bring past knowledge into play in order to find the correct solution.

The instructor linked new material with old. He showed that he had a larger view of his task than one or two lessons, and referred several times to materials previously encountered. (Example: [Stopping in the middle of a problem] "For this step, fortunately we have a formula for this. [Digression to recall and discuss the formula].) He also supplemented textbook information. At one point, he wrote a couple of formulas on the board, recalling that they were already familiar from past work. He recommended that they be used in conjunction (possibly as replacements) for formulas in the new materials introduced during this lesson. In general, the instructor was able to "talk through" problems in a convincing manner; he presented problems not as "things" to be digested as is, but "processes" to be discovered, thought about, explored. Rather than simply point students toward a solution, he knew how to go with them as they tried to reach it.

The instructor was constantly aware of the importance of setting a clear goal. Several times, he explicitly asked: (Given a problem like X) "What do we want to do"? Then, "How do we get there?" Overall, the instructor knew how to push his students to think for themselves.

Language. In many respects, this was an exemplary class--and the instructor should be commended for his accomplishment. No doubt, thanks to his leadership, all the other classes which will not be observed and commented on will be equally fine.

There is one outstanding flaw, however: the instructor's use of English. If only it were better!

Pronunciation is borderline most of the time, and often lapses into unacceptable. Talk is almost always somewhat hard to understand, and there are stretches which become incomprehensible. Fortunately, the speaker is very sensitive to feedback, and when comprehension fails, he senses the breakdown and immediately tries to repair (by paraphrase, repetition, chalkboard examples, etc.). Grammatical structures, too, are often quite ill-formed. At moments, only fragments come out, the effort to compose full sentences completely abandoned.

I suspect that with the weight of writing a dissertation and the pressure of daily life, attention to a language has slid down the list of personal priorities. In fact, it wouldn't surprise me to learn that actual sustained use of English had diminished, rather than increased, over the past few months. The remedy for this situation is simple: more practice. Of course, practicing a language more intensively may involve changes in personal routine and habits. And only a conscious decision by the individual concerned can bring about these changes.

Example 3.

This individual had a very good command of English and prior teaching experience. During the lesson, he had to work to overcome two disadvantages: his own voice (a clear tenor with poor carrying power), and the physical surroundings.

Class Description. Once-weekly discussion and quiz section. October 13, 1988, 2:00-2:50 p.m. 19 students: 8 women, 11 men.

Lesson Plan. Materials dealt with domain and range of functions and logic problems. About fifteen minutes before the end of the period, the instructor passed out the weekly quiz.

Class Atmosphere. The physical layout and conditions in the room made a major contribution to prevailing atmosphere. The room is large, wide, and low-ceilinged. It could probably hold fifty desks or more, but actually contains only about thirty or forty. Consequently, there is quite a lot of space in the room--all of it in the front. The first row of desks is much farther from the blackboard (and the teacher) than is usual. This arrangement has the effect of isolating the teacher; he (or she) seems to inhabit one zone, while the students watch from another. Students sitting near the back of the room seem very far, even isolated, from the instructor.

The room is also noisy. There is no carpeting or other sound-absorbing materials. Small noises are magnified almost to the point of being distracting; from the back of the room, the chalk hitting the board sounds a little like the crack of a small rifle. Ordinary classroom noise creates a steady background din with which the instructor must consciously or unconsciously contend all the time. External noise interferes, also; when the airconditioning equipment kicks in, it becomes even more difficult to catch the instructor's words.

Teaching. The instructor arrived early and began returning the previous quiz to those who were present. He did not seem to have to ask students' names. (During the lesson also, he showed that he knew the names of the members of his class.) At the beginning of the lesson, he carried out some administrative duties: announcing a test and giving the mean of the quiz he had just returned. He stressed the importance of the test grades in the final grade assignments and urged his class to prepare thoroughly.

The presentation was energetic and confident. The instructor had a clear idea of what problems needed review and where the weak points were likely to be. He tried to elicit the interest and participation of his class; on a couple of occasions, when the only response to his questions was silence, he called on students by name in an effort to draw the group into the discussion.

One student asked a question about a word problem which had occurred in an earlier section. Even though the question was somewhat irrelevant (and he had not expected it), he was able to demonstrate an appropriate approach and show how the equation was to be set up and solved.

Language. The instructor's English is generally very good. He expresses himself confidently, with a wide range of vocabulary and effective understanding of registers, idioms, and colloquialisms. His accent is noticeable, but by no means does it interfere with communication.

The instructor's speaking voice is not naturally powerful enough to override the continuous noise of this particular classroom. He might try slowing down a bit with crisper articulation. He should be aware of the way he comes across on the back rows and consciously attempt to project his voice to those sitting there. As for the din, he might ask his supervisor to schedule his class in a different room.

REFERENCES AND RESOURCES

1. Hatch, E., and M. Long. 1980. "Discourse analysis, What's that?" In M. Saville-Troike (ed.) *Discourse Analysis in Second Language Research*, pp. 1-40. Rowley, MA: Newbury House.

2. Rounds, P.L. 1985. "Talking the mathematics through: Disciplinary transactioon and socio-educational interaction." *Dissertation Abstracts International, 46*, 3338A. (University Microfilms No. 86-00, 543).

3. Rounds, P.L. 1987. "Characterizing successful classroom discourse for NNS teaching assistant training." *TESOL Quarterly, 21*, 643-671.

TOEFL/TSE

For a sample tape and user's manual:

TOEFL/TSE
Educational Testing Service
Princeton, NJ 08541-6151

Bulletin requests, phone: 609/771-7900.

Purchasing SPEAK

Initial kit: $300.00
Includes training materials and practice
tests, and Form 1 of the SPEAK/TSE.

Additional forms: $75.00 each

A Basic Self-Help Tape Library

(Note: This list is purposely truncated. There
are many other products on the market to choose
from. In general, materials emphasizing grammar
drills and repetition data from an earlier time when
such work was thought very useful in language
teaching. Publishers bring out new pronunciation
and listening materials regularly, so these items are
more up-to-date. Finally, recordings cannot
replace live contact and personal involvement.)

Church, Nancy and Anne Moss. 1983. *How to
Survive in the U.S.A.* New York: Cambridge UP.
Student's book: $7.50; Cassette: $13.96.

Dixson, Robert J. 1963. *Oral Patterns drills in
Fundamental English.* New York: Regents
(Prentice-Hall). Textbook: $5.75; Cassettes:
$90.00.
Gilbert, Judy. 1985. *Clear Speech.* New
York: Cambridge UP. Student's book: $5.95;
Cassettes: $24.95.
Orion, Gertrude. 1987. *Pronouncing American
English.* New York: Newbury House (Harper &
Row). Textbook: $18.50. Cassettes: $150.00.
Rost, Michael A. and Robert K. Stratton. 1978.
Listening in the Real World. Tempe, AZ: Lingual
House. Textbook: $6.50; Cassettes: $42.50.

Manuals for International Students

Althen, Gary. *Manual for Foreign Teaching
Assistants.* Iowa City, IA: University of Iowa,
1988.
Cohen, R. and R. Robin, ed. *Teaching at
Berkeley: A Guide for Foreign Teaching
Assistants.* Berkeley, CA: Graduate Assembly,
University of California Berkeley, 1985. (A newer
edition is available, edited by Vassilis
Panoskalpsis; it is titled *International Teaching
Assistants' Handbook: A Guide for Foreign TAs
at University of California, Berkeley*).

Media Attention

Some of the many recent articles related to
international teaching assistants are listed at the end
of Chapter 1.

Chapter 5

PART-TIME AND TEMPORARY INSTRUCTORS

This project began because of concern for the effectiveness in college mathematics classrooms of TAs and nontraditional teaching personnel. In graduate departments, teaching by graduate teaching assistants and some "post-docs" is traditional. Statistical and anecdotal evidence indicates that much teaching is done by other than regular faculty and those in these traditional apprentice classes.

Who are these teachers and what titles do they hold? How many are there? A section of Chapter 2, titled TEACHING PERSONNEL, describes the 1987 survey categories. Because of cost, the survey population had to be limited to graduate mathematics departments. Part-time and temporary instructors are seen to be filling a deficiency in the ability to staff classes in these departments. (Graduate departments are denoted GP1, GP2, GP3, and master's nondoctoral departments are denoted MAT and MAP.)

This chapter includes some information about the more general situation. There is a look at statements made through professional and accreditation organizations which have considered the academic and ethical questions raised by such staffing. These nontraditional teachers in graduate departments are also discussed in Chapter 3, MODELS OF PROGRAMS; some of the information and materials described there are applicable in departments without graduate students.

The practices included here are directly related to the selection, orientation, training, assignment, supervision, evaluation and retention of part-time and temporary instructors in two- and four-year colleges. Several of these colleges were included in the site visits made possible by the grant funding. Helpful information was obtained during enjoyable visits with faculty from:

Austin Community College, Texas	North Lake College, Texas
Brevard Community College, Florida	Richland College, Texas
Brookhaven College, Texas	St. Petersburg Junior College, Florida
Eastfield College, Texas	Tallahassee Community College, Florida
Manatee Community College, Florida	Towson State University, Maryland

WHO ARE THESE TEACHERS?

It is easier to describe who they are not. They are not tenured, tenure-track, continuing contract or other faculty with expectation of indefinite reappointment. They are not visiting professorial faculty or partially retired faculty, and in graduate departments they are not graduate mathematics students or post-docs who teach a reduced load in favor of research activities. Teachers from these traditional categories have a vested interest in mathematics, in general, and in the department which appoints them, in particular.

In the Chapter 2 data, graduate students seeking degrees in other departments and undergraduate students appointed for the same type of duties traditionally handled by graduate TAs are included in this nontraditional category. Although selection and retention processes for these out-of-department and undergraduate TAs differ significantly from graduate TAs, the recommended orientation, training, supervision and evaluation are similar; many of the suggestions in Chapter 3, *MODELS OF PROGRAMS*, are applicable.

Even in two- and four-year colleges where there are not TAs and post-docs, the definition of nontraditional faculty is not always clear. Terminology describing these teachers includes "on an eight-month contract" and the oxymoron, "full-time part-timer."

To describe more directly these nontraditional faculty members: They are usually called Lecturers or Instructors, often with a modifying term such as *Adjunct, Part-Time, Acting,* or *Visiting*. Usually they are appointed for what is considered a part-time load but which may, in fact, be as much teaching as is done by regular, full-time faculty. They may be appointed on a full-time basis, but without expectation of continued regular or full-time appointment. At the convenience of the department, they may be full-time in some terms and part-time during others.

Some frequently–held beliefs about the nature and intents of the part-time faculty in mathematics which are probably not true:

•Most teach one course. *Fact*: The average number of sections taught by a mathematics part-timer is reported in 1985 (See [2], p. 51.) as 1.54.

•Most of these faculty are high school teachers. *Fact*: The data shows that a minority of these faculty are high school teachers. In particular: 10%, GP1; 5%, GP2; 29%, GP3; 34%, master's departments. In [2] the percentage of part-time two-year college faculty who are high school teachers is given as 37%.

•Many faculty spouses do not want full-time teaching. (*Again*: note the biographies of Julia Robinson and Mary Ellen Rudin.)

•Mathematicians working in industry or government who had a good experience as TAs play a big role in the part-time pool. *Fact*: There are no GP1 department reports of such teachers, and the percentages in the other graduate department groups are approximately 15%. In [2] the "Industry" percentage for two-year college part-time faculty is 24%.

One- and three-year contracts which specifically preclude reappointment are often used. Only sometimes are there health, retirement and other benefits for these teachers when they are full-time, but rarely if part-time. These matters are discussed more fully in the last section of this chapter.

IS TEACHING BY NON TRADITIONAL FACULTY WIDESPREAD? IS THE PRACTICE INCREASING?

In the graduate departments responding to the 1987 survey, nontraditional faculty taught in virtually all except a few Group 1 departments (and in 75% of those). This percentage increased in all the categories of departments between 1985 and 1987 with the greatest increases occuring in the GP2 and MAP departments. In two-year colleges, the number of part-time faculty increased by 95% in 1975-1980, by 12% in 1980-1985, and in 1985 stood at 54% of the total faculty. Private four-year colleges have the lowest percentages of teaching by part-time faculty. The educational qualifications of

part-time two-year college faculty have declined since 1970. (See [2] for more data concerning mathematics faculties.)

Not only are more departments utilizing these nontraditional teaching personnel, but the total and average numbers of sections taught by them are also increasing. In the GP1 departments, approximately 10% of all individual classes are taught by these. In the GP2 and GP3 departments, this percentage is up to 21% and has increased between 1985 and 1987. This percentage is approximately 23% in the MAT category and is up to 35% of the individual classes being taught by nontraditional faculty in the MAP departments. This has increased from 1985 as well. These percentages reflect all individual classes. In the 1987 survey, departments were asked to report distribution of sections among teaching personnel by level: lower level, calculus, upper level. The percentage of lower level sections taught by nontraditional faculty is much higher: 29%, GP1; 40%, GP2; 34%, GP3; 44%, MAT; and 53% MAP. Anecdotal information leads to the conclusion that there are more people who are teaching one or two courses at each of several different institutions, up to a total equivalent to 18 or more contact hours of independent classes per week. There are instructor applicants with resumes reflecting three or fewer years teaching at each of a dozen or more schools.

Careful analysis of regular faculty teaching loads compared with the total amount of teaching led to the conclusion: "Perhaps the increase in the incidence of lecture sections in university departments and the hiring of more non-professorial faculty explain how expected professorial teaching loads can be kept low in the face of rising enrollments per FTE faculty member." (See [2], pp. 47-8.)

PRACTICES AND MATERIALS
IN TWO-YEAR COLLEGE DEPARTMENTS

In the two-year colleges setting, nontraditional teaching personnel include part-time instructors and those full-time instructors who are definitely regarded as temporary or who have duty assignments markedly different from the norm. (e.g., No committee duty but more classroom teaching; less than full-year contracts; reduced benefits...)

In some cases, advertisements are run in local newspapers that part-time instructors are needed in certain departments of a two-year college. The resulting candidates are interviewed and a file is generally put together with some recommendations and transcripts; it is unusual if any independent checking is done.

There are reports of departments with as many as three-quarters of the classes taught by part-time instructors, but the most recent data implies much lower averages. Some schools report much turn-over and others little. In a department with only a few part-timers and little turn-over, it may be appropriate if orientation is on an individual basis and if supervision is loosely organized. In many cases when the number of part-timers is rapidly growing and/or the base population from which they are usually hired is a fluid one, there is a need for a well organized program to insure quality teaching.

Generally there is not a concern about non-native speakers of English among two-year college part-time teachers because the numbers are few, and those who are employed almost always have previous teaching experience in this country. Some two-year colleges located near large universities find many applicants in their part-time pool who are present or past graduate students, often in engineering or science, and are not native English speakers. Generally they are not able to train adequately those for whom English is a real problem, and hence do not employ them—or do not reemploy them, if an error in judgment is made.

There are sometimes college-wide regulations on the full-time equivalent amount which one part-time instructor is allowed to teach; this may be as low as 25%. But there are horror stories of part-time instructors teaching six or more courses at one school, or one or more courses at several different schools. Generally part-time instructors do not serve on committees and the office hour requirements are different from those of regular faculty. Often this is justified by saying that the tutorial centers usually available on two-year campuses make up for lessened access to instructors.

As regards the orientation of new part-time instructors, St. Petersburg Junior College (FL) has a weekday night or Saturday college-wide orientation session. A packet of materials is distributed describing resources available to students and faculty and giving administrative information. One of the regular mathematics faculty members has some released time from teaching to coordinate the work of the part-time instructors. A number of the part-time faculty teach three or four courses, have offices, and function much like regular faculty. It is not unusual for a person to teach first one class, then several in a semester, then to apply for and be apointed to a vacancy or expansion position on the full-time faculty. At St. Pete, there are student evaluations and classroom observations of the part-time faculty; it is college policy that not more than 25% of the total number of mathematics classes may be taught by part-time faculty.

Sometimes extensive locally-written materials are prepared for part-time instructors. At Brevard Community College (FL), there are both a school-wide *Faculty Handbook for Part-Time Instructors* (see Appendix) and detailed instructional guides for each mathematics course.

Austin Community College (TX) provides extensive support with both written materials and a "live" contact person in the department. A Faculty Course Manual (see Appendix) includes general guidelines, instructor notes, outlines and syllabi for most of the mathematics courses taught. The editor states: "We hope that this manual will continue the long-standing tradition in the mathematics department of unity of purpose and commonality of standards and content among all sections of the same course, no matter where the section is taught within ACC." Additionally, each new instructor is assigned a mentor, a "contact person" in the department. This nonsupervisory person is a ready reference for new instructors, providing support, guidance, and information.

These materials provide a quick reference for both departmental and college policies. Many part-time instructors teach proportionately more evening classes than do the regular faculty. They are on campus at hours when the mathematics office staff is unavailable. These materials provide much needed information for getting course materials prepared.

The policy and practice at North Lake College (TX), where part-time instructors teach over 50% of the sections, recognizes that this teaching greatly influences the quality of the program. A description of the procedures and the motivation behind them follows.

"Standing alone and unexplained, even detailed printed course guidelines are ineffective.

Departmental exams, administered in the testing center, narrow the "gap" of variation in instructor interpretation of course goals and objectives. Student competency is then defined and the completion of course goals and objectives is insured. To assure consistent and meaningful assessment of student progress, instructors are provided with general assessment guidelines; all student exams are kept on file in the department. Since full- and part-time faculty largely do not communicate with each other, concrete information is provided to faculty about departmental philosophy, goals, and objectives. Part-time instructors need and appreciate information so that their grading practices and procedures are consistent with those of full-time faculty. They may even be relieved not to think they have to participate in a popularity contest, manipulating their grades to "prove" their effectiveness as an instructor. In a sequence of well defined skill building courses, students need accurate and meaningful assessment. Any objection to such philosophy or practices concerning grading policies as an infringement of academic freedom does not seem to be defensible in sequential courses; success in a present course is closely related to and perhaps ensured by success in prerequisite ones."

The sites visited on the project and described here have some exemplary activities and materials. Anecdotal evidence indicates that the most typical situation does not provide much help for and monitoring of new teachers. Most typically, new part-time instructors have a conference—twenty minutes to two hours—with either the department chair or a course coordinator and are given the syllabus, instructor's manuals, text, and administrative information. This is sometimes within a day of the first day of class. Student evaluations and classroom observations by the chair or a coordinator are less likely than for regular faculty. Departmental examinations are sometimes used in two-year colleges, but this coping device to insure consistency appears much less widespread than in graduate departments. (One difficulty reported with departmental examinations in the two-year college situation is that the large number of nontraditional and night class students preclude a common time examination for all sections of a course.) The pay rate is often very low; figures named are sometimes one-fifth of a prorated regular salary. The effectiveness of the teaching and the consistency with departmental standards vary.

Some dedicated part-time instructors seek out the information they need and their students are well served. In other cases the result is not so good. To make ends meet financially, instructors may be harried by the pressure of accepting more work than possible to do well. Others may be the result of last minute hiring without adequate screening of academic and teaching backgrounds. Institutions and departments owe to their students a good effort to help all instructors teach effectively.

RELATED QUESTIONS ABOUT ETHICS AND ACADEMIC INTEGRITY

Professional societies in other academic disciplines and organizations concerned with a broad spectrum of disciplines have long been concerned about potential academic integrity problems and ethical employment considerations imposed when a lot of teaching is done by nontraditional teaching personnel. In the past, less mathematics was taught by these faculty than the proportion in some other disciplines. But the gap has narrowed; in 1985, in two-year colleges, 59% of the total faculty are part-timers and 54% of the mathematics faculty are part-timers. [2] Legislation has been introduced in Congress which would affect the benefit status of part-time teachers. Court cases have established that a property right may exist for a part-timer employed for many years. The charges of inherent gender discrimination which have long been implied ([3], [5], [8]) are newly and prominently aired [1]. Examples follow of some concerns addressed by other organizations, which are appropriate for consideration by mathematics departments and their college administrations.

Regional Accreditation Organizations

Each of the regional accreditation organizations has faculty credentials standards and statements concerning part-time instructors. For an example, the Southern Association of Colleges and Schools, Commission on Colleges, provides as a minimum standard under the heading Professional and Scholarly Preparation, Baccalaureate, "...must have completed 18 graduate semester hours in their teaching field and hold a master's degree."

Specifically regarding part-time faculty, SACS requirements include:

Part-time faculty teaching courses for credit must meet the same requirements for professional, experiential and scholarly preparation as their full-time counterparts teaching in the same disciplines.

Each institution must establish and publish comprehensive policies concerning the employment of part-time faculty. The institution must also provide for appropriate orientation, supervision and evaluation of all part-time faculty.

In some departments, graduate students seeking degrees in other disciplines are one category of nontraditional mathematics teachers. In some cases they are not required to be taking mathematics courses; at other institutions, they must take some mathematics even though they are degree candidates in another discipline. Even so, they generally never acquire very many credits in graduate mathematics courses. With duties limited to conducting recitation sections, this does not seem to be a problem; teaching which includes independent classes, even at an elementary level, is questionable. This accreditation organization's statement about graduate students, although regional as regards enforcement, may be of general interest:

Graduate teaching assistants who have primary responsibility for teaching a course for credit and/or for assigning grades for such a course, and whose professional and scholarly preparation does not satisfy...above must be under the direct supervision of a faculty member experienced in the teaching field, receive regular in-service training, and be regularly evaluated. They must also have earned at least 18 graduate semester hours in their teaching field.

The above requirements do not apply to graduate teaching assistants who are engaged in assignments such as...attending...lectures, grading papers, keeping class records, and conducting discussion groups.

Modern Language Association: Statement on the Use of Part-Time Faculty

The following statement on the use of part-time faculty was developed by an ad hoc committee of the Association of Departments of English and adopted by the MLA Executive Council in May 1982:

The recent dramatic increase in the use of part-time teachers in many departments of English and foreign languages is already threatening departmental integrity, professional standards, and academic excellence. Although some part-time appointments add significant dimensions to curricula and some professionals prefer to accept only part-time academic appointments because of other commitments, most part-time appointments are not made for educationally sound reasons. Indeed, the primary motivation for many of these appointments has been to reduce the cost of instruction.

From the point of view of the departmental administrator, part-time teachers fall into two general groups. Most are clearly temporary members of a department. Others teach from year to year and become virtually permanent. Graduate students who serve as apprentice teachers enjoy a special status in their departments and are therefore distinct from these groups.

The very conditions under which most temporary and permanent part-time teachers are employed define them as nonprofessionals. Often they are hired quickly, as last-minute replacements, with only hasty review of their credentials. They receive little recognition or respect for their contributions to their departments; in many instances they are paid inequitably.

The potential damage to academic programs caused by the excessive use of part-time teachers cannot be calculated exactly, but some negative effects are unavoidable. Because part-time teachers are not treated as members of the departmental community, they often have a limited commitment to the institution and its students. Because part-time teachers rarely participate, as professionals should, in the development of courses, the continuity of sequential courses and the consistency of multisectioned courses suffer. Because part-time teachers are rarely available to advise students or, if available, may not be fully informed about institutional programs, inordinately heavy responsibility for advising falls to the full-time faculty. In addition, because of the low professional standing of part-time teachers, their frequent assignment to composition and introductory language courses diminishes the importance of basic courses at a time when society recognizes a need for special attention to this part of the curriculum.

In the face of present conditions and concern about the decline in quality of humanities programs, the MLA urges college and university administrations to make new and concerted efforts to eliminate the excessive use of part-time teachers, to improve the conditions under which part-time teachers are employed, and to recognize the professional status and important contributions of such teachers. Continuation of excessive, unplanned use of part-time teachers can only exacerbate administrative difficulties, invite student dissatisfaction, and threaten the quality of education.

The MLA offers the following guidelines for the employment of part-time teachers.

Guidelines

1. Each department should develop a long-range plan that clarifies the use of both temporary and permanent part-time teachers in terms of departmental needs and goals. This plan should establish an appropriate limit on how many part-time teachers may be hired in relation to the number of full-time faculty and graduate students who serve as apprentice teachers.

2. All part-time teachers should be treated as professionals. They should be hired and reviewed according to processes broadly comparable to those established for full-time faculty. They should be given mailboxes, office space, and clerical support. They should receive adequate introduction to their teaching assignments, departments, and institutions. They should either be paid a pro rata salary or receive a just salary that accurately reflects their teaching duties and an additional stipend for any duties outside the classroom they are asked to assume. When appropriate and in accordance with well-thought-out policies, part-time teachers should participate in determining departmental policies and in planning the courses they teach.

3. If there is a recurrent need for the services of part-time teachers, departments should consider establishing a cadre of permanent part-time teachers. In addition to the privileges outlined in item 2, above, these teachers should receive apporpriate fringe benefits outlined in item 2, above, these teachers should receive appropriate fringe benefits and incentives that foster professional development, for example, merit raises and access to research and travel funds.

Florida AAUP: Quotations from *Dialectic*

The Summer, 1988, edition of this Florida Newsletter of the American Association of University Professors Newsletter was devoted to "The Role of Adjuncts on Florida Campuses." Reference was made to the situation of an adjunct instructor at Daytona Beach Community College which had been covered by the *Daytona Beach News-Journal*. The Editor's note states: "In recent years adjunct faculty have become increasingly prominent on college campuses. As the situation at DBCC seems something of a paradigm for the national situation, *Dialectic* feels that allowing its pages to become a forum for a discussion... is... appropriate..." The issue ends with a "highly opinionated essay" by Steve Glassman, (Embry-Riddle Aeronautical University):

Remember a couple of years ago when the blue ribbon panels warned us about the dire perils of our grade and high schools? Those same panels made soothing noises about the state of higher education in this country. Well, I've got some bad news for you. America's colleges and universities are fast going the way of her primary and secondary systems. If we don't wise up, and soon, it won't be long before the mark of the solid middle-class family will be a foreign car in the garage and the kids studying at a university overseas.

For the root causes of this mess we may as well blame the usual culprits, the Vietnam War and the social upheaval it caused. That is probably too glib and fashionable an answer to hold the entire truth but, for our purposes, it will do. You remember that during the Vietnam era, the baby boomers were crowding the schools and colleges. In order to help potential draftees keep their 2-A deferments, teachers passed out grades like popcorn. And universities cut back on scholastic requirements in order to generate enough professors to teach the swollen college ranks. After a while, the baby boom crested and the war ended. The colleges were no longer overcrowded. In the meantime, everyone and his pet dog who wanted one had gotten a graduate degree.

Now in spite of what academics sometimes like to think, colleges are like any other institution. They respond to the pressure of supply and demand. Suddenly, we were in the seventies. Inflation became a household word. Colleges' real-dollar budgets were shrinking. And the market was glutted with potential teachers. On the other hand, the ranks of the college student were badly eroding.

College administrators responded "creatively." Instead of rolling requirements back to pre-Vietnam standards, they lowered them to attract the so-called nontraditional student. Looked at in the best light, this movement could well be American education's finest hour. No longer were only the top third or half or even three-quarters of the nation's high school graduate college bound. Now anyone of any age or background could go to college and, for a while at least, Uncle Sam would pick up a hefty portion of the tab.

But there was an unhappy side to this phenomenon. Remember that huge pool of graduate degrees. Unlike most other things in life, a college degree will not go away if it is not used. So college administrators, seeing an excellent chance to save a few bucks, tapped that pool. The way it

was supposed to work was that professionals in their fields were given a chance to become part-time professors. Unhappily though, in many more cases, underemployed persons with graduate degrees were simply hired at the last minute at a whopping savings to the school in salary and bennies. This practice has so gotten out of hand that no matter where you send your kids to college in this country, the chances are about half their first-year courses will be taught by part-timers, often paid less than a thousand dollars a section.

Another interesting practice that developed about this time was the use of the student evaluation of faculty. Many colleges, in an over-generous spirit of accommodation to the now-endangered student, began to rely on their evaluations of the teachers as the primary tool of faculty evaluation. On the face of it, this idea must sound pretty good, especially to administrators who have to make judgments on people whose work is done in sealed classrooms. But unhappily, we all know that variation of Heisenburg's uncertainty principle: it isn't the job you do that counts but the job you are perceived as doing.

Teachers who once decried their students' mad obsession with grades suddenly found themselves engaged in an even madder scramble for good student evaluations in order, frequently, just to keep their jobs. 'Fess up, what would you do if forced to choose between assigning a difficult, but important, task and risking bad marks on those all-important student evaluations? Until human nature undergoes a Pauline change, it is unfair and unwise to expect anyone to pick duty over self-preservation.

I have taught at six schools. The best known is a major southwestern university. Not long ago this school boasted of endowing thirty chairs at one million dollars each. What the press releases didn't tell you about were the lecturers in the English Department trenches, sweating it out for less than twenty thousand a year. These people were reappointed annually or by semester, wholly on the basis of their student evaluations. Likewise, you weren't told that a college over-sight agency, appalled by the treatment of the junior faculty, recommended de facto tenure for those who had taught seven years. The university's response was to slap a three-year ceiling on reappointments.

If this is the way a well-heeled school operates, you can figure how the really hard-up carry on. For instance, at one junior college at which I taught, only two permanent faculty have been added since 1969. All the slack over the past sixteen years has been taken up by part-timers. At another university, most of the freshman courses were taught by graduate students who would not have been qualified to teach in the primary or secondary schools.

Okay, so how do we fix this mess? For starters, we are going to have to demand colleges begin acting like responsible institutions again. We are going to have to draw the line in the dirt and say we have done the cheap thing and the easy thing too long. We are going to have to insist the preponderance of instructors be allowed to concentrate on their classes, rather than making ends meet. We are going to have to give the students their rightful say without handing over control of the classroom. Unless we get smart, and soon, a degree from an American college will some day be just another piece of paper, worth more than a 1920's German mark but less than a handful of Confederate dollars.

REFERENCES AND RESOURCES

Many of the references in Chapters 1, 2 and 3 have applicability to part-time and temporary instructors.

[1] Aisenberg, Nadya and Mora Harrington. "A 2-Tier Faculty System Reflects Old Social Rules that Restrict Women's Professional Development," *The Chronicle of Higher Educ.* xxxv, No. 9 (October 26, 1988) A56.

[2] Albers, D.J., R.D. Anderson and D.O. Loftsgaarden. *Undergraduate Programs in the Mathematical and Computer Sciences: The 1985-1986 Survey.* MAA Notes No. 7. Mathematical Association of America, 1987.

[3] Apkarian, Alice. "Another Aspect of Part-Time Teaching," *The NEA Higher Education Journal* III: 1 (1987) 117-118.

[4] Batell, Mark. "Viewpoint: On Part-Time College Teaching," *The NEA Higher Education Journal* III:1 (1987) 113-116.

[5] Clausen, Christopher. "Part-Timers in the English Department: Some Problems and Some Solutions," *Assn. Dept. English Bulletin*: 90 (Fall 1988) 4-6.

[6] Franklin, Phyllis, David Laurence, and Robert D. Denham. "When Solutions Become Problems: Taking a Stand on Part-Time Employment," *Academe* (May-June 1988) 15-19.

[7] Taylor, Vernon. *Teaching Tips for Part-Time Teachers*. Key Productions, 1974. (P.O. Box 525, Chatham, IL 62629. $5.95.)

[8] Wallace, M. Elizabeth. *Part-time academic employment in the humanities*. New York: Modern Language Association of America, 1984.

Chapter 6

PERCEPTION SURVEYS

It is important for each mathematics department to monitor the success of its efforts to help teaching assistants and part-time instructors teach effectively. Some helpful information comes from students: complaints or praise or their lack. Classroom observations and the student evaluations often conducted on an institution-wide basis provide structured information. More specific information from students and the perceptions of the new teachers who have participated in programs can help in an ongoing process of improvement through modification in training and assignment.

The Committee on Teaching Assistants and Part-Time Instructors designed two surveys to gain additional information about the success of TA programs. The agency providing much of the project funding has a strong commitment to learner-centered programs and the grant negotiator encouraged this activity. One questionnaire, designed to be completed by TAs, involves their perceptions of the helpfulness of various orientation and training activities. The other questionnaire is a modification of a typical "student evaluation" form. (The survey forms follow.)

During Spring, 1988, each survey was administered in a few departments to collect a sample of this kind of information. The information from these surveys must be considered as anecdotal. The TA and student responses were voluntary. The sampling of schools is neither random nor stratified although the departments represent a variety of public and private institutions and Ph.D. and master's-only departments from various parts of the country. Relatively new international TAs and their students were over-represented in the sample.

THE TA SURVEY

The purpose of the survey was to elicit attitudes concerning the value of the early training and orientation activities which TAs receive to prepare them to teach. Two hundred thirty-two TAs from seven departments responded to the survey. The top section of the survey requested some demographic information. The item "Undergraduate degree from..." was answered by 214 of the 232 TAs. In addition to those with

undergraduate degrees outside the United States, some of those reporting undergraduate degrees in the United States are international students. The item *Language of childhood home: (English/Other)* should be added to this demographic information.

The information reported was:

TA: Mathematics graduate student	178
Graduate student: other department	34
Undergraduate student teaching assistant	11
Nonstudent instructor	2
Year as graduate student:	
1st	62
2nd	72
3rd or beyond	78

Highest degree you intend to complete at this school:
      Master's   68
      Ph.D.   148

Undergraduate degree from:
      USA   115
      Outside USA   99

The tables below reflects responses to the nine statements to which the TAs were asked to respond. They were told to omit any items which did not apply to the training at their school; however, it appears that some TAs used the "No opinion" category for that purpose.

The first table shows the percent of the 232 TAs who agreed (sum of strongly agree and agree) with each item and the percent of those who disagreed (sum of strongly disagree and disagree) with each item. The remainder either omitted the item or marked no opinion.

The two items, 6 and 7, regarding English training do not provide useful information since native English speakers could not be excluded in analyzing the results. (See above suggestions about an added item.) On item 9, it may be interesting to look at the range of answers by school. (See the second table, **Helpfulness of Training**.)

TA Survey Responses

		Agree	Disagree
1.	The TA orientation sessions before classes were helpful.	69%	11%
2.	Practice teaching sessions were helpful.	59%	5%
3.	The video taping with review was helpful.	39%	6%
4.	The reading materials about teaching were helpful.	45%	13%
5.	More training should be done before teaching duties are assigned.	30%	30%
6.	The spoken English class helped improve my communication ability.	16%	10%
7.	More time should be spent in spoken English training.	22%	12%
8.	The TA training interfered with my academic progress.	13%	50%
9.	Overall, TA training helps me perform my teaching duties.	65%	11%

Helpfulness of Training

School	Number Responding	Agree	Disagree
A	32	69%	22%
B	44	80%	16%
C	125	54%	8%
D	5	100%	0%
E	7	71%	14%
F	14	86%	0%
G	5	80%	0%

TAs were asked in an "open" item to suggest how their TA training might be improved. Many TAs responded, and their comments covered a wide range of opinions. Some suggestions seemed to occur rather often. These included doing more video taping, dealing with specifics rather than generalities, going over the teaching of particular topics which might prove difficult, and providing more feedback to the TA during the semester. One TA suggested that training should "teach them how to be funny. Serious teachers are usually boring."

THE SURVEY OF STUDENT PERCEPTIONS

This survey was written to find out how students feel about their TA instructors. Six of the seven mathematics departments from the TA survey participated. Since the survey was not administered to a random population, definitive conclusions should not be drawn. Several classes taught by United States citizen and international TAs were surveyed. The survey was completed by 508 students of which 277 had U.S. and 231, international TA instructors. In particular, each department was asked to include the students of at least one international TA who had some difficulties and some new TAs. To be sure to get constructive suggestions, the departments probably included more TAs known to be having trouble than their proportion in the total group of TAs. For example, the percentage of undergraduates in this survey who had an international TA is 45%; this is a little greater than the overall percentage of international TAs, 41%, reported in the 1987 survey. (See Chapter 2, Table 33, and related discussion.)

Because the competence of international TAs is a serious issue, the total reponses are shown, and, separately, the response concerning U.S. and international TAs. It is interesting to note that 81% of the students responding felt that their TA was an effective teacher, and that the difference on this important overall criterion between U.S. and international TAs is not significant.

Responses to item 6 ("understands and responds effectively to communication outside the class hour") seem to indicate that many mathematics TAs are generally perceived as somewhat lacking in some communication skills; again, the difference between U.S. and international TAs is not very great--especially when the sampling bias toward new international TAs is remembered. The major differences fell in items 3 and 5, the questions where variation is most natural.

It is encouraging to note that the overwhelming majority of these students feel that their TA instructor is thoroughly competent and is concerned with whether students learn the material. No significant difference was noted between U.S. and international TAs (items 8 and 9).

In presenting these results, there is some concern by the Committee that the numbers will be quoted out of the context of the biases of the survey sample. However, because of the potential value of the information, it is presented with the hope that it not be improperly used.

Student Survey

	Percentage marking *Strongly Agree* or *Agree*			Percentage marking *Strongly Disagree* or *Disagree*		
	All TAs	U.S. TAs	Int'l. TAs	All TAs	U.S. TAs	Int'l TAs
1. The instructor writes clearly and legibly on the blackboard.	84	84	84	10	10	9
2. The instructor organizes material well on the blackboard.	81	84	77	8	7	11
3. The instructor communicates ideas effectively.	72	81	61	14	11	17
4. The instructor understands and responds to questions about content.	83	89	76	6	4	10
5. The instructor speaks clearly and audibly.	75	94	53	14	3	29
6. The instructor understands and responds effectively to communication outside the class hour.	50	57	42	3	3	2
7. The instructor usually holds the attention of the class.	77	84	68	11	6	16
8. The instructor seems to be concerned with whether the students learn the material.	84	84	83	4	4	5
9. The instructor appears to be thoroughly competent in the course content.	88	89	87	5	2	9
10. In general, the instructor is an effective teacher.	81	85	75	9	6	13
11. I would recommend this instructor to a friend who is required to take this course.	69	77	60	14	9	20

GENERAL PURPOSE SURVEY FORM

TA–Math Grad——— ○	1) Year as graduate ○	1st———————— ○	MARK ONLY ONE RESPONSE
Grad–other dept.— ○	student at this school:	2nd———————— ○	PER ITEM.
Undergraduate——— ○	○	3rd or beyond—— ○	
Non-student ○	○	○	DO **NOT** STAPLE OR TEAR.
Instructor———— ○	2) Highest degree I ○	○	
Language of your ○	intend to complete at	○	USE NO. 2 PENCIL ONLY
childhood home: ○	this school: ○	Masters———— ○	
English———— ○	○	Doctorate——— ○	IMPROPER MARKS / PROPER MARK
Other———— ○	3) Undergraduate degree	USA———————— ○	
○	from: ○	Outside USA——— ○	

In the above section darken the correct responses. Mark
the items below, responding only to those items which apply to
you or the training you received at your institution. Training
is any activity such as orientations, seminars, workshops, or
video taping intended to help you teach effectively.

E. Strongly Disagree
D. Disagree
C. No Opinion
B. Agree
A. Strongly Agree

1. The TA orientation sessions before classes began were helpful. 1 Ⓐ Ⓑ Ⓒ Ⓓ Ⓔ

2. Practice teaching sessions were helpful. 2 Ⓐ Ⓑ Ⓒ Ⓓ Ⓔ

3. The video taping with review was helpful. 3 Ⓐ Ⓑ Ⓒ Ⓓ Ⓔ

4. The reading materials about teaching were helpful. 4 Ⓐ Ⓑ Ⓒ Ⓓ Ⓔ

5. More training should be done before teaching duties are assigned. 5 Ⓐ Ⓑ Ⓒ Ⓓ Ⓔ

6. The spoken English class helped improve my communication ability. 6 Ⓐ Ⓑ Ⓒ Ⓓ Ⓔ

7. More time should be spent in spoken English training. 7 Ⓐ Ⓑ Ⓒ Ⓓ Ⓔ

8. The TA training interfered with my academic progress. 8 Ⓐ Ⓑ Ⓒ Ⓓ Ⓔ

9. Overall, TA training helps me perform my teaching duties. 9 Ⓐ Ⓑ Ⓒ Ⓓ Ⓔ

10 Ⓐ Ⓑ Ⓒ Ⓓ Ⓔ

11 Ⓐ Ⓑ Ⓒ Ⓓ Ⓔ

What would you suggest that might have improved your TA training?
(Write your answer below)

12 Ⓐ Ⓑ Ⓒ Ⓓ Ⓔ

13 Ⓐ Ⓑ Ⓒ Ⓓ Ⓔ

14 Ⓐ Ⓑ Ⓒ Ⓓ Ⓔ

15 Ⓐ Ⓑ Ⓒ Ⓓ Ⓔ

16 Ⓐ Ⓑ Ⓒ Ⓓ Ⓔ

17 Ⓐ Ⓑ Ⓒ Ⓓ Ⓔ

TA Survey

18 Ⓐ Ⓑ Ⓒ Ⓓ Ⓔ

19 Ⓐ Ⓑ Ⓒ Ⓓ Ⓔ

20 Ⓐ Ⓑ Ⓒ Ⓓ Ⓔ

FOLD HERE

FOLD HERE

GENERAL PURPOSE SURVEY FORM

MARK ONLY ONE RESPONSE PER ITEM.

DO **NOT** STAPLE OR TEAR.

USE NO. 2 PENCIL ONLY

IMPROPER MARKS

PROPER MARK

This form enables you to rate your instructor on several characteristics. Please respond as accurately and honestly as you can. Respond to each item according to the key at the right.

E. Strongly Disagree
D. Disagree
C. No Opinion
B. Agree
A. Strongly Agree

Instructor's Name_____

1. The instructor writes clearly and legibly on the blackboard.

2. The instructor organizes material well on the blackboard.

3. The instructor communicates ideas effectively.

4. The instructor understands and responds to questions about content.

5. The instructor speaks clearly and audibly.

6. The instructor understands and responds effectively to communication outside the class hour.

7. The instructor usually holds the attention of the class.

8. The instructor seems to be concerned with whether the students learn the material.

9. The instructor appears to be thoroughly competent in the course content.

10. In general, the instructor is an effective teacher.

11. I would recommend this instructor to a friend who is required to take this course.

How could your instructor become a more effective teacher?
(Write your answer below.)

Student Survey

1 A B C D E
2 A B C D E
3 A B C D E
4 A B C D E
5 A B C D E
6 A B C D E
7 A B C D E
8 A B C D E
9 A B C D E
10 A B C D E
11 A B C D E
12 A B C D E
13 A B C D E
14 A B C D E
15 A B C D E
16 A B C D E
17 A B C D E
18 A B C D E
19 A B C D E
20 A B C D E

FOLD HERE

APPENDIX

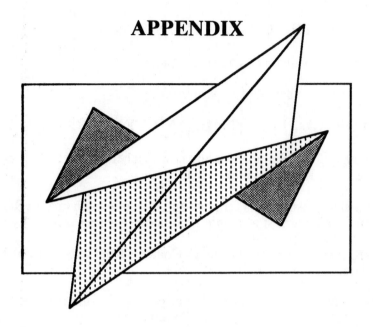

APPENDIX CONTENTS

THE COMMITTEE SURVEYS

1985 Survey (Sent to all graduate mathematics departments.) 94

1987 Survey (Sent to all graduate mathematics departments.) 95

GUIDES FOR TAs, PART-TIME AND TEMPORARY INSTRUCTORS

Chalking It Up—Advice to a new TA. (Suitable for any new teacher but written for new TAs assigned individual classes.) Bruce A. Reznick 99

Faculty Course Manual. (General instructions for part- and full-time mathematics faculty. Includes notes and syllabi for all courses; a sample is copied.) Austin Community College, TX . 114

Faculty Handbook for Part-Time Instructors. (Table of Contents from two-year college guide.) Brevard Community College, FL 122

Faculty Mentors: New Roles, New Relationships. (Description of organization of plan where regular faculty work with part-time faculty.) Austin Community College, TX 123

For the Newest MA 111 TA on the Harrelson Circle. (Course guide.) North Carolina State University 124

Helpful Hints to Good Teaching. (Departmental guide.) University of Wisconsin—Madison . . 129

How Not to Teach Mathematics—An Example. (Commentary on a videotape viewed by new TAs.) North Carolina State University 140

How to Teach Math 151. (Course guide.) Ohio State University 144

Recitation Instructor and Lecturer Responsibilities. (Generic checklists to supplement more specific information.) Florida State University 150

The Torch or the Firehose. (Advice about recitation sections in mathematics and physics.) Massachusetts Institute of Technology 153

ORIENTATION SCHEDULE AND COURSE INFORMATION

Advance Information. (Contents of packet of material sent to new TAs two months before their expected arrival.) University of Maryland 191

Class Notes, MAT 5941: Internship in the Teaching of College Mathematics (The 6-week summer departmental workshop activities and requirements.) Florida State University . . . 192

Course Guideline for the TA Workshop (Math 300): Manual for the Instructor. (Essential reading for the professor conducting a TA workshop; other departments have successfully adapted these ideas. Features short department and university orientations followed by weekly sessions.) University of California, Berkeley. 198

Ethics and Professionalism in College Teaching. (A collection of statements.) 212

Fall TA Orientation. (Schedule for graduate school orientation with first 6 days for international TAs, then 6 days for all followed by 1 day of departmental activities.) Syracuse University . . 214

Introduction Activity. (A "game" to help a new group get to know each other and provide an indication of language capability.) University of Maryland 217

Orientation and Workshop for New Graduate Teaching Assistants and New and Visiting Faculty. (Departmental; new faculty are also invited.) Oregon State University 218

Orientation for New Graduate Students and Teaching Assistants. (5 days, departmental.) University of Wisconsin—Madison 219

TA Orientation Schedule for Faculty/Mentors/Staff. (Schedule for those conducting one week TA orientation. TAs also go to one graduate school meeting; nine weekly sessions follow, completing the topics in *Class Notes, MAT 5941*.) Florida State University 220

Teaching Assistants' Orientation. (A 2-day small department program.) Villanova University . 222

EVALUATION AND OBSERVATION FORMS

Classroom Observation Checklist. (See last two pages of MIT's *Torch or Firehose*, listed above .) 189

Evaluation/Observation of Recitation Instructor. (To be completed by Lecturer; concerns both classroom observations and general performance of duties.) Ohio State University . . . 223

Report on One Class Visit. (Mathematical Sciences Department form.) Clemson University . . 224

Student Evaluation. (University-wide.) Clarkson University 225

Student Evaluation. (Departmental.) Ohio State University 227

TA Workshop Evaluation. (TAs indicate which activities were helpful and how others could be more helpful.) North Carolina State University. 228

INTERNATIONAL TA CONCERNS

Manual for Foreign Teaching Assistants. (Addresses problems and concerns encountered inside and outside the classroom.) Gary Althen 229

Report on Mathematical Exposition Seminar. (Set up for international TAs but the activity could be used for a general group.) Duke University 244

Teaching at Berkeley: A Guide for Foreign Teaching Assistants. (Selected articles are copied. A newer edition is available, edited by Vassilis Panoskalpsis; it is titled International Teaching Assistants' Handbook: A Guide for Foreign TAs at University of California, Berkeley.) . . 246

THE "BOTTOM LINE"

Letter of Recommendation for... University of Wisconsin—Madison 266

The MATHEMATICAL ASSOCIATION of AMERICA
INCORPORATED

1529 Eighteenth Street, N.W.
Washington, D.C. 20036
202 — 387-5200

Edgar H. Vaughn Building
DeCare Mathematical Center

Alfred B. Willcox, Executive Director

5/1/85　　Survey on Teaching Assistants and Part-time or Temporary Instructors

Institution _____

Department _____

All information provided will be treated as strictly confidential; the report will contain only summary information without institutional identification.

This questionnaire concerns teaching assistants and other teaching personnel in your department whose functions are those traditionally carried out by TA's. Currently, many departments do not find enough TA's among their own graduate students, and consequently employ graduate students from other departments, undergraduates or other part-time or temporary teaching personnel. We refer below to all such individuals as Teaching Assistants (TA) and Part-time or Temporary Instructors (PTI), or "TA/PTI".

In a number of questions below, separate responses are requested (1) for TA's who are graduate students in your department and (2) for all other TA/PTI. Please include in (2) those people assigned to lower level courses who hold such titles as lecturer, instructor, out-of-department teaching assistant, or adjunct. The following kinds of individuals should not be included: tenured or tenure track faculty, even if only part-time in your department; Visiting Assistant Professors; Research Associates or Postdoctoral Fellowship holders who are teaching part-time; partially retired professors from your institution.

How many of your "TA/PTI" were:

A.　Load Taught by TA/PTI (Fall term, 1984)

1. TA's who are graduate students in your department. _____

2. Other TA/PTI (e.g. students in other departments, PTI who are not students, undergraduate TA's). _____

3. Total no. TA/PTI. _____ (line 1 plus line 2).

How many "sections" (number of individual classes plus number of large lectures plus number of recitation sections) were taught by your department Fall 1984? _____ (4)

	Number sections taught by your TA's from line 1	Number sections taught by other TA/PTI from line 2	Number Sections by All TA/PTI	Number sections taught IN DEPARTMENT
Individual classes	(5)	(6)	(7) = (5)+(6)	(8)
Recitation sections	(9)	(10)	(11) = (9)+(10)	(12)
Large Lectures	(13)	(14)	(15) = (13)+(14)	(16)
				(8)+(12)+(16)=(4)

Load of TA/PTI in hours per week

Indicate the distribution of the actual numbers of hours worked per week (contact teaching in individual class, large lecture or recitation sections, grading, proctoring, office hours, etc.) by your TA/PTI. (NOTE: In many institutions approximately 3 contact teaching hours are considered as corresponding to 10 clock hours--25% full-time-equivalent--spent on all duties.)

Number of TA/PTI Working Indicated Number of Hours Per Week

Your graduate student TA's (from line 1)	Other TA/PTI (from line 2)
1 - 10 hours _____	1 - 10 hours _____
11 - 15 hours _____	11 - 15 hours _____
16 - 20 hours _____	16 - 20 hours _____
> 20 hours _____	> 20 hours _____
Total (= line 1) _____	Total (= line 2) _____

Distribution of class sizes of sections taught by TA/PTI

(If it does not fit your situation, please comment below on your class sizes.) Among the sections reported in lines (7), (11), and (15), (those taught by TA/PTI) give the number of sections with enrollments in each of the five indicated size ranges.

A table is provided to yield uniform data.

	< 20	21-30	31-40	41-65	> 65	
Number of Individual Classes						sum of row is (7)
Number of Recitation Sections	N.A.	N.A.	N.A.	N.A.	N.A.	sum of row is (11)
Number of Large Lectures	N.A.	N.A.	N.A.	N.A.	N.A.	sum of row is (15)

Comments on class sizes:

Prohibitions at your institution

Does your institution have a policy against using teaching assistants in individual classes or as lecturer in large lectures?　Yes _____　No _____

Does your institution have a policy against using undergraduates as TA's?　Yes _____　No _____

If "no", how many undergraduates are included on line (2) above? _____

If you employ undergraduates, describe duties:

B.　Training of TA/PTI

General Training

Does your department have a formal training or orientation program prior to or concurrent with first employment for new TA/PTI?　Yes _____　No _____　If yes, check appropriate boxes:

Number of Hours Spent in Training/Orientation Program

	1-4	5-10	11-20	> 20
By new TA's who are graduate students in your department (from line 1)				
Other new TA/PTI (from line 2)				

Check all of the following which apply:

Program is conducted before selection for hiring _____
Program is conducted before first teaching duties _____
Program is conducted concurrently with first teaching duties _____
Experienced TA's help with this program _____
Program involves simulated classes _____
Program involves videotapes _____
Program uses the MAA publication "How to Teach Mathematics" _____
Program uses other publications _____

Briefly describe your training/orientation for new TA/PTI. How is it administered? List or attach materials used, program outline, etc.

Competency/training in English

Some non-native speakers of English (those for whom English was not the language of the childhood home) use the language in an awkward manner which may cause difficulty for some students.

Number of non-native speakers included in line 1 above _____ (Number included in line 2 above _____)

Is a test to assess fluency in English required for TA's who are non-native speakers?　Yes _____　No _____
If "yes", which test (and cutoff if appropriate) is used?

_____ TOEFL _____ Test of spoken English (ETS) _____ Foreign Service Interview
_____ Other (describe)

Is an English language instructional program required for new TA's not considered sufficiently fluent?　Yes _____　No _____

Further comments:

C.　Selection, Supervision and Evaluation of TA/PTI

Please describe your procedures. A check-list is provided of frequently found situations. Please check each of the following which is applicable for the indicated category, and then comment on your special features.

	Your TAs (line 1)	Other TA/PTI (line 2)

Supervision TA/PTI is solely the responsibility of one faculty member. _____

(Optional: The supervisor's title is _____.)
General coordination of the duties and supervision of TA/PTI is the responsibility of one faculty member, but other faculty are also involved. _____

(Optional: Coordinator's title _____.)
There is supervision of small groups of TA/PTI by a number of faculty members. _____
Faculty members work with at most one of the TA/PTI at a time. _____
There is classroom observation by a regular faculty member of TA/PTI with the results of these observations discussed with the individuals. _____
Student evaluations are administered, reviewed by a regular faculty member, and discussed with each TA/PTI. _____
Are TA/PTI either renewed from the classroom or not renewed as a result of the evaluation procedure? _____

Please describe techniques of selection and other matters concerning supervision and evaluation and attach copies of any pertinent forms.

Thank you very much for your help. We welcome any additional comments or information you may care to provide.

(over)

The MATHEMATICAL ASSOCIATION of AMERICA

1529 Eighteenth Street, NW   ■   Washington, D.C. 20036   ■   Telephone (202) 387-5200

1986 SURVEY
TEACHING ASSISTANTS AND PART-TIME INSTRUCTORS

All information provided will be treated as strictly confidential; the report will contain only summary information without institutional identification.

Respondent's Name/Title _____

These questions concern teaching assistants and other teaching personnel in your department whose functions are, traditionally, those carried out by TAs. Currently, many departments do not find enough TAs among their own graduate students and, consequently, employ graduate students from other departments, undergraduates or other part-time or temporary teaching personnel.

Terminology

In a number of questions below, separate responses are requested for different categories. Please use the following descriptions.

<u>Lower Level</u> refers to non-calculus based courses such as College Algebra, Trigonometry, Analytic Geometry, Finite Math, Liberal Arts Math, Remedial Math, elementary Statistics; exclude computer science.

<u>Calculus</u> refers to any course which is primarily elementary calculus of functions of one or more variables.

<u>Upper Level</u> refers to <u>all</u> other undergraduate mathematics courses (linear algebra, differential equations, discrete mathematics, advanced calculus, statistics using calculus); exclude computer science.

<u>Mathematics Instruction</u> (MI) refers to teaching individual classes or holding recitation (tutorial) sections from large lectures with grading responsibility. It does not include those who work exclusively in a "mathematics laboratory" setting or those who grade without classroom or tutorial section teaching responsibility.

A. Teaching Personnel (Fall term, 1986)

Please use Fall 1986 data to give the numbers in each of the following categories for your department.

_____ <u>Professor</u> refers to all faculty in professorial ranks, whether or not tenured, including visiting, research and partially-retired professors.

_____ <u>Full-time doctoral</u> refers to non professorial personnel who hold a doctorate in the mathematical sciences.

_____ <u>Full-time non-doctoral</u> refers to non professorial MI personnel who do not hold a doctorate in the mathematical sciences.

_____ <u>Math TA</u> refers to mathematics graduate students involved in MI.

_____ <u>Other Graduate TA</u> refers to graduate students, studying in departments other than mathematics, involved in MI.

_____ <u>Undergraduate TA</u> refers to any student without a bachelor's degree involved in MI. Exclude undergraduates who work exclusively in a "mathematics laboratory" setting or grade under the direction of regular faculty.

_____ <u>Part-Time Instructor</u> refers to all other persons involved in MI.

Secondary School teacher: MA or higher_____   other _____

Government or Industry based: MA or higher _____   other _____

Other part-time instructor: MA or higher _____   other _____

Please enter the total number of graduate students supported by your department or institution _____.

Some of the mathematics graduate students supported by your department or institution may not do Mathematics Instruction as defined above.

How many have non service fellowships? _____

How many are research assistants? _____

How many do grading or other activities which do not involve classroom or tutorial instruction? _____

B. Class Size and Teaching Load Distribution

How many undergraduates were taught mathematics at all undergraduate levels in Fall, 1986? _____(1) (Include major courses, service courses and evening or off-campus courses for which your department provides staffing; do not include mathematics education or computer science courses.)

For the table below, use these conventions about Large Lectures: If each student from a Large Lecture is assigned to one of several "regular size" problem/quiz sections (perhaps called recitations or tutorials), use the first two columns under the appropriate level heading. (For example 4 Lower Level Large Lectures with 8 recitation sections each would report as a "4" for item (2) and a "32" for item (3).) Large Lectures (classes over 65) which do not have associated tutorials are reported in item (5).

	LOWER LEVEL				CALCULUS				UPPER LEVEL			
	Large Lecture	Reci-tation	Class ≤ 65	Class > 65	Large Lecture	Reci-tation	Class ≤ 65	Class > 65	Large Lecture	Reci-tation	Class ≤ 65	Class > 65
Total Number of Sections	(2)	(3)	(4)	(5)	(6)	(7)	(8)	(9)	(10)	(11)	(12)	(13)
Average Class Size												
Avg. Class Hours/Week												
Of the total number of sections above, indicate distribution among:												
Professor												
Full-Time Doctoral												
Full-time non-doctoral												
Math TA												
Other Graduate TA												
Undergraduate TA												
Part-Time Instructor												

Night or Extension Courses: If your department is responsible for mathematics courses taught in a night or evening division for which data is NOT included above, enter it separately below. Are these students included in line (1) above? (Yes/No)

Professor												
Full-Time Doctoral												
Full-Time non-Doctoral												
Math TA												
Other Graduate TA												
Undergraduate TA												
Part-Time Instructor												

Do you teach self-paced courses or have students not otherwise accounted for in the "sections" on lines (2)-(13) above? (Yes/No) If so, how many students are taught that way? _____

If you use the lecture-recitation mode, please complete the following.

Course Name	Credit Hours	Number of Lecture Hrs./Wk.	Number of Recitation Hrs./Wk.	Course Name	Credit Hours	Number of Lecture Hrs./Wk.	Number of Recitation Hrs./Wk.

C. Prohibitions

Does your department/institution have a policy against using graduate TAs in individual classes? (Yes/No)

Does your department/institution have a policy against using TAs as lecturers in large lectures? (Yes/No)

Does your department/institution have a policy against using undergraduates as TAs? (Yes/No)

If you employ undergraduates in any way, check as many as apply:

_____ grade for regular faculty

_____ grade for TAs

_____ conduct tutorials which are not associated with particular sections

_____ conduct specific recitation or tutorial sections including grading

_____ have responsibility for teaching and grading individual sections

_____ Other (explain)

D. Work Load of Graduate Students

Indicate the general number of hours per week expected of graduate TAs seeking degrees in mathematics. Indicate the number of hours, presuming the person was assigned only one type of duty. (Most, if not all of your TAs may be assigned a mixture of duties, for example 2 recitations plus open tutorials. But if in your department, the only duty assigned were recitations and 3 would be a full assignment, then complete the chart using this "hypothetical" TA.)

The answers will be presumed to hold for graduate students seeking mathematics degrees. Check here if they are applicable for graduate TAs from other disciplines _____ or undergraduate TAs _____.

		Contact Hours, Individual Classes	Contact Hours, Recitations	Grading Only	"Math Lab" or Tutorial Room	Office Hours, etc.	Total Hours Per Week
First Year TA	full assistantship						
	reduced assistantship						
Beyond the first Year	full assistantship						
	reduced assistantship						

E. International TA/PTI

	FIRST YEAR DUTIES			DUTIES BEYOND FIRST YEAR		
	Mathematics Instruction	Supported Non-MI	Not Supported	Mathematics Instruction	Supported Non-MI	Not Supported
(A) Number of Math TAs						
(B) Number of above who are International Students						
Number from B with Undergraduate education outside the U.S.						

Please give numbers from the other part-time MI categories for whom English was not the language of the childhood home: Other Graduate TAs: _____ Undergraduate TAs _____ Other Part-Time Instructors _____.

Competency/training in English

Some nonnative speakers of English (those for whom English was not the language of the childhood home) use the language in a way which may cause difficulty for some students.

Have you had complaints from students about "foreign accents"? (Yes/No)
Does your institution have a program to promote acceptance by students of instructors with these accents? (Yes/No) (If so, please attach a description and comments.)

Before appointment, does your Institution require examinations for nonnative speakers of English? (Yes/No)
If yes, please give minimum acceptable scores.
TOEFL _____ Michigan _____ TSE _____ Foreign Service Interview _____ Other (Specify) _____

Before appointment, does your Department require examinations for nonnative speakers of English? (Yes/No)
If yes, please give minimum acceptable scores.
TOEFL _____ Michigan _____ TSE _____ Foreign Service Interview _____ Other (specify) _____

After the student arrives on campus, is there additional language screening or testing? (Yes/No) How?
_____Individual interviewing.
_____Testing by Language expert.
_____Testing by Language expert jointly with mathematics department faculty.
_____Testing by mathematics department faculty.
_____Other (explain)

Please attach materials which describe your testing program.

If a student does not have sufficient fluency for teaching, what kind of instruction or training is available?

Is this instruction or training required? (Yes/No)

What are the duties of nonnative speakers of English, supported by your department, who do not teach recitations or individual classes?

Indicate the number of clock hours per week expected on these duties for a full stipend. _____ Comments?

F. Training

Does your department/institution have a formal training or orientation program prior to or concurrent with first teaching duties? (Yes/No) Is it required? (Yes/No)

How many hours of training is required for each category?   Mathematics Graduate TAs _____ International TAs, if different from above _____ Other Graduate TAs _____ Undergraduate TAs _____ Other Part-time Instructors _____

Check <u>all</u> of the following which apply:
_____ Program is conducted before selection.
_____ Program is conducted before first teaching duties.
_____ Program is conducted concurently with first teaching duties.
_____ Program uses experienced TAs (if so, how many? _____.)
_____ Program involves simulated classes.
_____ Program involves videotapes.
_____ Program uses the MAA publication, "How to Teach Mathematics".
_____ Program uses other publications (if so, please list titles, cost, ordering information).
_____ Program involves group discussions with faculty and advanced graduate students.

Personnel involved with the training of TAs include: (check all which apply) _____ professorial faculty _____ non professorial faculty _____ professionals from other departments _____ non faculty professionals

Please attach a description of your training/orientation for new TA/PTI and the name of the person who sets policy related to it. List or attach materials used, such as program outline, etc.

G. Supervision, Evaluation, Selection, Scheduling of TA/PTI

Policy matters arise concerning the selection, training, assignment of duties and supervision of TAs. Give the name and title of person(s) with responsibility for directing implementation.

_____

Determination of duties for particular TAs is done by _____.

To help you describe your procedures, a check-list is provided of frequently found situations. Please check each of the following which is applicable; then comment on your special features.
_____ Supervision of TA/PTI is solely the responsibility of one faculty member/administrator.
_____ General coordination of the duties and supervision of TA/PTI are the responsibility of one faculty member/administrator, but other faculty are also involved. These are: _____ Professionals in other departments. _____ Professorial faculty. _____ Non professorial faculty.
_____ TA/PTI are divided into small groups. One faculty member supervises each group, no overall coordination.
_____ Faculty members work with at most one of the TA/PTI at a time.
_____ There is a classroom observation by a faculty member or specialist member of the TA/PTI with the results of these observations discussed with the individual. First term teaching _____ Each term _____
_____ Student evaluations are required. First term teaching _____ Each term _____
_____ Student evaluations are reviewed and discussed as appropriate with each TA/PTI.
_____ TA/PTI are selected by Graduate Coordinator or Department Chair.
_____ TA/PTI are selected by a committee.
_____ Departmental course outlines are provided (name courses).
_____ Suggested departmental problem lists are provided (name courses).
_____ Uniform final exams are used (name courses).
_____ Uniform final exams are group graded (name courses).
_____ Uniform scale for grades of all sections is used (name courses).

Are TA/PTI removed from the classroom or not renewed as a result of the evaluation procedure? (Yes/No)

Please attach copies of any pertinent forms for supervision, evaluation, selection, scheduling of TA/PTI.

_____

The Committee on Teaching Assistants and Part-Time Instructors is preparing a <u>Resource Manual</u>. What specific information would you like included?

A Workshop about TA/PTI is planned for the January, 1988 Joint Meetings. What types of activities should be included?

Advice to a New TA

by Bruce A. Reznick,   University of Illinois

Introduction: page 100   Preparing: page 102   Lecturing: page 104

Grading: page 107   Coping: page 111

This material has been made available courtesy of:

The Random House Series for the Mathematical Sciences

This series is a recent undertaking of Random House and will feature textbooks for beginning courses in the mathematical sciences. It is a series intended not merely to meet immediate, existing textbook needs but also to anticipate new developments with strong, innovative textbooks. Random House, whose publications (including the Alfred A. Knopf imprint) have made it one of the best- known and most- respected names in North American publishing, was founded in 1925 and publishes college textbooks, trade and reference books, juvenile books, and school texts. Random House has published *Chalking It Up* and is making it available to the mathematical community as a service.

For more information about the Random House Series, or to obtain copies of this booklet, or permission to reproduce it, write to:

Random House Publishing
215 First Street
Cambridge, MA 02142

BRUCE REZNICK is an associate professor of mathematics at the University of Illinois at Urbana-Champagne. He received his bachelor's degree in 1973 from Caltech and his Ph.D. in 1976 from Stanford, where he was a Danforth Graduate Fellow. His research interests range promiscuously among analysis, number theory, algebra, geometry, and combinatorics.

Introduction

"The secret to education is respecting the pupil."
RALPH WALDO EMERSON

Facing your first class is a frightening experience. Think of it! Thirty human beings have paid good money to a fine university for their education, and they have been entrusted to *you.* You represent three percent of their college instruction, and you have never taught before. What can you do? Plunge ahead. Erase the board, pronounce your name and spell it in chalk, give your office address and phone number, the course name and section number, hold up the textbook and describe the syllabus, announce your office hours and intended grading scheme. By the time you call the roll, you are already a teacher. Congratulations!

This is a practical essay, rather than a theoretical article, but a few underlying beliefs will echo throughout. *Remember that your students are just like you.* You have probably been a student for most of your sentient life, so you know more about teaching than you may think. Recall what you didn't like as a student and don't do it as a teacher. (Be prepared to find a new respect for your old teachers!) *Remember also that your students are not at all like you.* You are interested enough in mathematics to pursue an advanced degree; your first students are probably taking a remedial high school course. Not only do you know a lot more than they do, but are probably much more talented mathematically. *Your students are not stupid people!* Be patient with them. Try to remember how it felt to take your least favorite required course and keep in mind that half your students will feel that way toward your course, no matter how well you teach. Many students will always look forward to a math class the way they look forward to a dental appointment.

Teaching can be extremely rewarding, but that is not why you have been given an assistantship: your department has a lot of teaching to do and it needs your help. *Your first job as a teacher is to convey as much of the material as you can to as many students as possible, not merely to have a good time.* Any reasonable technique for increasing this transmission of knowledge is, by definition, a good teaching technique. By knowledge, I mean algorithms and viewpoints as well as facts and theorems.

Since you have other obligations as a graduate student, I recommend the frequent use of a kind of educational judo. Students are prepared to devote a certain amount of energy and work to a course, just to get by. If you can channel this energy to a constructive purpose, the class will largely teach itself. The class must participate actively in learning; knowledge cannot be given, it must be taken.

A good teacher must be manipulative (in a positive sense). Almost everyone has the mental capabilities to understand algebra, trigonometry, and even calculus; the barriers to learning are environmental and psychological. *A teacher should try to persuade students that they really can learn the material.* Motivation can be found everywhere. We are bombarded these days with the need to understand math, science, and technology. Your students know of this need, and also fear falling behind in their studies. *Try to present yourself as an alpine guide to your class of climbers, rather than as a part of the mountain.* Create an atmosphere of common purpose rather than a battleground of wills. Encourage your class. Make them feel that, with time and effort, they *can* succeed. Most people want to please those around them, and your students have a dozen years' experience pleasing their teachers. Use this to an educational purpose. Make it clear that you *want* them to learn, that it matters to *you* whether they understand your lectures and can do the homework. Make the class an individualized experience, and not just another videotaped hologram at the Big U. If you really do care and you really do try and you succeed even moderately, you will find that teaching can be a very rewarding experience.

Please meet now two caricatures of imaginary teaching assistants. Morrie must assume that his students are miniature versions of himself: brilliant (of course), interested, and able to assimilate the material like a vacuum cleaner. He doesn't bother to prepare his lectures, because the course is so trivial that one reading of the text is enough, and besides, his students should see how a *mathematician* approaches a new situation. He cuts off those stupid questions in class and is forever annoyed that his stupid and lazy students don't make the minimal effort to understand him. He writes an "easy" test, but half the class flunks. Morrie wonders how these people will ever find gainful employment after school. He should wonder how he'll support himself next year when they take away his assistantship. Lester, on the other hand, doesn't know how those poor children would make it through the term without his help. He has given out his home telephone number as a 24-hour hot line for student problems. He gives two review sessions before each test and encourages students to ask for make-up exams if they are just not ready. He takes it so personally when a student misses a problem that he practically reads the exam during the review session. Les gives out door prizes to encourage attendance. He neglects his own studies and will also be looking for another job next year. Morrie and Lester are extreme cases, but they *do* exist. They come in various intensities and genders and a little bit of them lives in every teacher.

The rest of this essay is divided into four sections—preparing, lecturing, grading, and coping. Many of my suggestions may strike you as obvious or ridiculous. Feel free to accept as many as you like; you must develop your own style. If a particular technique makes you uncomfortable, don't use it. Your discomfort will distract from your teaching. To paraphrase Tolstoy, all bad teachers are alike, good teachers are good in their own way. You should know that there are quite a few articles and books around which also give advice to new teachers; keep them in mind.

I happily acknowledge the suggestions of the many people who have read drafts of this essay. In particular, Barry Cipra, Nancy Diamond, Dar-Veig Ho, Diem Kratzke, Tom Kratzke, John Martindale, Ray McEachin, Beverly Michael, Larry Riddle, Robin Sahner, and Korin Spongberg may recognize some of their comments. I bear, of course, full culpability for the contents. If you would like to see your name in this paragraph of next year's edition, drop me a note with your suggestions. This guide was written for use in graduate student orientation at the University of Illinois. The generous response it elicited encouraged me to seek a wider audience.

1/ Preparing

"Mere proof won't convince me."
from a Thurber cartoon

When you teach mathematics, you have an enormous advantage over the teacher of most other subjects: you can make statements in complete confidence of their truth. The value of a definite integral does not depend on political or religious beliefs. Consequently, you have a special obligation in preparing your course material. *You must know what you are talking about, and you must know when you don't know.* It is far better to admit ignorance than to feign knowledge. Expect this to occur; the old cliché that you never learn a subject until you teach it happens to be a *true* cliché. After you tell your students "I don't know," you should prepare a full explanation for the next class. Among other things, this shows your students that you work for them, and encourages them to work for you.

Your syllabus should give you an idea of how much material to cover in a lecture. Most beginning teachers, in their nervousness, talk too fast, cover too much, and do not give enough opportunity for questions. It is advisable to run through your lectures before you give them, writing down as your notes what you expect to put on the board. *Work out the calculations in detail, and make sure you know the correct definitions.* If you are running a recitation section, do the homework! And do it in the way that your students are expected to do it. Emphasize the algebra of the algorithms, not the arithmetic. There is no bigger waste of class time than that spent fumbling through an example with errors, false starts, etc. Of course, you should involve the class in the lecture, but this can be done in a premeditated way; there is nothing unethical about asking a class for advice when you already know what you want to do. Another old and true cliché is that mathematics is not a spectator sport.

Arthur C. Clarke has said that a smoothly running technology would be indistinguishable from magic. Formal mathematical proofs, stripped of tangible intuitive underpinnings, fall into the same category. You should always prepare some sort of explanation for what you do, even if, as in the case of trig identities, all you can say is, "It turns out that...".

I have heard Morrie's rationalization for not preparing examples (that it's good for students to see how a mathematician approaches a problem) from new TAs and from world-renowned mathematicians. It is self-serving rubbish. There is no reason why you cannot prepare your examples while indicating your thought processes. Further, for most of your students, learning how a mathematician thinks is of secondary importance to competency in the subject. Finally, your behavior in the classroom is a model for your students, a small and limited utopian society. If your students see you trying to bluff your way through a lecture, you are telling them, in effect, that it's okay for them to bluff their way through the course. Such is not the path to knowledge. I should also reemphasize that your students' lack of expertise in mathematics does not make them fools; they can tell when a teacher is unprepared.

Having said all this, I should warn against overpreparation. Your lecture is not being recorded for posterity; it is simply one of your several tools toward the larger goals of the course. *Even the most brilliant lecture will leave many students befuddled; do not take this personally.* You should leave room in your preparation for questions and unplanned tangents. (If you are uncomfortable in English, communication must take precedence over spontaneity, but you should never ignore questions.) You must also think of your other responsibilities as a graduate student; two or three hours out of class for each hour in class is sufficient in most cases.

In a broader sense, you should spend the first week of the term familiarizing yourself with the entire course. You must learn the particular notations of the text, even if they seem arbitrary or ungainly, since your students will be asking you questions in that language. Mathematics in a consistent set of notations is confusing enough for the non-mathematically oriented student. If you are running a recitation section, make sure you know what the lecturer has covered, especially where it diverges from the text. You should always have a rough idea of what you will be doing the next day or week. If you announce a major topic in advance of covering it, you will enhance the sense of accomplishment when you are done. Every topic lives in four tenses: "soon we shall consider...," "now we consider...," "now that we understand...," and "you are responsible for...".

There are literally hundreds of textbooks in print for any course you are likely to teach as a TA. These provide an excellent resource for the preparation of examples and test questions. Your department probably has a collection of unused complimentary textbooks sent by publishers. Shop around. (It is dangerous to write your tests *entirely* from another textbook; one of your students might catch on.)

I will say this many times, but it is important to remember: when in doubt, ask a more senior colleague. You do not need Zorn's Lemma to prove that every teacher has been a first-time teacher. If you are really ambitious, you may be able to find professional journals in your library which are dedicated to the teaching of college mathematics; but I'd hold off on these until you are more experienced.

2/ lecturing

"No dark sarcasm in the classroom."
PINK FLOYD

So now that you have prepared for the class, you must face your students. According to pollsters, people are more afraid of public speaking than anything else. Two ways to reduce this fear are to be confident in your knowledge of the material and to have a congenial class atmosphere. Even if you shake visibly, remember that students are much more scared of a trembling teacher than you can be of them. It *does* get easier with time.

I recommend that you arrive at the classroom as early as possible, and on time under all circumstances. The interval before the bell is very useful. You can personally distribute the corrected homework (and so learn names and faces) and you can also talk to the class in a more relaxed way. If you establish yourself as a person before the bell, it is easier to assume the role of the lecturer after the bell. (By lecturing, I include answering questions and solving problems as well as presenting new material.) Once the bell rings, you should start immediately on organizational material (test dates, etc.) and on old business. Review the last ten minutes of the previous class and ask for questions. *Do not cater to late arrivals by waiting for their attention.* Time is precious, and just a minute a day, three times a week for a semester, adds up to a full class period.

The mechanics of lecturing are straightforward, but hard to implement consistently. Write in a large, distinct, and legible hand so the people in back can see. Speak clearly and slowly enough to be understandable. Try to vary your pitch and tone, so the class will know when you're making an important point and won't be lulled to sleep by your monotone. Establish eye contact with students throughout the room. This also keeps the class alert, and provides you with honest and immediate feedback. Urge your students to complain if they cannot see or hear you.

Always ask for questions. *Never say that a question is stupid!* For every student who doesn't understand and asks, there are probably many others who don't understand, but are too timid to ask. *You may despair of their ignorance, but remember that your job is to teach these people.* When you answer a question, be patient and informative, and maintain eye contact to see if you've gotten through. It is proper to defer an answer to later in the hour or week, if new material will address the issue. If you become hopelessly bogged down, you should ask for a truce, move ahead, and pick up the topic in the next class. A reaction which suggests, "You are an idiot, you are wasting my time, get out of my life," will reduce the number of questions and the amount of learning. Many people have been turned off to mathematics by just one such arrogant teacher. I do not like to interrogate specific class members. This is college, not high school, and if students dread coming to class, they may not show up. (Maybe you can interrogate better than I can.)

Be sparing in your use of such words as "clear," "trivial," and "easy"; they provide no reinforcement to the good student and discourage the bad student. *Never be afraid to repeat yourself.* Someone might not have been paying attention to you, and some ideas only make sense the third or fourth time around. Students are always reassured to hear their teacher say something they already know. Never be afraid to repeat yourself.

After a dozen or more years of schooling, the student establishes a direct link from eye to hand, bypassing the brain. Accordingly, you should never intentionally write something incorrect on the board, even if you *say* it is wrong. Someone won't be listening and will copy it down as valid. Another tip on board work is to label the source of your examples if they come from the text. Students have told me that it helps them study.

I have avoided specific discussion of course material, because there is such a variety in courses taught by TAs, but I have a few general suggestions. The syllabus is there for a reason;

stick to it. Going too fast or too slow is one of the biggest problems for a new teacher. Ask for help from your colleagues if you are having trouble with the syllabus. Try to emphasize the unity and reasonableness of the material you are teaching. Present solutions in an algorithmic way, rather than as a bag of tricks. Avoid shortcuts unless they are widely applicable. The level of rigor in the text may distress you, but do not try to raise it. Explanations are probably better than proofs at this level. Appreciating proofs is an acquired taste, and must be held secondary to competency in the material. Perhaps the most important part of a college education is learning how to think independently. Keep in mind that most of the low-level mathematics you will be teaching is valuable as a tool in other disciplines, rather than an end in itself. No one wants a course consisting of 57 varieties of problems, all solved by rote, but attention must be paid to the fundamentals. Teach your students how to *read* a problem and establish guidelines for what constitutes an acceptable answer.

The same rules of courtesy apply to the end of the class as to the beginning. *Stop on time or, on rare occasions, one or two minutes late.* By the time the bell rings, a student is already thinking about lunch or the next class and will not give your lecture the attention it deserves. It is especially bad to start a new topic late in the hour, since you will probably have to repeat everything the next time. Do not be afraid to stop lecturing a *few* minutes early, if you can think of no constructive use for the time and no one has any questions, but don't do it often. If at all possible, stay a few minutes after class to answer questions. This is an efficient alternative to a short office visit for all concerned.

I have found at least five distinct media through which a student learns: the textbook, the lecture, the homework, the exams, and out-of-class work with other students. Each medium is best suited for a different kind of learning. The most brilliant lecture series makes for a superficial or incomplete textbook, and the best textbook, if read aloud, would be a ponderously dull course. The text is best used as a reference, to learn material which requires reflection and careful study, and for detailed calculations and lengthy and precise definitions. The best use of the lecture is "real-time" interaction with the students: answering questions and presenting examples which show how things work. Handy hints and suggestive analogies which would be out of place in print are ideal for the lecture. The lecture is most effective when the student has already made some attempt to learn the material; I encourage my classes to read the appropriate sections of the book both before and after my lecture. Some even do it. The homework is essential for the assimilation of skills; problems should not be assigned until similar problems have been done in class. Exams, or, more accurately, the studying for exams, provide you the opportunity to focus class attention on the crucial parts of the syllabus. (The next section is devoted to these two media.) Out-of-class interaction is largely beyond your control, but it is often the way your students will learn the most. You should certainly encourage it, and even offer to match students who have no one to work with.

New teachers often wonder how they can make their lectures more interesting. The best way is to show an interest in the material yourself. Don't just copy down the book's examples; change the numbers around. (This also increases the student's supply of worked examples.) Look for opportunities to pursue student questions, especially if they are in the direction you want to go.

If you like to tell jokes, tell jokes. *Avoid jokes which might possibly offend anyone, and never make a student the object of humor.* Remember that you are an authority figure; it is inappropriate to smirk about how hard the next exam is going to be. *If you don't like to tell jokes, please don't tell jokes.* Nobody expects a trigonometry class to be Monty Python's Flying Circus. Showmanship in education is only justified when it increases attendance or helps maintain class interest.

3/ Grading

"You knew the job was dangerous when you took it, Fred"
SUPERCHICKEN

In the ideal academic groove, there would be no testing and grading, just lecturing and the correction of assignments. Students would not need certification and would care enough about the content of their courses that the stimulus of grades would be unnecessary. We are not in an ideal groove. Virtually everyone agrees on the need for some hour-long exams during the term and a final at the end. Your university requires a final course grade in any case. You will probably be teaching one section of a multisection course, so it is important that your grading schemes be consistent with your colleagues'. I would now like to discuss the various components of a student's grade: homework, hour-long exams, and the final. Much of this section represents my personal taste, and should not be considered as pedagogical dogma.

Many teachers use weekly quizzes as a substitute for graded daily homework; I think this is a bad idea. I must add that I am in the minority on this. *Frequent homework, collected, corrected, and returned promptly, can be an extremely valuable tool for learning.* (By frequent, I mean two or three times a week.) It gives the students a chance to test their skills, and gives you an accurate idea of what your class has understood. It also encourages students to work steadily throughout the semester rather than pull all-nighters just before the tests. The former style is more conducive to long-term learning. I realize that this much grading is unpleasant and time-consuming for the teacher, but it works. (Remember, though, the advice of Superchicken to Fred the Lion, quoted above.) *I have found a way to reduce my work load, while increasing the effectiveness of daily homework: I distribute worked solutions to the homework on the day it is due.* This kills at least five birds with one stone. I can correct assignments without having to repeatedly write down the correct solution. I can plan my lectures with more confidence. (After presenting one or two problems on the board, I ask my students to read the solutions and bring up other questions the next time.) I can forbid late homework, the bane of every grader and a bad habit of many students. I provide the class with, in effect, a supplementary text of worked problems. Finally, I am forced to psych up in advance for the class by preparing the solutions. You will only recognize this problem when you are teaching the same class for the third or fourth time.

Homework assignments ought to require at least an hour's work per class for the average student, beyond reading in the text and reviewing class notes. I do not assign homework in advance, waiting for the end of the class period to write the assignment on the board. This is not an especially popular practice, but it allows me to tailor the assignment to the particular material covered that day, and subtly encourages class attendance. You should stick to the text's problems until you are quite familiar with the syllabus. Another labor-saving device is to divide the homework into "hand in" and "also do" problems. I write up solutions to both, but only grade the "hand in" problems. The "also do" problems are advertised as good candidates for the hour-long exam. *I try to return homework by the next class period, so the students get rapid feedback, and I get to know what they haven't mastered.* I prefer to grade homework on a fairly crude numerical scale, on the basis of 3 or 4 points per assignment. I do this so I do not have to agonize over partial credit or regrade many assignments. *Students will work together on the homework.* If you encourage this, it is inevitable that some students will copy a better student's homework without fully understanding it; I know I did this in college physics. If you forbid collaboration, the same students will still copy homework, but you will now have created a large group of scofflaws, hurting the social contract even more than copying. I allow collaboration on homework.

I am not a fan of short quizzes. They are too often a substitute for the kind of homework I have described above, which gives students a chance to work out their skills without serious time pressure. Quizzes tend to distort the entire class period in which they are given. When comprehensive, they do not give the student enough time to think about the problems, and reward the rapid regurgitator. When really short, they make testing too hit-or-miss for my tastes. I concede that quizzes can be useful, if given in conjunction with homework. They are good for students, in reducing the terror of test-taking, and they are good for teachers, in reducing the terror of test-making. I repeat that they are a poor substitute for homework.

Even after you have done your best to explain the material, corrected the homework, and provided many examples, your students will disappoint you with their lapses of knowledge. This is inevitable. It is a fact of life that for many students, the only part of the class that matters is the final grade. *Students will study the material they believe they will be tested on.* You can use this leverage to your advantage by testing the most important parts of the course. Let your students know the skills you consider most important, base the bulk of the homework on them, and then write boring and straightforward exam questions. I do not like to see difficult questions on marginal topics or trick questions on exams; these aren't prelims. Students should concentrate on hitting fastballs before they worry about curves, and a little batting practice won't hurt. An exam should be a dipstick into the crankcase of knowledge and not just a shaft.

In writing a test, you should first identify the skills you want to test and then write problems to fit them. A good test should take a good student about forty-five minutes (of a fifty minute class) to complete, leaving some time to check the answers. You should always take the test yourself before giving it to your class. Many innocent-looking problems require fiendish computations. If a fifty-minute test takes *you* forty-five minutes, it is too long.

You will want a range of difficulties, with an A-level problem to challenge the better students. Any problem which is only solved by one or two students is necessarily a bad problem for that particular class; aside from frustrating most of them, it will ruin the scores of good students who work on it to the neglect of the rest of the test. Make the first question relatively easy, to encourage your students and reduce the frequency of clutching. Cute story questions are often very tempting; I have found that the level of student literacy is such that many students cannot translate them into mathematics. In all cases, you should indicate the point value of each problem. Try to balance your allocation of points so that no single question is weighted too heavily. Multipart questions are also tempting, but should be used with extreme caution. You don't want a student's exam ruined by an error in the first part.

No matter what sort of students you teach, everyone becomes a pre-law major before an exam, looking for a loophole in the syllabus. It is merciful to tell a class that you have no intention of testing them on a particularly tricky offshoot of the main discussion. You want them to concentrate their limited studying time on the most important topics. On the other hand, you shouldn't give away the store. I have given up on review sessions, unless the mood of the class is unusually panicky before an exam. Review sessions perpetuate the idea that exams require extraordinary preparation. I prefer to make them a natural extension of the rest of the course. You may want to provide review questions, especially if you have taught the course before. This way you minimize the advantage of students who have access to fraternity and sorority files of previous tests.

I suggest generosity in partial credit towards those who know how to do a problem, but make silly errors or miss a relatively minor point. Care in working is certainly an important skill, but most exams force students to work at almost a reckless pace. *I like to grade an exam problem-by-problem to standardize partial credit.* It's a good idea to correct a few papers without assigning points to them in order to get a sense of how you want to award the partial credit. You should clearly mark where the student went wrong, but I find it instructive to avoid giving solutions in full detail. The students should try to work out their own errors. *A successful exam is one which accurately measures the abilities of the students, not necessarily one which neatly puts them into five piles.* It is very important to return exams individually to the students, in order to ensure the privacy of their grade. This rule may be waived when you are giving back a perfect exam. If you are grading part of a large exam with other instructors, consistency is crucial. Do not make unilateral policy decisions.

By the end of the term, both you and your students should have a pretty accurate conception of the final. I like to give about a two-hour exam in the three-hour time slot. The final should cover the course material in a balanced way, including variations on the highlights of the course. It is quite appropriate to be less generous with partial credit on the final, since the material is being tested for a second time.

As far as course grades go, your conscience must be your guide. When in doubt, ask for advice or bow to peer pressure. *Your grading scheme should be consistent, immutable, and well publicized from the start.* I prefer grading on a numerical basis during the semester, assigning a letter grade only at the end. (If students ask you how they are doing, it is best to give a range of grades. This method is more robust to small changes in partial credit.) My scheme is generally this: the homework counts 10%, three hour-long exams count 20% each, and the final is 40%. This adds up to 110%. I drop the 10% with the lowest percentage score, so that one bad test can be discounted. Class involvement is too subjective a criterion for me to weigh. I give semi-flexible percentages separating A/B, B/C, etc.; that is, 90% *will* be an A, but 88% *might* be an A, depending on how the class does. Thus, students have specific goals to shoot for, but there is still a curve. You will find that your students clump very naturally into the various grades, except for a few inevitable borderline cases, which will give you heartburn. Again, if you are in a more structured grading environment, you must go with the flow.

Do not be talked into changing your grade by a hard-luck story. You would be chagrined to learn how many of them aren't true. In any case, one grade in one course does not get a student into trouble. Any university has an extensive apparatus for handling students in hot water. (No school likes to admit that its admissions office goofed.) This apparatus relies on objective evaluations from the professional staff. That means *you*. If you feel that a course grade does not represent the full story, give it anyway, and get in touch with the appropriate dean. As always, consult your more senior colleagues if you are uncertain.

Teaching provides many opportunities for self-delusion. A good grade or a bad grade on a poorly designed test may be close to meaningless. Morrie and Lester would undoubtedly have radically differing opinions of the same students. Maybe your trig class is stupid, or maybe you're giving them two problems too many on their tests. When that median score is 93%, let someone look over your test and grading policies. Maybe you are a brilliant teacher and they are a brilliant class, but maybe you're not expecting enough of them.

4/ Coping

> "I was thrown out of NYU for cheating on my metaphysics final. The professor caught me looking into the soul of the boy sitting next to me."
>
> WOODY ALLEN

This last section is a miscellany of advice for dealing with some difficult situations. I claim no originality and make no warranty for my suggestions. Since the particulars vary, you should ask more experienced teachers for advice as the circumstances warrant.

Cheating is a real problem, and *the truly dedicated cheater is probably close to undetectable.* The best way to avoid cheating is to create an atmosphere of common purpose in which cheating is clearly antisocial and destructive — an offense against the majority of honest students and an assault on the integrity of the educational system. Peer pressure can be an effective tool. It never hurts to remind your class not to cheat and to proctor exams with care. Move around the classroom, including the back if possible, so the cheater never knows what direction you'll be facing. You can minimize the opportunities for out-of-class cheating by being circumspect with test materials. Write the exam at home, after the last class meeting, and don't make up a formal key until the test has been given. When grading a test, be suspicious of non-sequiturs in reasoning, or oddly missing steps in a calculation. Be suspicious of friends who sit together at test time and make the same errors. However, cheating is a very serious charge to make, and friends who study together often make the same mistakes. You are well within your prerogatives to ask them not to sit together, but do it tactfully. You should let someone in authority know if you have caught a cheater dead to rights. Most universities have rather complicated regulations on cheating.

Students often complain about the grading of their homework and exams. They may point out something the grader missed, or claim that their work deserved more credit. In either case, you should listen carefully. When this happens to me, if there is clearly no merit in the claim, I tell them so. Otherwise, I take the paper back and regrade it on my own time, away from their scrutiny. On questions of judgment, I am reluctant to change my mind. You should be too. You are a professional with a far richer perspective on the course material than the student. Still, all teachers make mistakes, and we must own up to them. In cases of a serious or intractable disagreement, you may want to consult a more experienced teacher for advice. I must also mention that cheaters have been known to write in tests after they have been graded and demand extra points. If you suspect someone of doing this, photocopy future tests before you return them. You should always make a mark on a blank examination page, to discourage after-hours inspiration.

Teachers at the University of California are required by state law to post their office hours (it seems that an assemblyman's son could never find his prof), and it's an excellent idea. You ought to be available two or three fixed hours per week to answer questions, talk to absent students, etc. You should also be available for appointments, but setting regular office hours in advance is a good time-management technique. Do not expect a lot of activity until just before a test. Office hours are a good time to write lectures and grade homework and are a desirable resource for students, but they are not a very efficient teaching medium. If two people ask you the same question during office hours, you are well advised to cover it in class.

Many mathematicians are naturally shy people who find it hard to maintain the self-image of authority during class. *Always remember that you have the superior knowledge and the power to assign grades.* When you see a student learn as the result of your efforts, your own self-confidence in life will benefit. Mine did. On a related subject, you will sometimes have to deal with unruly students. It is hard to give an algorithm for dealing with them; books have been written on the subject. I start with polite requests and try to get the class on my side. If necessary, you can ask the offending party to leave. In case the class as a whole becomes too boisterous, silence is the best weapon. Stand in front of the class, look annoyed, and say nothing. Don't speak until you have regained their attention. Five seconds is usually all it takes.

Gilbert Highet once noted that a teacher must be friendly without becoming a friend. However relaxed your classroom atmosphere, you must retain a distance from your students. You must not let personal relationships interfere with your professional responsibilities. Dating a student is *a priori* an unprofessional act. In cases of sudden true love, I urge patience or a change of section. Universities have begun to enforce rigorously their rules on sexual harassment; avoid any behavior which might be taken the wrong way.

Finally, as a new teacher, you will be bombarded with advice from everybody. You cannot possibly accept it all. Teaching gets easier as you become more experienced. You will learn that many ideas that you think will work won't work, and vice versa. Before too long, you'll be giving other people advice. You've been very good to read this far — I think I'll let you out a few pages early.

MATHEMATICS

* * *

Austin Community College

* * *

FACULTY

COURSE MANUAL

* * *

1988 – 1989

Edition

* * *

AUSTIN COMMUNITY COLLEGE 1973

**produced by the mathematics task force
for college-wide use**

Table of Contents

	General Guideline to Classroom Policy	1
MTH 1423	Intermediate Algebra	12
MTH 1513	Mathematics: Its Spirit and Use	18
MTH 1523	Modern Mathematics I	25
MTH 1533	Modern Mathematics II	28
MTH 1563	Elementary Statistics	31
MTH 1643	Mathematics for Business and Economics	42
MTH 1674	Business Calculus I	46
MTH 1684	Business Calculus II	50
MTH 1743	College Algebra	54
MTH 1753	Trigonometry	60
MTH 1764	Precalculus	66
MTH 1854	Calculus I	70
MTH 1864	Calculus II	
	Fall 1988	73
	Spring 1989	75
MTH 2054	Linear Algebra	78
MTH 2154	Calculus III	
	Fall/Spring 1988-1989	80
	Summer 1989	81
MTH 2164	Differential Equations	84
MTH 2254	Calculus IV (Vector Calculus)	87

This manual in its 1988-1989 version contains information about 17 of the 24 courses currently offered within the mathematics department at Austin Community College. The manual will be revised each summer. Later editions will include information about the remaining courses. With one exception, all of the courses not now in the manual regard vocational/technical mathematics, a curriculum which is currently in a state of change and development.

Mathematics

Course Manual

Austin Community College

with

Notes for Instructors

and

Information for Students

to be used by

Full-Time and Part-Time

Faculty

1988-1989

INTRODUCTION
Stephen Rodi

This manual contains teaching suggestions and other useful information for 17 of the 24 courses offered in the mathematics department at Austin Community College. The material in the manual is written by full-time faculty members in mathematics at Austin Community College. The manual was mandated by the college-wide mathematics Task Force in spring, 1988, for use by all mathematics instructors in all locations of the college.

The manual was put together during the spring and summer, 1988, for use beginning in September, 1988. I assumed the responsibility for editing all the text material and trying to assure that there was commonality among the various course documents as regards issues like attendance, withdrawals, and incompletes. The faculty is greatly indebted to division secretary James Harper for the original typing and numerous rounds of redrafting that went into the final version of the manual.

Users will find that the material for each course occurs in two parts. The first is a short description (generally one to three pages) of "Notes for Instructors." These were written by faculty members for the use of instructors only. These pages are a combination of practical guide for the instructor, including directions of what instructors must do to prepare for a course, as well as teaching hints on specific topics within the course. The second part of course material is a combination course outline/syllabus which is entitled "Information for Students." Our intention is that this course outline will be distributed by the instructor to all students together with an additional page produced by the instructor giving specific information about that instructor's grading, testing, and homework scheme. Indeed, in most instances, the mathematics department will prepare forty copies of the "Information for Students" sheet for the instructor to distribute to the class.

Our intention is to upgrade this manual every summer. Hence, all the documents are currently dated 1988-1989. A complete new version of the manual will be produced in summer, 1989, for the following academic year. Therefore, please be sure to relay to the mathematics faculty any suggestions or ideas you might have for the improvement of this manual. In particular, the mathematics task force (curriculum committee) will review this first edition of the manual during 1988-1989 in order to fine-tune it for future editions.

At the front of the manual, the instructor will find many general observations about teaching mathematics at Austin Community College. Most of these cut across all courses and should be read by all instructors. In the past, much of this information has been given to instructors at the part-time orientation session which precedes each semester. It is our hope that this sort of general organizational information can now be read by all instructors in this manual. This will allow us in the future to try to use part-time orientation sessions for more academic topics that directly pertain to the content and teaching of various courses.

We hope that this manual will continue the long-standing tradition in the mathematics department of unity of purpose and commonality of standards and content among all sections of the same course, no matter where the section is taught within Austin Community College. As the college continues to grow and spread in different locations with multiple administrative structures, this unity (which is critical to our credibility as a mathematics faculty) becomes more and more difficult to achieve. Fortunately, over the years, the combined efforts of full- and part-time faculty in mathematics have given us a great deal of success in this common endeavor. If you are a new part-time faculty member, we welcome you to this enterprise and hope that this manual gives you much support as you begin your work with us.

A General Guide
to
Classroom Policy
for
Mathematics Instructors

1. When you are assigned to teach a mathematics class at Austin Community College, please begin by reading the items in these general guidelines. Later in the manual you will find a short set of "Notes for Instructors" for the course you are teaching. Read these notes next. You also are provided with a one-page handout which contains (1) the name of the text for the course, (2) a syllabus which must be covered, and (3) a suggested timetable in which to cover the material. In most instances, you will be provided with enough copies of this last information page to give one to each student. Some of these information handouts run more than one page, depending on the complexity of the course and the amount of detail the handout includes. Most of these handouts do not contain a testing and grading format for you to follow. Hence, it is your responsibility to produce an additional handout for all students which gives precise details on how you will test and grade your section of the course. This handout, produced by you, also should indicate how you plan to deal with homework and what your standards will be as regards attendance, withdrawals, and incompletes. On these latter three points, there is a departmental policy which in some cases is included in the "Notes for Instructors" and in other cases is included in the "Information for Students." Please follow these departmental policies.

Testing and Grading

2. Your grading-testing schedule should be consistent with the following guidelines. If you wish to deviate significantly from these guidelines, please consult the division chair or department head.

    a. Generally, have between 4 and 6 major tests in a semester.

    b. Incorporate into the course and into the grade some procedure for dealing with homework. A variety of procedures are possible. Here are some ideas.

    - assign homework problems in each class and pick them up next class for "spot" checking
    - ask students to keep a systematic homework notebook with the solutions of homework problems
    - once a week (or more often) give brief homework quizzes in class (perhaps allowing the use of the homework notebooks)
    - occasionally check (perhaps during a major in-class test) and award credit for the homework notebook
    - have the sum total of the homework grades count for a major test

    c. Use of the testing center for some major test can free up additional classroom time for instruction. If you elect to use the testing center, be sure to obtain the testing center regulations and follow them. If tests are given in the testing center, you must have multiple versions of the test available and a procedure worked out for assigning a given student to a particular test. Testing center use is optional. Indeed, it is possible the testing center will become unavailable for regular class tests. You will have to check the most recent testing center policy on this matter. When tests are given in the testing center, class must meet at the scheduled time. Using the testing center cannot replace a class meeting.

    d. Always give the final examination in class during the last class period. (The testing center should not be used in the last week of the semester.) Classes should meet through the end of the semester.

    e. If your grading scheme provides for dropping one grade, never allow the last grade to be dropped. (Some students will simply stop attending class during the last weeks, if you allow the last grade to be dropped.) The same problem can exist if other test grades are dropped. Hence, many instructors prefer not to drop any grades since they want students to study all material equally seriously.

    f. Use a numerical grading range that is consistent with what most other instructors do. Thus, generally, A=90-100; B=80-89; C=70-79; D=60-69.

    g. Some instructors allow re-testing on some or all tests. This is an acceptable, but not required, procedure. If you allow re-testing, it is suggested: (1) that not all students be allowed to re-test, and (2) that there be an upper limit on the grade a student can earn as a result of re-testing so that students who take a test twice are not able to do significantly better than students who take a test only once. (We want to encourage students to do well on the first testing.)

    One possible re-testing procedure is as follow:

    - allow only students who score below 70 on the initial test to re-test with no re-tests on the last test
    - allow only one re-test during the semester
    - allow the maximum possible grade one can earn on the re-test to be 70 (some instructors average the two testings or weight the two tests so that the final grade is the weighted average of the two testings or 70, whichever is less; this encourages good performance on the first test and keeps 70 as a cap)

    This suggestion for re-testing is based on the philosophy that a re-test exists for students who did not earn a minimum C on the first testing and need one additional chance to earn a minimum C grade.

3. Students should receive a timetable of testing during the first class when the grading-testing procedure is distributed.

4. Make it very clear to students that is is the STUDENT'S responsibility to submit the necessary paperwork by the required date in order to withdraw from class and receive a grade of W. Do not assume this responsibility for students as one of your usual activities. You may, or course, withdraw a student who has stopped attending class. College policy gives the instructor the right to withdraw students who are not progressing satisfactorily toward completion of the course objectives as announced at the beginning of the course. You may (but are not required to) withdraw students who simply disappear from class rather than assign these students a grade of F at the end of the semester. But this is different from promising students that under all conditions you will assume responsibility for submitting withdrawal forms. You should include on your first day handout the statement that you "may" withdraw students with more than 4 absences (to emphasize your right to do so). But use the word "may," not the word "will".

5. Students should expect at the end of the semester to receive a grade of A, B, C, D, or F. A grade of I should be given VERY rarely. Generally, a grade of I only should be assigned to students who are otherwise passing a course, have done all required work and have a personal tragedy or accident that prevents them from taking the last examination. Very few students should fall into this category.

6. It is important that you stand by the grading and testing procedure which you announce on the first class day. If disputes about such matters come to the department chair or the division head, among the first questions to be resolved will be if the instructor was acting in accordance with this policy announced for the class. So, be sure to give this policy careful consideration, to state it clearly, and then to maintain it.

7. While the above points on testing and grading may sound somewhat stern, you should also remember that the "spirit" and "style" of the handout you prepare for grading-testing should convey to the students that we are here to be of service to them and are planning all of our efforts to help them maximize their chance of success. Reasonable regulations reasonably expressed only improve those chances.

Syllabus, Attendance, Attrition, Withdrawals

8. You have the right to be sure that all students meet the prerequisites for the course. We attempt during the registration process to make sure that students are not allowed to register who do not meet prerequisites. However, as an open door institution, we do not have the advantage of screening student applications through a pre-admission procedure. Also, it is not unusual for students at registration to give us a rosier picture of their current ability than is accurate. If you discover that students do not have prerequisites, you may withdraw them from the course. Ideally, one should try to make this discovery during the first or second class and direct the student either to late registration or to the department head/division chair's office where class adjustments can be made through the 9th day, if space is available.

9. You are expected to finish the syllabus which is proposed for the course you are teaching. Since in almost all cases mathematics courses are sequential, you have an important responsibility to make sure that students who have passed the course have seen all the material which instructors in succeeding courses will assume the student knows. Hence, students who are not able to keep up with the regular pace of the course should be directed to withdraw and drop back to a preparatory course. The syllabus for the course should not be truncated in order to accommodate less than adequately prepared students.

10. Most of our syllabi provide very little time for "review." It is our assumption that students come prepared for the course they have registered for. As we all know, some brief review is necessary and helpful during a course. Hence, it is recommended that you do not set aside blocks of time specifically for review, but rather "review" necessary material as it becomes appropriate in the context of a course. Thus, for example, one should not spend time at the beginning of a course reviewing all the properties of logarithms, but rather touch on that topic when the logarithmic function is introduced later in the course.

11. Attendance is required in all classes. Instructors should be very reluctant to give students permission not to attend class since it is very unusual for students to perform adequately who are frequently absent from class. The first day handout, which the instructor provides the students, should include the observation that the instructor may drop a student who misses more than 4 classes. In unusual circumstances, you may want to require that a student attend class regularly and successfully pass the first test before you will consider special attendance arrangements.

12. It is not unusual for attrition to be high in mathematics classes. On average, attrition in mathematics runs between 45% and 50%. In some courses, attrition can even exceed 50%. We would like to do everything that we possibly can to reduce the attrition rate. But, for your own peace of mind, you need to know that such attrition occurs regularly at other community colleges and at 4-year institutions. In particular, attrition is influenced by many factors over which you have little control. Among these are the student's prior preparation, too optimistic projections by the student of how much mathematics they remember from previous courses, not enough outside study time scheduled by students (frequently as the result of overly demanding school and work schedules), unrealistic expectations by students of how much outside study time is required in mathematics classes, and poor student attendance. We hope that you will take all reasonable steps to reduce the attrition in your class. But it is possible that high attrition will occur even in the presence of very strong and effective teaching.

13. The inappropriate way to deal with the attrition problem is to reduce our standards. Since so many of our students transfer to four-year colleges and universities, it is important that we maintain our credibility and continue to graduate from our courses students who are prepared to perform successfully wherever they take their next mathematics course. In general, our standards in a particular course should be the same as those of other colleges and universities with whom we interact.

We should be proud if students say "it is easier" at Austin Community College because we have small classes, concerned teachers, good textbooks, walk-in tutoring, faculty available during office hours, and a well structured curriculum. We should be embarrassed (and immediately correct the situation) if students say "it is easier" because we offer significantly less material in a given course than neighboring colleges and universities or because our standards for performance are significantly lower. Maintaining the careful balance between standards and compassion is one of our most important and delicate responsibilities both for the department and for individual instructors.

14. You are no longer required by the college to complete attendance forms. However, it is strongly recommended that you keep a record of class attendance. If you have a student receiving VA (Veterans Administration) benefits, you must notify the Admissions Office as soon as the student has excessive absences. (Such students are indicated by an asterisk on your official class roster.) For other students, an attendance record gives helpful information about the student's general performance.

15. If you drop a student from class, be sure to keep your copy of the official drop form which should have been stamped and dated by the Admissions Office. Occasionally, these forms do not get processed properly. When this happens, it will be necessary for you to "prove" that the drop was processed before the deadline date by producing your copy of the drop form. (When withdrawals are properly processed prior to the deadline date, the student will receive a pre-printed W on the final grade report. If such a pre-printed W does not occur but you dropped the student properly prior to the deadline, you must have a drop form in order to assign the W grade.)

16. Be sure to check the withdrawal date each semester. It is listed in the catalog. As you will see observed frequently in this manual, you should not promise students that you will withdraw them. You may elect to do so when students are not meeting course requirements, but you should not assume this responsibility for students.

17. At the end of the semester, you may not assign a grade of W to a student. (In general, students should expect to receive a grade of A, B, C, D, or F. In very rare cases, as has been previously observed, you might assign a grade of I. See number 5.)

18. At the end of the semester, be sure to note the date by which the Dean's Office will require that you turn in final grades. Please take this deadline very seriously. When you turn in your grades, your department head/division chair will want you to follow a specific procedure. That procedure will be very similar to the one outlined at the end of these guidelines.

Useful Information

19. In almost every instance, at your teaching location, you will have a personal mailbox. Once the semester starts, be sure to check that mailbox regularly (at least every time you come to teach class) for information.

20. Be sure to follow testing center rules and regulations in whatever location you are teaching. It is possible that the testing center will become unavailable for regular class testing. If this should occur, all tests will have to be given in class. If the testing center remains available for regular class testing, you will probably have to submit your examinations at least two days in advance. There are other comments about the testing center in number 2 above.

21. The Parallel Studies Division (PAR) supports mathematics classes in two important ways. "Lab classes" are available to support almost every mathematics course. These "lab classes" are entirely optional but are recommended for many students. Students must sign up for these courses during regular or late registration or by special arrangement in the Parallel Studies Division before the 9th class day. Such classes meet two hours per week in a small group with a tutor available to answer questions. They are sometimes described as regular organized homework sessions. The other service provided by Parallel Studies is walk-in tutoring at the Parallel Studies laboratory. This tutoring is available most hours of most days according to a schedule posted in Parallel Studies.

22. Our mathematics courses form a complicated interwoven network. For more details, consult the chart which immediately follows these guidelines.

23. It is expected that all classes will meet as scheduled. In short, the instructor does not have the authority to "give the students a walk." If you anticipate missing a class, it is your responsibility to obtain a substitute. Please have your department head/division chair approve this substitute. Payment of a substitute can be handled in two ways. There is a formal college procedure for deducting money from your paycheck and transfering it to a substitute. In many instances, it is easier for you and the substitute to work out a personal arrangement of payment or trading classes. Substitutions of this type should be relatively rare during the semester. If you have a last minute emergency which requires that you miss class, please call the division office or the Dean's Office at the campus where you are teaching so that the students can be told that you will be absent. One such cancelled class in a semester will not effect your paycheck. Additional cancelled classes will result in a paycheck deduction.

24. The mathematics faculty from all locations at Austin Community College is organized in a college-wide committee called the Mathematics Task Force. The task force serves as a curriculum and policy committee for all matters mathematical. The task force usually will meet at least once each semester. You should receive a notice of task force meetings and we hope that you will try to attend. With so many part-time faculty members in our system, we find that much of our most valuable experience comes from the part-time faculty.

25. Austin Community College now has a plan which requires many entering students to be assessed. (Some students are exempted from assessment based on their SAT scores, ACT scores, or previous college record.) A student may register for one semester without going through this assessment process but will not be allowed to re-register for a second semester until assessment is completed. These facts are mentioned for your information in case a student ever asks you about it.

END OF SEMESTER DETAILS

1. Instructors at most satellite campuses may turn in their grades at that location, if they choose to. If they do so, they should MAIL THE FOLLOWING VIA CAMPUS MAIL to their department head/division chair:

   (a) the instructor's copy of the grade sheet and any "I" forms
   (b) textbooks, if the instructor is not teaching the same course the following semester
   (c) withdrawal slips described in (e) below

2. ALL OTHER INSTRUCTORS should go to their Divisional Office to turn in the following materials. **Always go to the Division Office first.** Only go to the Dean's Office if the Divisional Office is locked. Turn in the following at the Divisional Office:

   (a) instructor's copy of grade sheet
   (b) textbook and supplement according to guide below
   (c) incomplete form precisely and thoroughly filled out, if you choose to give an I grade (which should be rarely)
   (d) any college keys, if the instructor is not returning
   (e) a properly dated withdrawal slip for any student who does not have a pre-printed W on the grade sheet but who according to your records should have a W. (A grade of W cannot be assigned without such a properly dated and stamped withdrawal form--the form verifies an error by the registrar in not pre-printing a W on the grade sheet.)

   After you have turned in these items at the Division Office, turn in the audit copy of your grades and get your paycheck at the Dean's Office.

3. Instructors not teaching the same course the following semester will be asked to return all textbooks and supplements as a condition of receiving the final paycheck. This should be done at the Division Office.

4. The Dean's Office will set the date and time each semester by which final grades must be turned in.

26. Students are required to attend the class section for which they are registered. A student who is not on your class roll and who does not have an official class change from the Admissions Office should not be allowed to remain in your class. Do not say to students, no matter how much space might appear available in your class, that the student may remain in your class. The individual instructor at ACC does not have the authority to make this promise. Such promises, no matter how well intentioned, only make it harder for division chairs/department heads to do their jobs.

27. No changes of any kind in class registration will be made by the division chair/department head before the end of late registration. Up to that time, the student has direct control of his/her own registration through the normal registration process. From the end of late registration to the 9th class day, if space is available, the division chair/department head can arrange to change students who are in the wrong course (e.g., change from college algebra to intermediate algebra). This process may not be used to change sections of the same course. Occasionally, because of verifiable work changes imposed on a student by an employer after late registration, a student may change sections of a course. This "Informal Class Transfer" must begin with the division chair/department head and is not allowed unless all instructors and division chairs approve. Instructors may not implement such transfers on their own. Generally, even these transfers will not be approved after the first 25% of the course is completed. Finally, such transfers are never approved for reasons of personal convenience involving class time, location, transportation or personal preference about the instructor.

28. Occasionally, every instructor has a discipline problem in class dealing with cheating. When you have a case of scholastic dishonesty, you should obtain from the Student Services Office the appropriate form. Directions of how to proceed are included on that form. In short, you complete the form and impose the penalty which you think is reasonable. If the student signs the form accepting the penalty, the matter is completed. If the student challenges the penalty, a formal hearing committee is called into session to deal with the case. The latter process is complicated and is not used very often. Generally, an instructor can deal with such matters directly with the student. Penalties imposed by instructors range from 0 on the assignment or test to an F in the course, depending on the seriousness of the offense and other factors.

29. If you are a new part-time faculty member, you will have assigned a "mentor" during your first semester of teaching. This is an experienced faculty member (sometimes full-time, sometimes part-time) who is available to you for questions and guidance. Of course, your department head/division chair can also be of assistance. The mentor is paid a small stipend for this work, so you should not be shy about contacting your mentor when you need assistance.

30. Austin Community College also has an active Staff Development Office which can assist you. This office publishes a part-time faculty manual with general information about the college and suggestions about teaching.

MTH 1684
Business Calculus II

Notes for Instructors
Stephen Rodi
1988-1989

Students who register for this course need to have completed a prerequisite course that has treated both differentiation and integration of polynomial, rational, log, and exponential functions. Students without these skills, with a grade of less than C in the preceding course, or who have taken the preceding course more than 18 months ago are not likely to do well in MTH 1684. Such students should be strongly urged to go to late registration and sign up for MTH 1674.

All students in MTH 1684 can benefit from the "Business Calculus II" lab taught in the Parallel Studies Department. If a student has not registered for this lab, suggest the student do so at late registration. The lab meets about 2 hours weekly in small groups of 8-12 to discuss homework problems. Also remind students that walk-in tutoring is available in Parallel Studies.

The text on which the following comments are based is Applied Calculus, third edition, by Barnett and Ziegler, Dellen/Macmillan Publishers.

Chapter 5

The preceding course at ACC ends at Section 5.2 and this course starts at Section 5.3. This is done deliberately to give students some chance to review integration early in MTH 1684. This also allows improper integrals to be taught in the same semester as Chapter 10 (continuous probability theory) where such integrals are used to find cumulative distribution functions.

Chapter 6

This is a long chapter with a great variety of material. Hence, the syllabus suggests a test after Section 6.3 (covering both the Chapter 5 material and the first part of Chapter 6). It is important in 6.1 to have the student understand what a function of two variables is and to give the student a sense of graphing in 3-space since this will be used in 6.2 to interpret the partial derivative. However, do not go into great detail about such graphing. There should be two points emphasized in 6.2: the mechanics of taking partial derivatives and the geometric and practical interpretation of the partial derivative. The figure on page 464 is important.

In Section 6.3, remind the student this extrema test is similar to the second derivative test in one-variable calculus. The Lagrange multiplier technique in 6.4 should not be omitted.

Section 6.5 is one of the most important in the course. It is a nice application of maxima-minima techniques from 6.4 and shows the students (almost all of whom will be required to take a statistics course) why the coefficient formulas in linear regression have the form they do. This is not a statistics course. Hence, the presentation should go beyond just plugging in the formulas on page 495. On a test, I always require students to do a problem which imitates the step by step development of equations (1) and (2) on page 495. In other problems, I recommend they use (1) and (2) which come right from the data rather than the more contorted (3) and (4).

In Sections 6.6 and 6.7, problems involving changing the order of integration help discover if students really understand the various processes.

Chapter 7

Do not overly emphasize the graphing in Section 7.1. You can omit logistic growth in Section 7.2. In 7.3, I think it is far preferable to emphasize the logical development of the use of the integrating factor and discourage students from merely memorizing the box on page 564.

Chapter 8

This material is the least important in the syllabus. If you need to omit a section, here is the place to do it. The point of this chapter is to have students see how a Taylor polynomial at $a=0$ can be used for approximation of a function value (how to get a polynomial to approximate a non-polynomial function). Do not do expansions at $a\neq0$, that is, do Maclaurin expansion at $x=0$ only.

Chapter 10

An extremely important chapter, along with Chapter 6 the other "central" chapter of the course. You need to spend time on finite probability. The binomial distribution with the associated histograms makes a good transition to continuous probability. Students will have a tendency to confuse density functions and cumulative distribution functions. To emphasize the difference, I suggest you always say "cumulative distribution." In the continuous case, be sure to relate the measure of probability to area under a curve.

Supplemental Handout

The focus of this material is geometric linear programming in Section 2.2. Chapter 1 develops various ways of solving systems of equations (a useful topic in itself for students who have never had matrices). These are used later in finding corner points. Chapter 2 shifts the emphasis to systems of inequalities and linear programming.

First Day Handout

In addition to the sheet Information for Students provided for your use, you must prepare for students a first day handout which explains the testing, grading, and homework procedures for your section of this course.

MTH 1684
Business Calculus II

Information for Students
Regular Semester
1988-1989

Text: (1) Applied Calculus, 3rd edition, by Barnett & Ziegler, Dellen/Macmillan Publishers.

(2) Supplemental Handout on Linear Programming (excerpted with permission from Finite Mathematics, 4th edition, by Barnett & Ziegler, Dellen/Macmillan, pp. 85-203). This handout is available from the bookstore for about $3.50.

Week	Chapter	Sections
Week 1	Chapter 5 (Applied Calculus)	Sections 3, 4
Week 2	Chapter 5	Section 1
	Chapter 6	Section 1
Week 3	Chapter 6	Sections 2, 3; Test
Week 4	Chapter 6	Sections 4, 5
Week 5	Chapter 6	Section 5
Week 6	Chapter 6	Section 7; Test
Week 7	Chapter 7	Sections 1, 2
Week 7	Chapter 7	Sections 3 (Omit Logistic)
Week 8	Chapter 8	Sections 1, 2
Week 9	Chapter 8	Sections 3, 4; Test
Week 10	Chapter 10	Sections 1, 2
Week 11	Chapter 10	Sections 3, 4 (Omit Beta)
Week 12	Chapter 10	Section 5; Test
Week 13	Chapter 1 (Special Supplement)	Sections 1, 2, 3, 4
Week 14	Chapter 1	Sections 5, 6, 7
Week 15	Chapter 1	
Week 16	Chapter 2	Section 2; Test

Notes:

1. Business applications will be emphasized throughout the texts.

2. Instructors may introduce supplemental materials as needed to enhance and clarify topics covered in the text.

3. One possible testing scheme instructors may use has been suggested above. Your instructor may use a different scheme. Your instructor will give you an additional handout with details of testing, grading, and homework for your section of this course.

4. Attendance is required in this course. Students who miss more than 4 classes may be withdrawn. After the withdrawal date each semester, neither the student nor the instructor may initiate a withdrawal. It is the student's responsibility to initiate all withdrawals in this course. The instructor may withdraw students for excessive absences (4) or failure to meet course objectives but makes no commitment to do this for the student.

5. Incomplete grades (I) will be given only in very rare circumstances. Generally, to receive a grade of I, a student must have taken all examinations, be passing, and have a personal tragedy occur within the final three weeks of the course which prevents course completion.

MTH 1684
Business Calculus II

Information for Students
Summer Session
1988-1989

Text: (1) Applied Calculus, 3rd edition, by Barnett & Ziegler, Dellen/Macmillan Publishers.

(2) Supplemental Handout on Linear Programming (excerpted with permission from Finite Mathematics, 4th edition, by Barnett & Ziegler, Dellen/Macmillan, pp. 85-203). This handout is available from the bookstore for about $3.50.

Chapter 5	Sections 3-5	1 week
Chapter 6	Sections 1-7	3 weeks
Chapter 7	Sections 1-3 (Omit Logistic)	1.5 weeks
Chapter 8	Sections 1-4	1.5 weeks
Chapter 10	Sections 1-5 (Omit Beta)	2.5 weeks
Supplemental Handout	Sections 1-7 & 1-2	1.5 weeks
Tests	After Chapter 6 Section 3	
	After Chapter 6 Section 7	
	After Chapter 8 Section 4	
	After Chapter 10 Section 5	
	At the end of the Supplement	

Notes:

1. Business applications will be emphasized throughout the texts.

2. Instructors may introduce supplemental materials as needed to enhance and clarify topics covered in the text.

3. One possible testing scheme instructors may use has been suggested above. Your instructor may use a different scheme. Your instructor will give you an additional handout with details on testing, grading, and homework for your section of this course.

4. Attendance is required in this course. Students who miss more than 4 classes may be withdrawn. After the withdrawal date each semester, neither the student nor the instructor may initiate a withdrawal. It is the student's responsibility to initiate all withdrawals in this course. The instructor may withdraw students for excessive absences (4) or failure to meet course objectives but makes no commitment to do this for the student.

5. Incomplete grades (I) will be given only in very rare circumstances. Generally, to receive a grade of I, a student must have taken all examinations, be passing, and have a personal tragedy occur within the final three weeks of the course which prevents course completion.

BREVARD COMMUNITY COLLEGE

Faculty Handbook for Part-Time Instructors

August 1986

A PLACE TO GROW

TABLE OF CONTENTS

	Page
INTRODUCTION	i
WHAT IS INSTRUCTION?	1
A. Preparing for the Class	2
B. Meeting the First Class	5
C. Availability of Students	8
D. Student Evaluation	8
E. Methods of Instruction	11
F. Student Evaluation of Instruction	16
PERSONNEL SERVICES	18
Application	18
Contracts	18
BUSINESS SERVICES	18
Pay	18
Time Sheets and Attendance Reports	19
Legal Assistance	19
INSTRUCTIONAL SERVICES	19
Information and/or Assistance	19
Desk Copies of Textbooks	19
Class Cancellations	19
Schedule Changes and Substitute Teachers	20
Class Hours and Breaks	20
Class Rolls and Attendance	20
Teaching Aids and Equipment	21
Faculty Seminars	21
STUDENT SERVICES	21
Student Supplies	21
Emergencies - Student	22
Emergencies Other Than Student	22
Withdrawals	23
Final Roll and Grades	24
PHYSICAL PLANT	24
Buildings and Grounds	24
Security	24
Keys	24
APPENDIX 1 - ADMINISTRATIVE WITHDRAWALS - VA STUDENTS	25
APPENDIX 2 - INSTRUCTIONAL PERSONNEL CHECKLIST, PART-TIME	26
APPENDIX 3 - SCHEDULED PAY DATES FOR CONTACT HOUR INSTRUCTORS	27
SCHEDULED PAY DATES (FULL TIME AND PART-TIME CREDIT HOUR INSTRUCTORS)	28
APPENDIX 4 - TELEPHONE NUMBERS/GENERAL DIALING INFORMATION	29
APPENDIX 5 - 1986-87 ACADEMIC CALENDAR	30
APPENDIX 6 - STUDENT EVALUATION OF INSTRUCTION	32

ii

INNOVATION ABSTRACTS VOL. IX NO. 14

Published by the National Institute for Staff and Organizational Development
With support from the W. K. Kellogg Foundation and Sid W. Richardson Foundation

FACULTY MENTORS: NEW ROLES, NEW RELATIONSHIPS

Part-time faculty are a valuable resource for two-year colleges. Their growing numbers and responsibilities at Austin Community College led us to look for ways to facilitate their orientation and adaptation to the role of community college instructors. With these thoughts in mind, we designed and implemented the Mentor Program.

Typically, an experienced faculty member acts as a mentor to new part-time instructors. However, other faculty may be assigned to the program: e.g., an instructor teaching a course for the first time, new full-time instructors, or instructors who need additional assistance in the delivery of instruction.

The problem this program addresses is significant. Most participants assigned to a mentor are new part-time instructors. For most part-time faculty, especially those teaching at night and at off-campus centers, contact with other ACC staff is very limited. Instructors may, through an entire semester, have contact only with Division Chairpersons, Department Heads, Campus Deans and Site Managers. Contact with other faculty, especially those teaching in the same department, may be virtually non-existent. It is this concern, as well as others, that led to the implementation of the Mentor Program.

The Mentor Program facilitates the orientation and adaptation of new faculty members by teaming them with an experienced faculty member, from the same department or division, in a *non-supervisory* relationship. Faculty mentors are full-time instructors or experienced part-time instructors, with special knowledge or skills, willing to assume this important support role.

The mentoring relationship is designed to extend over one semester and is characterized by flexibility and mutually agreed upon objectives. The Mentor Program is *not* supervisory or evaluative. Guidelines for the program include (1) a planning meeting during which time the new faculty member and his/her mentor establish mutual agreement on the objectives of the program and (2) a work plan for accomplishing them. The objectives should reflect the following outcomes:

1.   understanding the administrative requirements of his/her job (i.e., grade and attendance reports);
2.   discussing the use of a course syllabus;
3.   learning about instructional resources (i.e., LRS, Student Services, Faculty Development);
4.   modeling a positive example of the teaching role for the new instructor;
5.   becoming familiar with departmental student evaluation systems.

In addition to setting objectives, a schedule for additional contacts is agreed upon. While each relationship will be different and is structured to meet the needs of the individuals involved, there are certain suggested program elements: e.g., visits to observe each other's classes, periodic telephone conferences, other personal conferences, tours of college facilities, and scheduled meetings with other staff.

Mentoring brings added duties and responsibilities to these selected ACC faculty, both part-time and full-time. A nominal honorarium is offered to faculty mentors. And while it certainly does not repay the time and effort mentors invest, it does symbolize the value the institution places on the mentor's willingness to act in this capacity.

While there is always room for improvement, the Mentor Program has proven to be one of increasing value to our growing institution. It helps a multi-campus, multi-site college function with ever-increasing numbers of part-time faculty. Program policies demonstrate concern for the new faculty member's professional life and offer recognition to the faculty who serve as linchpins for the success of the program— the mentors.

Mimi Valek
Instructional & Staff Development

For further information, contact the author at Austin Community College, P.O. Box 2285, Austin, TX 78768.

TABLE OF CONTENTS

A WORD OR TWO. 1

Departmental Administration. 2

Handout to MA 111 Students 3

Comments on: "Handout to MA 111 Students" . .5

    Recordkeeping 6

    Tutorials 6

    Using the MA 111 Supplementary Sheets . . .6

    General Suggestion. 7

    A. Preparation for Class. 7

    B. In the Classroom 7

    C. Office hours 8

    D. Testing 8

    E. Determining the Final Average 9

    F. Questions for the Failing Student . .9

FOR THE NEWEST MA 111 TA

on the

HARRELSON CIRCLE

Marilyn S. McCollum
Department of Mathematics
North Carolina State University

Acknowledgments

    I would like to express my appreciation to the following persons for their suggestions and comments related to this booklet:

    Dr. J. B. Wilson

    Dr. Jerry Pietenpol

    Professor R. G. Savage

    Ms. R. L. Cranfill

    Appreciation is also extended to Christie Lehmann who wrote Everything You Ever Wanted to Know About Being a TA, But Didn't Know Who to Ask, which I used as my guide for this handout.

Marilyn McCollum

A Word or Two to the New TA. . . . .

In a typical year, more than half of the approximately fifteen thousand registrations in mathematics courses are for 100-level, first-year courses, and more than half of the teaching of those courses is done by Graduate Teaching Assistants (TA's). Moreover, nearly sixty sections for courses above the 100-level are taught each year as lecture sections (approximately 90 students), to which a TA is assigned to assist the lecturer by keeping records, having office hours to help students, and grading quizzes and examinations. So it is that, whether you are assigned to teach a course or to assist a lecturer, you are an essential and valued part of our teaching force and have an important share in this department's commitment to effective teaching in every classroom.

In pursuit of that effective teaching, I urge you to --

-- study and make good use of all the materials and advice which have been prepared to assist you with teaching techniques and with understanding important aspects of the course you are assigned. No matter how much any of us knows about a course, we are very unwise to enter a classroom without careful preparation and expect to teach a good lesson.

-- enter into our teaching program and into each day's teaching with enthusiasm. Enthusiasm is contagious, and a motivated student is apt to be a good student. On the other hand, if we appear bored, we cannot expect our students to be motivated or to think that this course is important.

-- discuss your teaching and grading -- and any attendant problems-- with colleagues and faculty members. Senior TA's in your office, your course coordinator (Ms. McCollum for MA 111), the faculty counselor assigned to you (or the lecturer you might be assigned to assist), and I are specific persons to whom I hope you will come freely for this purpose.

-- be assured that your part in this big task is seen by the Mathematics Department as very important and will be appreciated.

D. E. Garoutte
Director of Undergraduate Instruction

V. General Suggestions (These suggestions come from years of experience and many mistakes!)

A. Suggestions in Preparation for Class:

1. Practice talking through your lesson with a tape-recorder. Listen to yourself. Check your mathematical vocabulary, your organization, your "um's",etc.

2. Work the assigned homework before class or you might get embarrassed.

3. Ask questions of colleagues or faculty members when you need help. We have all been there.

B. Suggestions in the Classroom:

1. Be careful using the words "always" and "never" in mathematical statements.

2. If a student asks you a question that you are not sure of the answer, say "Let me think about that and I'll tell you tomorrow". And do tell them the next day.

3. Do not waste class time working a student's problem that was not a class assignment. Tell the student you will be glad to work it during your office hours.

4. Organize board work. Also, write out important comments. Student hear what you say but do not record it unless you write it on the board.

5. "Remain calm". If a student is disrupting class, ask if he/she has a question or a comment; if the student is a chronic disturber, tell him/her to see you in your office, and there explain that the disturbance must stop or he/she will have to be removed.

6. Keep class moving. Free time may allow for an unruly class.

7. Do not try to teach while students are talking. Explain that they are wasting valuable time.

8. Your students are adults, not children. Expect good work from them and you will get it. Be professional.

9. At the beginning of class, write down the H.W. assignment (which includes the textbook reading) and briefly state the objectives.

10. Class is not cancelled the day before holidays or breaks. You are expected to follow the syllabus. Expect your students to be there and they will come.

11. Attitude makes the biggest difference. "Convince" your students that they are having fun. Be enthusiastic and show them that you _care_ about their learning this material.

12. Keep on schedule.

13. Don't try to impress the students with what you know. Impress them by being a good teacher.

14. Don't get off the subject.

15. Try to return major test within 2 days; pop quizzes within 1.

16. Begin and end class on time. There is 50 minutes of work to be presented. Students like to leave class early until they fail the course. Then, they will blame their failures on your "short" lectures. Remember, give the student his/her money's worth.

17. _Never accuse your students of not studying._

C. Office Hours

1. Encourage students to organize (or make a list of) their questions before they come for help.

2. Don't work H.W. questions for the student who has not put out the effort. You can best help the student by asking him/her the questions that lead to solving the problem.

D. Testing:

1. Be "fair and firm" grading test. You are not doing your students any favor "giving" them grades. They will be over their heads the next semester and recovery will be much harder.

2. When you return a major test, work one or two of the questions that _many_ students missed. After that, tell your students to correct the test themselves using their class examples and homework. If they can't find their mistakes, they should come during office hours and you _guide_ them through the problem, not work it for them.

3. If a student does poorly on a test, write a note on the test requesting them to come by your office to discuss the test.

4. Always work the test before you give it to your students. This will alert you to overlooked typing errors or to vague questions.

E. Suggested factors to weigh in determining whether or not to give the 1/2 (or less) point needed on the final average in order to receive the next higher letter grades:

1. Did the student attend class regularly?

2. Did the student actively participate in class (i.e. work the inclass practice problems, ask questions, show enthusiasm, etc)?

3. Did the student come during office hours with specific questions that indicated his/her hard work?

4. Did the student turn in homework (if required) that indicated a real effort (not just answers)?

5. Were 4 of the 5 major test grades consistent?

If you get 3 or more "yes'", the student probably deserves the 1/2 point. Remember, you have already "inflated" the average by dropping the lowest of the first four major tests.

F. Suggested questions you might ask a student who is failing (or not doing as well as he/she wishes) but insists that he/she is working hard:

1. Do you attend class regularly?

2. Do you seek help regularly during office hours?

3. Do you rework (not reread) the inclass examples before you attempt your homework?

4. Do you read the text and work (not read) the illustrated examples?

5. Do you view the video taped lectures in the Audio-Visual Tutorial Center?

6. Do you ask the "tutor" on duty in the Audio-Visual Tutorial Center to help lead you through the problems, not work them for you?

7. Do you concentrate at least 1 to 1 1/2 hours per day outside of class on MA 111?

8. Do you have an outside job? If so, how many hours per week do you work?

If the responses are positive to the above questions, the student probably needs to change into MA 115.

HANDOUT TO MA 111 STUDENTS
NORTH CAROLINA STATE UNIVERSITY
DEPARTMENT OF MATHEMATICS

1. Instructor   : _____
   Office   : _____
   Office Hours : _____
   Telephone Number : _____

2. Textbook: <u>Algebra and Trigonometry with Applications</u>, 2nd ed., Munem and Foulis.
   Text Coverage: 1.1 - 1.7: 2.1 - 2.7: 3.1 - 3.9: 4.1 - 4.6;
                    5.1 - 5.5: 6.1 - 6.6: 7.1 - 7.7 (omit 7.5);
                    8.1 - 8.3: 9.1: 9.6: 9:7: 11.2

3. Classes will meet Monday, Tuesday, Wednesday, Thursday, Friday.

4. Major test days: _____

   _____

5. Final exam day and hour: _____

   Final Average: 60% Major test average

                   30% Final Exam

                   10% Homework and/or Pop quiz average

   A student who has 5 or fewer absences may drop the lowest of the five major test grades in determining the major test average. A student who has more than 5 absences may not drop any major test grades. <u>No distinction is made between excused and unexcused absences.</u>

6. Grading scale: $90 \leq A \leq 100$
                   $80 \leq B < 90$
                   $70 \leq C < 80$ No Curves!
                   $60 \leq D < 70$ No retest!
                   $0 \leq NC < 60$

7. No exam exemptions.

8. Calculators are allowed, but no instructions will be given.

9. "Make-up" test policy: Approval for making up a missed test should be given prior to the test. THE MAKE-UP TEST MUST BE TAKEN WITH THE <u>TWO</u> WEEKS FOLLOWING THE SCHEDULED TEST, assuming that the absence from the test is excused.

10. Homework policy: _____

    _____

11. Pop quiz policy: _____

    _____

12. Absences: <u>Attendance is expected in all classes unless extreme circumstances.</u> No distinction will be made between excused and unexcused absences.

13. "Corrections to the instructor's grading": If the student feels that an error was made in the grading of the test, she/he should explain the error on the outside of the test (or attach another sheet) and return to the instructor within <u>3</u> class periods after the test is returned.

(OVER)

I. Comments on "Handout to MA 111 Student"

Below are suggested options from which you may choose your policie homework, pop quizzes, and attendance. Once you decide on your policy, WRITE it on the board for your class and STICK TO IT. You may change it the next semester.

A. #10 Homework Policy [Students should plan on 7 - 9 hours of homework each week]. OPTIONS:

   1. Take up and grade all.

   2. Take up and grade only 2 or 3 selected problems.

   3. Take up none and make homework the student's responsibility.

B. #11 Pop quiz policy [No make-up ever of any pop quizzes] OPTIONS:

   1. [If you elect to not take up any homework, this policy is strongly recommended.] Give 12 pop quizzes during the semester and drop the two lowest grades. They are generally graded on the basis of 10 points (i.e., 9@95, 10100, 220). Questions for pop quizzes (generally 2 or 3 total questions) are exact homework problems and/or examples worked in class.

   2. Periodic 10 minute pop quizzes, no definite number given.

   3. Give no pop quizzes.

*Note: If you elect Homework Policy A.3, you will need to elect Pop Quiz policy B.1. If you elect Pop Quiz policy B.2 or B.3, you will need to elect either Homework Policy A.1 or a.2.

II. Record Keeping

A. Take roll everyday. At the end of the semester, you will be asked to turn in each student's attendance report along with their grade. A seating chart is very helpful.

B. Keep pop quiz scores and/or homework grades; and major test grades two places:

   one set in your grade book and one set at home. Keep your grade book with you at all times.

C. (Optional) Keep a record of the students who come for help during office hours.

III. Where you students may seek "free" help:

A. A student should always seek help first from his instructor during office hours.

B. The Mathematics Audio-Visual Tutorial Center - HA 244.

C. The Learning Assistance Center - Poe 528.

IV. How to use the MA 111 Supplementary Sheet (options)

A. Have your students work and turn in the day before "Review Day" problems from the MA 111 Supplementary Sheet that are applicable to the upcoming test. Grade 2 - 3 "selected" problems and record the grade as either pop quiz or homework. Return the graded work on "Review Day".

B. Begin each class with a 10 minute review of the previous lesson. Select the problems for the review from the MA 111 Supplementary Sheet and allow the students of his/her mastery of the previous lesson and alerts you to topics that need further explanations. Make sure that all supplementary questions applicable to the up-coming MA 111 test have been worked before the major test.

HELPFUL HINTS TO GOOD TEACHING

Department of Mathematics
University of Wisconsin
MADISON

Original Editions by
Leroy J. Dickey and Kenneth M. Hunter 1966, 1967
Revised and Expanded by Kim Bruce 1974
Updated by Charles Steinhorn 1977
Updated by Steve Bauman 1983

TABLE OF CONTENTS

0.   Introduction

I.   Departmental Resources

II.   Good Beginnings

III.   You and Your Students

IV.   Teaching Techniques

V.   Teaching Routine Facts and Skills

VI.   Blackboard Techniques

VII.   Lecture - Discussion Courses

VIII. Lecture-Recitation Courses

IX.   Teaching Your Own Course

X.   Homework Assignments & Quizzes

XI.   Exams, Grading & Cheating

XII.   Feedback

XIII. New TAs

XIV.   Coordinators for New TAs

XV.   Postscript: How Not To Win Friends and Influence People

XVI.   Suggesting Reading

XVII. Footnotes

Appendix A: "Some Techniques for the Mathematics Classroom" (Reprint of a Monthly article).

Appendix B: A Primer on Dealing with Some Problems That Your Students Might Have.

INTRODUCTION

The purpose of this pamphlet is to provide some ideas and encourage you to think about how to be a good teacher. Contained inside are many ideas and techniques that other teachers have tried and found useful. Rather than trying to be a how-to book on good teaching, this pamphlet presents (sometimes contradictory) suggestions that you might want to consider. No one will agree with everything that is suggested. Depending on how you see your role as a teacher, how you remember feeling about teachers as an undergraduate, and your own personal style, you will probably elect to use some of the ideas presented and reject others. No pamphlet or book could possibly list all the different ideas and techniques different teachers have used successfully and we make no pretense of doing that here. For example, no mention is made of the discovery method of teaching math. If this pamphlet encourages you to think about the different aspects of teaching, it will have been successful. The authors and the math department would be grateful for any comments, criticisms, or suggestions concerning this pamphlet.

I. DEPARTMENTAL RESOURCES

Every TA will be given a list of teaching supplies available. These include textbooks, pens, pencils, paper, ditto supplies, etc. Aside from these everyday things you will need, there is a collection of supplementary texts in room 214 that you may want to look at to get examples of different approaches to topics, good examples, and extra problems to put on quizzes or exams. Two other things that you should be aware of are the video-tape recorder and the availability of math films. The video-tape recorder can be used to record your teaching and then be played back later to give you a good idea of what your teaching looks and sounds like to your students. More about this in section IX. A catalog of films available through the university is located in room 214. Be sure to preview the film yourself or talk to someone else who has used it before you show it to your class. It can be very embarrassing to show a fourth grade level film to a Trig class of freshmen and sophomores. All you have to do is call BAVI at 2-1644 to arrange to preview a film. If you decide you want to order it you must get the approval of the faculty supervisor of your course and then order it through Sherry in room 214 (the math department must pay for the film). A Warning: start this whole procedure at least a week before you actually want to use the film! The bureaucracy moves slowly and the film you want may be reserved for the day you need it.

Reminder: Be sure to get a building permit from the receptionist. Otherwise you may be kicked out of the building by the Protection and Security guard if you are in the building after normal business hours.

II. GOOD BEGINNINGS

Getting off to a good start can save you a lot of trouble later on in the course. The first thing you need to do is become familiar with the book you are going to use. If you are aware of what is coming you'll be able to prepare your students by knowing what to emphasize and can alert them to what to expect. If possible get hold of a syllabus for the course--use one from the previous semester if you can't get a current one. You may find that you don't like the way the book handles certain topics and may want to change the order of or combine some topics. While you're teaching you should keep in mind that it is almost impossible to keep exactly within a syllabus. Its purpose is merely to show about where you should be at any given time so that all classes will have covered the same material at exam time.

By the first day of class you should have a pretty good idea of how you want to run your section, how often you will give quizzes, how much (if any) of the grade will be based on homework, quizzes, class participation, 6- and 12-week exams, and the final exam. See section XI for some ideas on this.

The first day of class you should give your students an idea of who you are, what you expect of them, and what they should expect to learn in the course. Write your name, office number, and phone numbers on the board. (You would be surprised at the number of students who have no idea what their TA's name is.) Put your tentative office hours on the board (you'll probably want to modify them slightly once you find out your students' schedules), tell them about your grading system and other policies such as on make-up exams. You may also want to tell your students a little about yourself.

Different teachers like to do different things on the first day of class, some possibilities are listed below:

(1) Give the students a rough schedule of the topics to be covered in the course. Often this is just a week-by-week summary of the syllabus.

(2) Have each student fill out a 3x5 card with their name, phone number, major, reason for taking the course, math background, etc. This will help you get to know your students' backgrounds and needs. The phone number is also very useful in tracking down students who miss exams or just drop out of sight.

(3) List some of the types of problems you expect them to be able to solve at the end of the course. This will give your students some idea of what they are working toward.

(4) Use the first class period to tell students about applications of the material they will be studying. It should be explained that the material you and they will be studying is worthwhile even though in some courses, such as matrix theory, some applications cannot be fully understood until the last few weeks of the course. In such ways students (especially engineering majors) can be prevented from giving up "because mathematics is too abstract."

Students are impressed by a teaching assistant who gets to know them quickly. You should take roll at least until you know them all. Returning assignments to individual students at the beginning of each class period, although time consuming, is one way of giving students some personal attention. During the first day of class, when roll call is taken, you may wish to note where each student is seated. At the next class meeting most of the students will be in approximately the same location.

Students often show an initial burst of enthusiasm. The first two weeks may best be used to dive deeply and quickly into a topic (1) which is new to most of them, (2) which seems useful to them, and (3) which uses a new skill (or new skill). The initial material is usually learned well, so a topic that will frequently re-appear through the course should be taught first. It is an unfortunate practice to teach conic sections at the beginning of a second-semester calculus course.

Most teachers seem to feel that beginnings must be slow. If the initial work is too slow, there will not be adequate time for more difficult material at the end of the course. Rapid beginnings will accustom the student to giving a relatively great amount of her time to your course (most of her other courses will begin at a slow leisurely pace). If enough difficult material is covered in the first few weeks, the 6-week exam can actually indicate how well the student is doing (so that the poor students will be able to drop the course).

III. YOU AND YOUR STUDENTS

Probably the most important trait of an outstanding teacher is the ability to respect the student as the student respects her. Sometimes it seems that educators look down on the student and condemn her for being so stupid. But the distinguished educator brings herself down to the student's level and then helps the student-teacher relationship. It is this teacher who succeeds in teaching everyone, not just the smart pupils (who can understand the material without the teacher's help anyway!).

These teachers always seem to gain more respect instead of less respect as many people would think. If a teacher could respect the student as the student respects the teacher, the relationship between them would result in the students tending to work harder, to learn faster, and to get better grades.

Exercise. Enumerate traits (good and bad) of teachers you have had. Try to determine what makes "the supreme teacher."

The greatest favor you can do for your students is to put yourself in their place and try to see the problems from their point of view. To do this you need to know a little bit about the kind of students you will have in your sections. First of all you should understand that you will rarely have a prospective math major in any of your sections. This is especially true in courses numbered 212 and below, but still holds true for the 221 calculus sequence. There are only about 200 undergraduate math majors at the UW compared to the 9000 students who take math courses each semester, so don't expect your students to act like budding mathematicians. In fact in the courses numbered 212 and below only a minority of the students are in sciences or engineering. As you begin to know your students and what they are and are not capable of you'll begin to be able to anticipate their problems and concentrate your efforts in these areas. One of the most difficult (and probably unsolvable) problems in teaching is "Who do you teach for?" If you teach for the brightest students you will lose the rest of your class. Somehow you will have to find a middle ground between these two extremes and yet still try to hold the interest of your brightest students and not lose completely the poorest students.

Many students seem to have mental blocks and have convinced themselves that they cannot do mathematics. Part of this may be the result of the fact that mathematics is so sequential. To understand what you are doing today you must understand what was presented yesterday, last month, and even last year. If a major topic is not learned,

the student will not understand most of what follows that topic. In the lower level courses there is a lot of review of what was supposed to be learned in high school. At this time you can try to patch up the gaps in the students' backgrounds. Encourage your students that if they fill in the gaps in their knowledge and keep up, they will be able to do mathematics and it will make sense. Anything you can do to encourage their success will really help. Some teachers have the attitude that you should "let the students win one" every once in a while. That is, give a quiz or exam that even your worst students can pass. For some students this taste of success will encourage them to work harder and succeed in the course. However, don't make your first quiz or exam too easy. Your students will feel that not too much is expected from them and will put in less time and effort.

One advantage that TAs have over older more experienced teachers is that they are closer in age and experience to the undergraduates. You can use this fact to your advantage if you choose to. First of all, having just received your undergraduate degree you probably remember the things that frustrated and pleased you about your life and teachers as an undergraduate. Your students have many of the same sort of problems and appreciate the same sort of things you did. (Keep in mind though that this appreciation does not extend to seeing mathematical proofs.) Sharing things that happened to you as an undergraduate makes you a more human and sympathetic person to your students. Another thing you can exploit is the fact that you too are a student. Let your students know about this and let them in on the way your classes are going and you'll find that soon they are sharing their problems and complaints about teachers and classes with you. This can be very enlightening, especially if you are open to comparing their criticisms of other teachers with the way you are teaching your class.

Try to set up office hours that will be convenient for as many of your students as possible and let them know that they can make appointments to see you at other times if they can't make it during office hours! You should plan on holding office hours at least three hours a week and be there when you say you will. A common complaint of students is TAs who are never around during office hours. If you have each of your students will out a schedule of when they are free you can try to pick hours that most of your students will be able to come to at least once a week. As a general rule avoid picking all three period in the same time slot, e.g. 11:00 MWF. Strongly encourage your students to come to office hours. Many students are afraid to go visit their teachers, so let them know they really expect them to come. (Note: You may wish to tell your class that you may be up in the 9th floor lounge during office hours. This way you will have a little more freedom of movement.)

Admit that you are human. Tell them you expect to make mistakes in writing on the board or saying the wrong thing and you would appreciate it if they would correct your errors. Do try to keep these mistakes to a minimum though, since if you are consistently making mistakes you will anger and frustrate your students and they will soon tune out.

Some other ideas that you might want to consider are listed below:

1. a) Call your students by their last names.
    b) Call your students by their first names and have them call you by your first name.

(At any rate call them by some name—any teacher who doesn't know her student's names by midterm doesn't show much concern for students.)

2. Encourage your students to work on homework together or help each other with their problems.

3. Make your students sit near the front of the room. Many of the rooms in Van Vleck hold 40 or more people and it is not unusual to have all the students sitting in the last two or three rows, with three empty rows between them and the teacher. It's hard to get your students to participate in the class if they are sitting 20 feet away.

4. Ask the students that do poorly on quizzes or exams to come talk to you outside of class. Often you can help these students understand the material they were supposed to know. Otherwise they may just give up and not even look over the exam except to see the grade at the top.

IV. TEACHING TECHNIQUES

You may have heard the old adage that a teacher should explain everything three times: First tell your students what you are going to tell them, tell them, and then tell them what you told them. This is still true and can be very effective if done right. This does not mean you should say the same thing three times, far from it. One way to do this might be to tell your students what sort of problem they are going to learn to solve and generally how you will approach it, then explain the new theory or technique to your students and work through an example. Finally go back and review the technique and how it applied to the example. In other words the idea is to let your students know what to look for in the new material, then explain it to them, and finally review it so that they all understand it and how it applies. (Note: We've now told you three times.)

Many teachers seem to forget that students know how to read. You may wish to lecture on everything mentioned in the book, but there is an alternative you might want to try instead. Have your students read in advance the material you are going to discuss each day. Then concentrate your efforts on the main points of the material. Go over the most important parts carefully and work examples, encouraging your students to ask questions on anything they don't understand. You should probably at least mention the other points covered in the text but you need not waste the class time copying definitions from the book and repeating things they have already read. Instead you'll have more time to answer questions, work through examples, and work on the more difficult points. Before you rely too heavily on a text, though, make sure your students can read it. In some cases the text may be hard to read and in others the students may not know how to read mathematics. In the latter case you may want to give your students some tips on reading mathematical texts (e.g. read with a pencil in your hand and a pad of paper by your side.).

Try to begin each class with a short discussion of data with which the students already feel comfortable, rather than plunging directly into new material. A brief resume of the previous lecture will help to provide continuity. By association of the new material with the old, the student begins to have solid contact with the new material; it seems real and stable. When presenting a new theorem, explain how it fits into the general theory of the course. A common complaint among students is that they understand this or that piece of information, but they are just not sure what it means.

A very useful teaching technique is to ask frequent questions of individual students. This has many advantages. When introducing new material it can stimulate the student's intuition. When reviewing or using material already covered, it can make the student aware of what she should know; it also gives the teacher the opportunity to rework a concept if the student is not sure of the answer. Time spent in this way is seldom wasted since there will probably be others with the same problem and it lets the students know what material you think is important. Another advantage is that it keeps students awake. Call on all students. If you call on just the bright ones, you will discourage the poor students. If you call on just the slow ones, you will give a false sense of security to the average students.

One view of teaching that has been suggested is that the teacher's primary function is not to teach facts, but to teach the students how to understand and remember the material. This does not mean you should spend your time thinking up mnemonic devices. Rather show your students how you think about the material, how it all fits together, and why it makes sense so that they don't just have to memorize facts but instead see that the facts fit together in a structure. Similarly when you are working problems don't just write down the solution. Explain (on their level) why you thought of doing each step and why it works. Thinking out loud as you work through problems can really help your students see how to attack a problem.

Encourage your students to ask questions. Sometimes it is very difficult to get a class to open up and ask questions or even respond to your questions. In this case you may have to work extra hard at first and encourage them to speak up. Generally if you get a class to open up at the beginning, they will stay that way the rest of the semester. Similarly once they get used to just sitting back and not asking or answering questions, you'll have a hard time ever getting a response from them. Never condemn a student for asking a stupid question. Once you stomp on a student in front of the class she may never ask another question for fear of embarrassment. Also when a student asks a question, repeat it so the rest of the class can hear it. The chances are that several other students have the same question.

Needless to say you should always be well prepared before you enter the classroom. One of the qualities students in low level math courses look for in a teachers is good preparation and organization in lectures. This is not to say that you should write out your lecture in advance (although the first few times you teach you may want to rehearse your lecture before class). Have a good outline of your lecture to take to class with you. It's very easy to accidentally skip saying something very important during a lecture, so have your outline ready to refer to. Work out examples and all pictures before you get to class so you won't have to play with them at the board to make them come out right.

On the other hand, don't read your lecture to the class. Know the material well enough so you only need to refer to your notes to know what topic you want to cover next or what example you are going to put up. Keep good eye contact with your students—don't always look at the blackboard or over your students' heads. Let your eyes rove over the class, reading your students' expressions so you will know what they are or are not understanding. With a little practice you will be able to know when they are confused, understand what you are doing, or don't care. You can then adjust your presentation to try and compensate.

To keep your students' interest in the subject you must be interested yourself. Care about your students and the fact that they are trying to learn. Figure out reasons why the material is worthwhile to them and show them. Don't be afraid to be excited about what you are teaching. Interest and excitement are contagious—almost as contagious as dullness and boredom. Don't be afraid to act. An actor uses her body and voice to communicate a feeling or idea. You can do the same thing on a lower level. You are trying to communicate something to your students so use every resource at your disposal. Use your voice to help you. There is nothing more boring than listening to a droning monotone voice for an hour. Raise your voice a little when you are making a point and then lower it to let the point sink in. Use your imagination and try to bring your students into what you are doing.

V. TEACHING ROUTINE FACTS AND SKILLS

Teaching Routine Skills

In any math course there are a number of routine problems which the student should learn to carry out. For example in an algebra course this might include solving quadratic equations; in calculus, differentiating polynomials; in matrix theory, finding the rank of a matrix, and so forth. Even if you prove that these techniques work, you should precisely and clearly state the algorithm and illustrate its use with several examples. Homework problems testing the students' ability to perform a new routine skill should always be given. Even the poorest student should learn to solve routine problems in a routine way—even if she doesn't understand why the technique works (this will give her the satisfaction of being able to do something in the course). You should personally make sure that everyone in the course learns to do the routine problems. The ability to perform routine manipulations sometimes gives a hopelessly lost student such a feeling of accomplishment that she may even try to catch up in the course.

Exercise. List the routine skills which all your students—even those who fail the course—will be able to carry out when the course is completed.

Teaching Routine Facts

It may be worthwhile to give a brief pre-announced quiz at the end of each week asking the students to reproduce exactly the important pieces of mathematical knowledge that you have indicated for the week (see section 6). Even the poorest student will thus learn some important basic facts which you have proved in class. Such quizzes are not a grading tool, but a teaching tool. Through them you can be sure that your students have a certain minimal knowledge (e.g., the integrals of some frequently encountered functions).

Exercise. List the routine facts which all your students—even those who fail the course—will know when the course is completed.

VI. ESSENTIAL BLACKBOARD TECHNIQUE

It is most important that you begin writing at the top of one blackboard panel, move down that panel, and then proceed to the next panel. You should not skip around the blackboard, placing equations haphazardly here and ther. Talk while writing. Some people suggest that if you are right-handed, you should begin at the rightmost front panel. When this panel is full, move to the next panel already filled (from which some students may still be taking notes) while a new panel is being written on. A gymnastic alternative is to face the class while writing. If you have more than two rows of students in the class do not write all the way to the bottom of the board because those in back will not be able to see. Keep the desk at the front of the class clear of any large object such as a lectern or a briefcase. Some teachers use an eraser to simplify a complicated mathematical expression. This technique is guaranteed to annoy students taking notes.

You should try to put complete statements on the board, not just a few symbols. Your students' notes usually will consist almost entirely of what you have put on the blackboard.[1] If a student doesn't understand some point presented in class, she would be able to turn to her class notes and recover the missing information. When you are solving a problem at the blackboard, you should put up a complete statement of the problem[2] to indicate where it can be found in the text. Define in writing on the board any variables used in the solution. Work the problem to completion so that your students will learn to recognize a solution. Important results that you want your students to memorize may be emphasized by outlining them in chalk. An example would be the equations for the derivatives of the trigonometric functions. A word of caution: do not make your course a memory course. Emphasize the importance of thinking a problem through from first principles or known results.

In using the triple-layer blackboards in the large lecture halls, use the middle-layer board, sandwiched between the front and back layers first. When done, push it up and pull the front board down, so that what was just written is still in sight. After using the front board, push it up, and use the back board (the front board will still be visible). Then move to the next triple-layer panel. One may wish to prove the proposition: There is no suitable technique for using all n layers of an n-layer blackboard if $n > 3$.

VII. LECTURE-DISCUSSION COURSES

A lecture-discussion course is a course with divided teaching responsibility. It has a lecturer who meets with the entire class (from 100 to 300 students) several days a week and a number of assistants who meet with small sections of the class (of 10 to 20 students each) at other times.[3]

When you teach a discussion section in a large lecture-discussion course, your primary responsibilities are to clarify and give examples of material already presented in the lecture and to do problems that have been assigned. It is relatively rare that you will actually be presenting brand new material in your section. Therefore most of your time will be divided up between working examples to illustrate the material in the lectures, answering questions on the lectures and homework, and sending students to the board to work problems (either individually or in groups). It

1. For as you undoubtedly know from experience, note-taking is a difficult skill to master. At best, many of your students will have acquired only the rudiments of this skill. To best serve the students, then, you must attempt to develop careful blackboard technique.

2. (Including the section of the text from which it comes and its number).

3. This is the situation in which most of you will find yourselves during your first semester of teaching.

is important that you attend the lectures in your course. Your students will be asking questions directly on what the professor said in lecture. Unless you were there and know what your students are talking about, you won't be much help to them.

You will meet very early in the course—maybe even before the first lecture—with the lecturer to decide on important course policies such as the number of exams to be given, the assignment of homework (i.e., whether the lecturer will assign problems to be the whole class or whether each assistant will assign problems to her own sections), and emphasis of the exams (such as mostly theory or mostly problems, etc.). Your lecturer will provide you with a syllabus for the course so that you will be able to adjust your pace in discussion sections to stay abreast of the lectures.[4]

Your lecturer will want to be informed about the problems your students encounter in the course. You are often the person best qualified to pass on this information. Your lecturer may want to visit your discussion section and must let you know in advance. At least one lecturer has been known to give assistants a vacation by taking their sections one at a time so that he can get direct student feedback that is impossible in a large lecture.

You will probably work with the lecturer in making up the examinations for the course and if you are lucky, she will work with you in grading them. The exams are usually group graded; you might grade problem number one on every exam while someone else grades problems two, and so on. The marking goes very quickly this way and is potentially the most uniform.[5]

VIII.   LECTURE-RECITATION COURSES

All of the courses in the remedial and intermediate algebra sequence are lecture-recitation courses. These courses are structured exactly as the lecture-discussion courses with respect to student allocation and contact hours. However, TA responsibilities in these courses are different.

As a TA in a lecture-recitation course, you are not expected to attend lectures or hold office hours. You do not have to make-up or grade exams, nor do you give quizzes and homework. You are expected to proctor exams (usually one per week), help in staffing a tutorial room (two hours per week in lieu of office hours), and conduct recitation sessions.

Recitation sessions are regularly scheduled periods when your students meet to go over problems from problem sheets that are given out during the lectures. Your duties during a recitation session are to assist the students in doing those problems. You are encouraged to have students solve problems at the board either individually or in small groups. During the recitation sessions you are also expected to go over the exams that your students have taken.

IX.   TEACHING YOUR OWN COURSE

In this type of course you have almost full responsibility for your sections. Your students will never see another teacher, and the only time all the students will be in the same room is for a major exam. This places much more of a burden on you as a teacher as well as offering you more freedom in what you will be doing. All the TAs for the course will get together to decide on a syllabus, create the 6-, 12-weeks and final exams, and decide policy on make-ups. It is up to you to decide on policies for attendance, homework, frequency of quizzes, and how the classroom part of your students' grades will be determined (see section XI on grading). Perhaps the hardest thing to get a feel for is the proper balance between lecturing, answering questions, and doing examples and homework problems. You will find that it is usually very easy to spend the whole hour lecturing. Unfortunately this probably is the worst thing you can do, and is almost guaranteed to put your students to sleep. Take advantage of the fact that you are teaching a small class of about 20 students. Communicate with them rather than talking at them. A typical class might proceed as follows:

1. Give a brief summary of what you did the previous session and perhaps work a typical example of homework problems. This should lead you directly into:

2. Answering questions about the previous day's material and the homework due that day. You may wish to send students to the board to work the problems others are having difficulty with. Encourage questions—your students will learn as much or more from your answers to their questions as from your lectures.

3. Once everyone (or almost everyone) has a good feeling for what has gone before, you are ready to dig into the new material. In fact if you are clever, your answers to their questions will sometimes lead right into the new material. For some topics (limits, continuity, convergence), everyone understanding of a topic is not essential for the things to come, don't waste too much time trying to get everyone to understand. Otherwise you will not have enough time available to cover the remaining material in the course. For ideas on how to present this new material see Section IV on teaching techniques.

4. Interspersed with your discussion of the new material you should have been giving lots of good examples. If you finish the discussion of the new material early you might want to work some typical homework problems or you might want to send the students to the board in groups of 2 or 3 to work on their homework.

Don't feel that you have to fill up every minute of your assigned 50 minute class period. Your students will resent you if they feel you're just trying to fill up the time. Let them out early if you have nothing to say or do.

It is almost impossible to do all of the things listed above in a 50 minute class period so you'll have to make some decisions on exactly what you want to try to fit into the class period. Get suggestions from your students on what is the most useful to them. Also be flexible in what you plan to say so that you can adjust your presentation if you are running out of time.

Teaching your own class can be a lot more rewarding than assisting in a large lecture-discussion course. It is here that you have the most potential for either expanding or destroying a student's knowledge and enjoyment of mathematics.

4. In most cases, though, a lecturer will allow you to conduct your own sections essentially as you please. This includes responsibility for deciding whether or not to collect homework, what type (if any) and how often you plan to give quizzes, etc. It is best to make these decisions in advance of your first class, so that when you first meet with your students you can inform them of what they may expect in terms of work during the semester.

5. (In the past, however, lecturers have provided detailed answer keys and partial credit suggestions, and had assistants grade their own exams. This experiment proved to be successful although it should be pointed out that most of those involved felt that some experience at grading on the part of the assistants is a necessary prerequisite for success.)

X. HOMEWORK ASSIGNMENTS

Some TAs give no assignments at all on the assumption that students will work on their own and ask questions about those things that they do not understand. Such a TA has probably forgotten her own undergraduate career. She is baffled when her students do not ask questions because she does not realize that most students do not understand enough to ask even the most basic questions.

One approach is to give short assignments daily, due the next class meeting, and to be returned by you at the next meeting after that. This method has certain advantages:

1. You remain aware of how your students are doing.

2. Short assignments can be corrected relatively quickly.

3. The student gets fast reinforcement, before she has lost interest. She knows continuously how well she is doing.

4. The assignments are not so long as to discourage students from studying the course material.

5. The student becomes accustomed to systematic regular study for the course (so that she won't wait until the day before the exam to begin studying).

6. The student gains the satisfaction of being able to do something for the course.

7. Returning assignments directly to your students helps you to learn their names.

8. The student finds out what sorts of problems she is expected to learn to solve in the course.

9. The student has a set of solved problems to review for the exam.

Some variations on this idea are:

1. Give a somewhat long (1/2 hour to hour) assignment but only have the students turn in two or three of the problems. This has the advantage that you can assign a mix of easy and hard problems so that all students will be able to successfully work at least some of the problems before they get stuck. Be prepared to answer questions on all of the problems, but you will only have to grade a small number of them.

2. At the beginning of each class period have each student turn in a slip of paper which contains a list of problems she was able to solve, those she tried to solve but failed, and those not attempted. Keep track in the grade book how well each student claims to be doing, and use the slips to call students to the board to work problems. This can work as a check on the student from claiming more than she knows.

3. Don't collect homework but give quizzes on the homework once a week. One devotee claims two advantages: First, it eliminates any advantage a student might get out of copying; Second, it lets the student look at his own work during discussion of homework with no danger of cheating.6

4. You might also consider suggesting interesting (perhaps unrelated) problems that some of your best students can work on.

It is usually a good idea to at least see how to solve all the problems before you assign them. Sometimes what looks like an easy problem is really very difficult. Also know how to solve every problem that you have assigned before you go to class the next day. One of the most frequent student complaints is that it took the TA half an hour to work a homework problem.

When you grade homework (or quizzes or exams for that matter) make your corrections helpful. You don't have to write out the entire solution, but show your students where they went wrong and which direction they should have gone. On complicated problems outline the way to reach a solution.

XI. EXAMINATIONS, GRADING & CHEATING

Many teachers use "blue books" for examinations. The advantage is being sure that there are no crib marks on the paper before the exam. Students generally dislike blue books and an alternative is to ditto examinations with one or two questions per page so that there is enough room for working the problem on the test paper. This can waste paper and trees if carelessly done. It will aid you in grading because you will not have to flip through many pages hunting for solutions. If you are teaching in a large lecture-discussion course, have a place for each student to put her name and her discussion leader's name on each page. The exams can then be separated so that each assistant can grade one problem on every test paper. (If you are unlucky enough to get a student with large fourth grade handwriting you can always supply her with extra paper.)

After the first examination have those students who have made D's and F's come to your office for individual consultation. Tell them that they are not doing as well as their classmates. Find out why. Most students will respond very favorably if they know that more is expected of them. If it is clear that the student is not prepared for the course perhaps she should drop the course. Ideally this situation should not arise because such a student should have been identified in the first two or three weeks by her performance on homework. This early identification gives the unprepared student the chance to drop back to a lower level course.

Your students will usually appreciate it if you give quizzes between the major exams. This can work to your benefit as well as theirs. Your students will have an opportunity to see how well they really know the material before taking an exam which may count 1/4 or more of their grade. In this way they will be able to identify those items they need help with and go back and review as well as ask questions. Quizzes can help you by showing how much of what you said, your students actually understood. If a large number of students miss the same problems you will probably want to go back and cover the corresponding material again. There is some question on whether quizzes should count toward the student's grade, and if so, how much. If the quizzes don't count the students can approach them more calmly and use them as the diagnostic tool

6. If you prefer not to collect homework (e.g., in a lecture-discussion course you might spend a substantial amount of class time answering questions about homework problems assigned by the lecturer, and thus might not require students to hand in the assigned problems), there are basically two varieties of quizzes that you might try. On the one hand, you can administer in-class quizzes of varying length, emphasis and difficulty. A variation on this standard would be to allow students the opportunity to recover points lost by permitting them to write up, at home, problems done incorrectly and handing them in at the beginning of the next class period. A problem thus done correctly might be worth say, half of the points lost originally. This gives an added incentive to learn techniques which they have not mastered completely. On the other hand, of course, you might try take-home quizzes. Many TAs prefer to give these since they serve to de-emphasize the "performance under pressure" situation of in-class exams or quizzes, and also permit the introduction of material to which a student otherwise might not have exposure.

they are meant to be. On the other hand, you may have a student who does well and then for some reason does poorly on the exam. The same sort of considerations apply to grading homework. One popular solution to this problem is telling students that you will make up a "classroom" grade based on homework, quizzes, and class participation. Based on this grade you can bump their final grade up or down half a point if they are on the borderline between two grades. Most TAs use this classroom grade only to raise grades, working on the premise that a student should not be penalized for learning from her mistakes on a quiz or homework. Other TAs have counted quizzes directly in the final grade, sometimes throwing out the worst two or three scores.

Your students will find it very helpful if you spend a little time in class reviewing before a major exam. Things you might do include:

(1) Giving the students a list of topics that will appear on the exam.

(2) Giving the students a set of problems to work on of the same type as those that will appear on the test. This is most often done by dittoing off a copy of the previous semester's exam. Your students will find this most helpful if you also provide a list of answers so they can check their work. Of course if you do this you must be sure not to repeat any problems from the exam on the new one. Also, in some courses, old exams are available to undergraduates in room 511 Memorial Union.

(3) Hold a review session, either in or out of class, where the students can ask questions about all the material covered in the exam. If you hold this outside of class time you must make a room reservation with one of the secretaries in room 214. Do this at least 48 hours before you need to announce the room number to your students.

Most major exams are made up jointly by all the TAs teaching the course. This has the advantage that it is easy to make up a massive curve from all students taking the exam. The TAs with the help of the supervising faculty member then decide the dividing lines for the grades. This usually turns out to be a relatively fair scale since the sample of students may be as large as 1000 or more. With all the grades together it is easier to decide whether an exam was particularly hard or easy, or whether it was just your students who did well or poorly. A curve based upon only 20 or 30 students is almost inherently unfair if it doesn't take into consideration that there may be an unusually high number of very good or very poor students. This kind of grading also tends to encourage the throat-slitting kind of competition that is all too common among students. It might be argued that the large scale curve encourages the same sort of competition, but in this case it is still possible for a large number of students in your classes to get A's without having to have an equal number of F's in your class to balance these out. If you feel very strongly about giving your own 6- or 12-week exams, this can sometimes be arranged with the faculty supervisor. However don't announce it to your students as fact before making sure it is possible.

As far as giving out final grades, no one enjoys giving students D's or F's and some semesters your students may all have learned enough so that you can pass them all. On the other hand most of the time you will have a couple of students who for one reason or another just haven't learned enough mathematics to pass the course. In

this case it's not fair to the student for you to tell her through a passing grade that she knows enough mathematics to take the next course in the sequence. By the way, because of the change in grading procedures, some students may feel it is more advantageous to get an F rather than a D, since if they take the course over the F will not count in their grade-point average. So don't be too surprised if you receive such a request. You should be aware that UW now uses a grading scale including the two intermediate grades AB and BC. Therefore the possible grades are A, AB, B, BC, C, D, and F.

When giving an examination there are some simple precautions that you can take to insure that what the student hands in is her own work. Stay in the room. Use alternate seating. Watch the students. Your students will not respect you if you give an implicit invitation to cheating. One teacher has suggested that if you suspect a student of cheating from a neighbor simply go over and whisper to the student that she should move over a seat to where she won't be as crowded. Another technique that has been tried is that if you notice a group of students cheating on a quiz, let them continue, and then announce the next class period that you noticed a group of them cheating, know who they are, and the next time you will call them on it. This has the possible advantage of worrying other students who may have been cheating without your noticing.

Whatever you do, be sure not to give your students the impression that you expect them to cheat.

If you suspect cheating by copying on an examination and cannot prove it, you can:

a) Have the students involved come to your office.

b) State clearly that you are not accusing them of cheating.

c) Have them observe the unusual similarity of their papers.

d) Inform them that you are keeping their exams on file to become "Exhibit A" if further evidence is found.

If you have a case of cheating and you think you can prove it, consult with the person in charge of the course you are teaching. The University has a definite policy on such matters. Some think that a minimum penalty for cheating should be failure in the course.

XII. FEEDBACK

It is through feedback that you can find out what you are doing right and what you are doing wrong. Feedback can come in many different ways. Some of them are described below.

(1) Perhaps the most obvious form of gaining feedback is through watching your students. Notice the way your students respond in class. Do they have a glazed look in their eyes when you are talking? Do they groan when you erase the board? Does

TA performance. The TA and undergraduate members of this committee are elected annually near the beginning of each school year. Please encourage your students to sign up for this committee, and if you are interested please sign up yourself. Near the end of the first semester the TA Review Committee asks all TAs to hand out the departmental evaluation forms to their students and then return these forms to the committee. The results are summarized and the TA Review Committee analyzes the data. The Review Committee uses the results of the questionnaires as a rough tool to classify teaching performance in four categories: outstanding, very good, satisfactory, and unsatisfactory. The vast majority of TAs fall into the satisfactory range while only about 10% fall into the outstanding category. The groupings are not considered to be extremely accurate and it is possible that a TA may go from a very good rating to satisfactory and then back again in successive evaluation. What the committee is really interested in are the two extremes of the range. Those at the top of the range may be nominated for campus-wide teaching awards or picked to be sponsors for new TAs, while those near the bottom will be called in to talk about their teaching. The committee tries to work with and help these TAs to improve their teaching. The committee has the authority to recommend to the department that a TA's assistantship be terminated but in the past has preferred to work with those TAs having problems and bring their teaching up to a satisfactory level.

XIII. NEW TAs

Learning how to teach is like learning how to do anything else. It might be compared to learning how to play tennis after having watched tennis matches on TV for years and then having read a book about it. You'd know a good teacher if you saw one but it takes practice before you become good at it yourself. You're bound to make some mistakes when you start but if you try to be aware of them and correct them as you notice them, it won't be long before you improve. See Section XII on feedback for some ways to become aware of your problems. The greatest teachers are those who continually work to improve their teaching.

Some suggestions for your first semester as a TA follow:

(1) Don't be too ambitious at first. Try to pick up the basics of good teaching before trying something radically different. Once you pick up these basic skills you will have plenty of time for experimentation.

(2) Work on developing your own style. Your favorite teacher may have a different personality than you do and what is highly successful for one person may be disastrous for someone else. Find a style that is comfortable for you and then work on improving it.

(3) Don't try making major modifications in the course you are teaching the first time around. Students find it much easier to follow a TA who sticks close to the book, and you may find later on that there was a reason for doing things in a certain order. It is also very tempting at times to throw in a lot of extra material that you may be interested in. This can be disastrous unless you know what you are doing. Your students also will resent you later when they find this material is not going to appear on the exams. This does not mean that you shouldn't bring in interesting examples as applications of the material you are teaching. Just don't introduce too much new theory.

their performance on quizzes show they are not understanding what you are saying? Keep an eye on your class while you are teaching and note their responses. They can tell you a lot about your teaching.

(2) Your students know what they do and do not like about your teaching and the way you are running the course. Ask them! Many TAs like to set aside a time after a couple of weeks to ask their students for suggestions on how to improve their teaching or the course. You should also make it clear that if they have any complaints at any time you would be glad to hear them. When you ask your students for suggestions, ask them some leading questions: "Do I stand in front of my writing?", "Are there too many quizzes?", "Is my lecturing clear?", "Why do you think you had trouble on the last exam? Is there anything I can do to help?"; for otherwise they won't have much to say. Another way to get feedback from your students is to hand out a questionnaire.

The math department has a questionnaire available in 214 Van Vleck or you may want to make up your own. If you do this by the middle of the semester you'll have a chance to correct any problems before it is too late to do any good. The TA Review Committee will ask you to hand out a questionnaire near the end of the fall semester. More about that in (5).

(3) Have other TAs or faculty members come into your class and observe you, and you go into their classes and observe them. Taken in a spirit of friendly constructive criticism this can be one of the best forms of feedback you can get. Other teachers can tell you things about your teaching that your students may not be able to tell you. Visiting classes run by other TAs or faculty members can give you ideas on how to teach your classes and examples of how other teachers communicate with their students (as well as serving as examples of things you may wish to avoid).

You can learn to be a pretty fair tennis player without ever watching players who are better than you, but you are likely to improve faster if you watch carefully the way they play and emulate their best points (keeping in mind that different players have different styles and what is useful for one is not necessarily useful for another). In the same way by watching better (and worse) teachers you will find some traits that you wish to emulate and others you wish to avoid.

It has often been said that teaching is an art, not a science, and many new (and old) teachers use this as a justification for not working on basic skills. You should keep in mind though, that there are very few great artists of any kind who have not become great by many years of hard work perfecting their talents as well as by studying the works of other artists who preceded them.

(4) The math department has a video tape recorder and monitor so that you can tape yourself in class and play it back later at your leisure. Use this tool! You may think you know what you look and sound like but it is always a surprise to see how you come across. Watching yourself on camera you can detect basic faults like standing in front of what you are writing, repeating yourself, using the same phrase 20 times per lecture, and so on. Besides, this is your big chance to see yourself on TV.

(5) On a slightly different tack, there is a departmental TA Review Committee made up of two faculty members, two TAs and two undergraduates whose task is to review

(4) Don't overestimate your students. The math background required to be admitted to UW is two years of high school math and many students don't have much more math background than that.

(5) Keep a teaching log. Jot down problems that you are having or things in your course that you now see should have been done differently. This log can serve two purposes: (a) it will keep you aware of problems you are having and keep you thinking about them, and (b) it can be a big help if you ever teach the course again.

(6) You can't avoid giving a bad lecture occasionally or getting stuck on a question. If you can't figure out the answer to a question relatively quickly, tell your students you'll figure it out and tell them tomorrow. As for bad lectures, just try to keep in mind this motto that appeared on a teacher's door: "NEVER HAVE TWO BAD ONES IN A ROW."

Don't take items (1) and (3) to mean that you should never try anything new. Once you are comfortable with teaching and have some good ideas for new ways of doing things, think them out carefully, talk to other people about them, and if they still seem workable go ahead and try them.

XIV. COORDINATORS FOR NEW TAs

Each new TA will be assigned an experienced TA who will serve as a coordinator or advisor. Each coordinator will be assigned approximately five new TAs. The coordinator's function is to help the new TA in her first semester of teaching. This help will take several forms, including many of the following:

(1) Meeting with new TAs during orientation to talk with them about teaching, to familiarize them with the department, and to arrange for the new TAs to give practice lectures before classes actually begin.

(2) Meeting with new TAs occasionally during the semester to talk about teaching problems. These meetings will take place at least every two weeks at the beginning of the semester and then taper off toward the end of the semester.

(3) Arranging for new TAs to visit the classes of other TAs and professors to get an idea of different teaching styles.

(4) Visiting the classes of the new TAs occasionally during the semester in order to observe and give constructive criticism. The coordinator will also encourage new TAs to visit each other's classrooms for the same purpose.

(5) Arranging for new TAs to videotape their sections at least once during the semester.

(6) Being available to discuss any problems that may come up at any time during the semester.

XV. POSTSCRIPT: HOW NOT TO WIN FRIENDS AND INFLUENCE PEOPLE

You may have noticed that only feminine personal pronouns were used in this pamphlet. While this may have seemed unusual to you, the previous version of this pamphlet used only masculine pronouns. It is very easy to alienate people by making them feel that you are not talking to or about them. If you are talking about people you are talking about men and women. Therefore it is wise to include an equal number of each as the characters that appear in problems or on exams.

In a similar vein don't slant all your examples toward physical science majors. try to hit the interest of all of your students at least once during the semester. As a result they will feel they have something to gain from the course and show more enthusiasm.

XVI. SUGGESTED READING

Professors As Teachers by Kenneth Eble
    See especially chapters 6 & 10.

Mathematical Discovery by George Polya (two volumes). Published 1962 and 1965 by John Wiley and Sons, Inc. See especially chapter 14 in Vol. 2.

Teaching Tips: A Guide Book for the College Teacher by Wm. J. McKeachitz (Livington, Mass: D.C. Heath & Co., 1969). Very practical, down-to-earth advice.

Improving College Teaching edited by C.B.T. Lee (Washington, D.C.: American Council on Education, 1967). See especially the articles titled: "The Future of Teaching," "Who Teaches the Teachers," "University Teaching and Excellence," "Innovations in College Teaching," "Research in Teaching: The Gap Between Theory and Practice," "College Teachers and Teaching: A Student's View," and "Less Teaching, More Conversation."

Do You Teach? Views on College Teaching by Hugh R. Skilling (N.Y.: Holt, Rinehart and Winston, Inc., 1969). Grossly sexist in parts but contains several good ideas. See especially chapters 4-11, 14, 15, and 21. Chapter 22 is interesting for historical reasons. Chapter 7 contains a lecture by George Polya.

The Importance of Teaching, A Memorandum To the New College Teacher, a report of the Committee on Undergraduate Teaching (New Haven, Conn.: The Hazed Foundation, 1969).

The Art of Teaching, by Gilbert Highet.

APPENDIX A: Some Techniques for the Mathematics Classroom
by Alan H. Schoenfeld

This note summarizes a longer paper with the same title, available upon request from the author. Briefly, the article addresses the following problem: that there are numerous obstacles between the dedicated teacher and his or her students in the classroom. The following is a list of some techniques my colleagues and I have found valuable for surmounting (or circumventing) those barriers.

A. Class evaluations. In the third or fourth week of the term you might give your students ten minutes to write a paragraph or two about their views of the class: pace and clarity of lectures, blackboard techniques, homework load, etc., are all fair game. At the next class meeting you might respond to these, explaining procedures they challenged and announcing changes (if any). At the very least, the students are convinced of your good intentions and you find out about dissatisfaction (if there is any) before it's too late.

B. Alternate class structure. If the material is appropriate or if you planned to do some review, you can hand out a list of problems, have the class break into small groups (say 4 or 5 students each) and have the groups work individually on the problems for half of the class hour while you circulate among the groups acting as a consultant. In the second half of the hour you can present brief solutions and tie the material together. This approach both prevents boredom (everybody participates) and stimulates participation.

C. Student-written class notes. At each class meeting one student is designated "official notetaker." That person's responsibility is to write up a complete set of class notes for that day, perhaps in consultation with you. The notes are then copies and distributed to the class. Freed from the responsibility of taking their own notes, all but one student can participate more easily in class; you obtain a complete set of notes and the opportunity to see a different side of the students.

D. In-class "experiments." Unlike the physical sciences, we do not have laboratories, but we can still perform "experiments" of our own. The simultaneous solution of the equations

$$y'' = -g; \quad y' = V_0\sin\theta; \quad y(0) = 0$$
$$x'' = 0; \quad x' = V_0\cos\theta; \quad x(0) = 0$$

and the effects of varying θ on the solutions can be demonstrated by tossing a piece of chalk through the classroom. a book crashing to the floor contradicts Zeno's paradox; Möbius strips fascinate students; and so on.

E. Quizzes without trauma. You might give brief, unannounced quizzes, to be taken anonymously. Examining the papers tells you what is or isn't getting across, at no "cost" to the students.

F. The class dummy. If you want participation and aren't getting it, you might assign or ask for a volunteer "dummy of the day." That person's job it to ask all the questions the others are afraid or ashamed to ask. When the students see that you welcome questions and respond reasonably, they start asking their own; and the "dummy" can be retired.

G. "Programmed" spontaneity. If you're doing routine problems, you can have the class participate in their construction. In a calculus class, for example, I ask the students to "give me some numbers." After they've called out "seven," "twelve," "pi," and "six point three," I ask them to differentiate $(7x^{12} + \pi)^{63}$. The break in routine can be entertaining.

H. "Treats." We're mathematicians because we like mathematics; our favorite problems or game can entertain students as well as ourselves. There are scores of mathematical amusements; why not share them with students if you have the time? (A particular favorite of mine is described in the larger paper.)

Seasame Group in Science and Mathematics Education, c/o Physics Department, University of California, Berkeley, Berkeley, CA 94720.

Reprinted from The American Mathematical Monthly, January, 1977.

APPENDIX B: On Dealing With Your Students' Problems: Academic and Otherwise.

At best, this section offers a basic common sense guide as to how you might assist students with problems. If a student has severe academic problems, or seems to be under great stress the easiest thing to do is simply to arrange to confer with the student. If you cannot make any real progress on your own, you might then consult with faculty members (e.g., lecturers or course supervisors) to assist you in suggesting an appropriate course of action to your student. Should you need further assistance:

(i) for academic problems, contact the Letters and Science Dean's Office, 104 South Hall, 262-2644.

(ii) for problems of a personal nature, the Dean of Students' Office, Bascom Hall, 263-5700.

N.b., detailed information regarding these and other services may be found in The Wheat and The Chaff: A Campus Sourcebook, which is available, if not from the secretarial staff in the Math Department, through the Dean of Students' Office, 117 Bascom Hall.

COMMENTS ON "HOW NOT TO TEACH MATHEMATICS - AN EXAMPLE"

The viewer might find it profitable to spend some time enumerating the many shortcomings of "How Not to Teach Mathematics." Aside from facilitating critical thinking about the teaching of mathematics, this activity could also produce items for consideration not covered in the discussion which follows.

Suppose we divide our criticisms into three categories - those pertaining to DELIVERY, those pertaining to CONTENT and those pertaining to MANAGEMENT. The flaws in the instructor's delivery are painfully and even comically evident. The flaws in the content of the presentation are more difficult to identify, especially for the student, who cannot know what issues the instructor needs to be addressing. The flaws in the instructor's management of his class are evident and generally result from this instructor's having shirked some of his responsibilites rather than from his flawed judgement. Let us discuss each of the categories is some detail.

Here are a number of items needing improvement in the delivery.

THE INSTRUCTOR CONSTANTLY STANDS IN FRONT OF HIS WORK, obscuring it from the view of the students. It is necessary to stand aside repeatedly to allow the students to assess the material presented and to make notes. IF THE INSTRUCTOR REMAINS CONSTANTLY AWARE THAT HIS OR HER OBJECTIVE IS TO COMMUNICATE THE MATERIAL to the students, he

or she will probably do this from instinct.

THE BLACKBOARD ORGANIZATION IS POOR. Try to organize your work in a systematic and readable fashion. Fragmented calculations are difficult to relate to the overall development. Whenever possible THINK AND WORK IN COMPLETE STATEMENTS AND EQUATIONS and encourage your students to do the same. The eye should be able to pick out the flow of a long computation without undue attention to details.

THE WRITING IS FAINT AND SLOPPY. Not only is faint writing difficult for students to read, it bespeaks a lack of assertiveness on the part of the instructor with regard to his subject matter. SHOW YOUR STUDENTS BY YOUR EXAMPLE WHAT PROFESSIONALISM AND CRAFTSMANSHIP MEAN WITHIN THE DISCIPLINE OF MATHEMATICS. If you give them a sense that you are in complete control of the subject matter you not only heighten their confidence in you but you contribute to their perception that mathematics can be mastered and comfortably applied, thereby increasing their confidence in their ability to cope with the subject matter.

THE INSTRUCTOR HAS ANNOYING AND DISTRACTING MANNERISMS. How many times did the instructor ask ""OK?" Undoubtedly you have had instructors who habitually insert an "..uh" or a "you know" into their speech. The "cool" language of the students is also out of place in the classroom. ALL OF YOUR ATTENTION AND ALL OF THE STUDENTS' ATTENTIONS MUST BE FOCUSED MOST OF THE TIME ON THE SERIOUS BUSINESS OF LEARNING. You need to be a confident leader in this activity, and it is not constructive to dilute this role by undue familiarity with the students either in your behavior towards them or in your choice of language. Did you notice that the instructor allowed his student to address him as "BOB?" How would you deal with this behavior?

THE INSTRUCTOR'S SPEECH IS MONOTONE AND FREQUENTLY INAUDIBLE. To repeat a point already made, REMAIN CONSTANTLY AWARE THAT YOUR OBJECTIVE IS TO COMMUNICATE THE MATERIAL to the students and not just to "hold class."

THE INSTRUCTOR ERASES THE BLACKBOARD TOO RAPIDLY. STUDENTS NEED TIME TO COPY MATERIALS FROM THE BOARD AND THEY NEED TIME TO ASSIMILATE THE MATERIAL. Of these two activities, the latter is far more important. In an elementary course, most if not all of the material covered will be available to the students in nearly identical form in their textbooks. While many students do not seem to recognize this fact, it is counter-productive for the student to devote all of his attentions to mindless copying of the materials presented. This ties in strongly with the next criticism.

THE INSTRUCTOR'S INTERACTION WITH THE CLASS IS INADEQUATE. The instructor seems to be talking to the blackboard or to himself. The entire unwritten message of his presentation seems to be that the class meeting is a scheduled chore to be discharged as expediently as possible. The instructor appears to be fairly disinterested in what is happening. His lack of interaction communicates to the students

that their participation is of minimal importance. It also reinforces a like attitude on the part of those students who couldn't care less about the material but just want to know what questions will be asked and how the instructor wants them answered. Even in his choice of content, which we shall discuss below, the instructor belittles both the students and the material under discussion. THE INSTRUCTOR NEEDS TO MAINTAIN CONTINUOUS CONTACT WITH THE STUDENTS TO KNOW WHETHER HE IS BEING UNDERSTOOD. The students need to know that whether or not they understand is of significance to the instructor and they need to feel comfortable about asking for clarification when it is needed. And perhaps most importantly of all, the instructor needs to be more than someone who reads the material from the book or from his notes. HE OR SHE CAN FACILITATE THE STUDENT'S ASSIMILATION AND GRASP OF THE SUBJECT MATTER THROUGH HIS OR HER INTERACTIONS WITH THEM. Confront the class from time to time with a key question. Let them have a chance to try a problem so that they can discover their questions or miscomprehensions. Repeat those points that need to be stressed. Make sure they are aware of what you are setting out to do in any explanation, share your thinking with them in the course of the explanation and then tell them what you have done when you have finished with the explanation.

THE INSTRUCTOR DOES NOT GIVE ADEQUATE EMPHASIS TO THE ASSIGNMENT. It is a good idea to WRITE THE ASSIGNMENT ON THE BOARD. At the very least it should have been repeated until there were no questions as to what was expected. This instructor didn't even bother to ask if anyone had any questions about the assignment.

Now let's turn to the content of the presentation. It is probably in this area that the greatest disservice is performed to the student and to the teaching profession. Here is where a teacher can most easily conceal his inadequacies and here is where the student has the most difficulty in identifying his or her problems with the teaching or with the subject matter.

THE INSTRUCTOR'S FOCUS IS ENTIRELY ON FORMULA MANIPULATION. This approach is grievously inadequate. When students adopt this approach to their subjects they cut themselves off from understanding the material and when the instructor adopts this approach he not only makes understanding difficult or impossible, he reinforces the disastrous inadequacies of many students' approach to mathematics. ALWAYS TEACH FROM A CAREFULLY PRESENTED CONCEPTUAL FRAMEWORK. In this presentation we have heard nothing from the instructor about quadratic equations or about solutions of equations. Indeed, a student could work the assigned problems and have no clear idea of what the point of the assignment was. Did you notice that the instructor reminded the students that the first ten problems were to have been solved using "yesterday's formula?" Perhaps the first ten problems did not contain quadratic equations. If so, this should have been pointed out.

THE INSTRUCTOR'S TERMINOLOGY IS INADEQUATE AND INACCURATE. Remember that PEOPLE THINK WITH WORDS. The verbal aspects of mathematics are consistently the most difficult for the students to

deal with. You need to use the correct terminology and to use it repeatedly. Encourage the students to use it and help a student to find the appropriate words to formulate his or her questions. MATHEMATICAL TERMS REPRESENT MATHEMATICAL IDEAS. The inadequacies of the content of this instructor's presentation can be identified quickly and surely through a list of the terms that he either did not introduce or did not clarify. Here are some of them: quadratic equation, coefficients, solutions, roots, real roots, distinct roots, complex roots, complex numbers, real numbers, discriminant, existence and uniqueness.

NO ATTENTION IS GIVEN TO THE ORIGINS, THE SIGNIFICANCE OR THE INTERPRETATION OF THE MATERIALS PRESENTED. Where did the quadratic formula come from? Even if a derivation is not given, some mention could be made of the nature of the derivation. Why learn the quadratic formula? Is it ever used? Frequently? Is there any reason for the sudden appearance of complex numbers? The instructor would do well to indicate how in some situations a parabola might not intersect the x axis, or might intersect it in just one point. These details not only heighten the students' understanding of the material, THEY ALSO CONTRIBUTE TO THE STUDENTS' INTEREST AND RETENTION OF THE MATERIAL. If time permitted, the instructor might even tell the students about how the Italian mathematicians of the middle ages held contests in finding the roots of polynomials. Show the students how to get numbers on their calculators and to check their answers by substituting these numbers into the given equation. This will reinforce their grasp of the meaning of a solution.

Finally, here are some criticisms as to the management of the class.

THERE IS RARELY A GOOD REASON FOR DISMISSING CLASS EARLY. The instructor is being paid to teach the entire class, the students are paying for an entire class and the syllabus is designed on the assumption that all available class time will be used. IF EXTRA TIME BECOMES AVAILABLE IT CAN BE USED PRODUCTIVELY FOR DISCUSSION, REVIEW OR DRILL.

THE INSTRUCTOR HAS KEPT A TEST UNGRADED FOR WEEKS. Students learn few lessons as permanently as those resulting from their mistakes on quizzes. Students doing well profit from prompt positive feedback. Students who are slack need prompt confrontation with the consequences of their inadequate preparation and ALL STUDENTS NEED AN OPPORTUNITY TO UNDERSTAND THEIR ERRORS BEFORE TOO MUCH NEW MATERIAL INTERCEDES.

THE INSTRUCTOR CANCELS HIS CLASS IN VIOLATION OF SPECIFIC DEPARTMENTAL DIRECTIVES TO THE CONTRARY. THIS INCREASES THE LIKELIHOOD THAT THOSE SAME STUDENTS WILL FAIL TO ATTEND THEIR OTHER CLASSES ON THAT DAY. No matter what day of the week a vacation begins on and no matter how long the vacation is, there will be students who will find reasons for leaving school early and for returning from vacation late. One way to increase students' attendance before a vacation is to schedule a quiz. WHEN A STUDENT

ENROLLS FOR A COURSE, HE UNDERTAKES A RESPONSIBILITY TO ATTEND THE CLASSES. IN LIKE MANNER, THE INSTRUCTOR UNDERTAKES A RESPONSIBILITY TO MEET THOSE CLASSES. In asking the students which of them planned to attend on the day before break, the instructor relinquishes the decision about holding class to the students. Many students would be uncomfortable about responding even if they did plan to attend and in any case, there should never be any question that their attendance is expected and required.

THE INSTRUCTOR DOES NOT TAKE ROLL. In elementary classes in particular, THE INSTRUCTOR SHOULD MAKE HIS CAREFUL ATTENTION TO ATTENDANCE RECORDS VERY EVIDENT to the students. This underscores your expectations of the students in regard to their attendance and in some cases can act as a deterrent to excessive absences.

HOW TO TEACH MATH 151

PREFACE TO THE THIRD EDITION

The new edition of the 151 orange book is a response to the adoption of the text of Edwards and Penney, and the introduction of a limited amount of departmental testing.

Like the previous editions, this one provides generic syllabi, sample tests and examinations, and grade distributions. It also includes my own opinions on the course, and on teaching mathematics here at Ohio State. I think that people new to teaching, or to mathematics, or to a student body like ours, might find in these pages some ideas worth considering.

Frank Carroll
January, 1987

SPECIAL ASPECTS OF 151

Math 151 is the first course in the main-line calculus sequence for engineers. Honors courses have skimmed off the cream, and the students remaining will learn best by proceeding from the concrete to the abstract. Many will never leave the concrete.

The course is taught in self-contained sections in Summer Quarter, at night, and on the regional campuses. At other times on the Columbus campus, it is taught in lecture sections of 120, subdividing into 4 recitation sections. Here is the recent GPA history of the course:

Year	Summer	Autumn	Winter	Spring
83/84		2.327	2.061	1.988
84/85	2.556	2.287	2.035	2.047
85/86	2.614	2.249	2.267	2.194
86/87	2.589			

CONTENTS

	Page
1. My letter to the lecturers, January, 1987	3
2. Sample letter from lecturer to students	4
3. The syllabus for Winter, 1987, to be regarded as a generic lecture-recitation syllabus	5
4. A generic syllabus for a self-contained class meeting five days a week	6
5. A syllabus for a course meeting two evenings a week	7
6. A sample Midterm 1	8
7. A sample Midterm 2	9
8. A sample Final Examination	10
9. Appendix. Advice on teaching mathematics at Ohio State.	

SAMPLE

To: My Math 151 students, Winter, 1987

From: Professor Popular

Welcome to Math 151, Differential Calculus.

It's a demanding course, and you'll need to stay current with homework. Sometimes I'll make explicit assignments. In case I don't, assume that you should do problems $1,5,\ldots,4n+1$, where n is approximately 10. Homework will be sampled and the samples graded.

I plan for the homework to count for a total of 50 points. There will be a departmental multiple-choice test on Thursday, March 5. I shall count that for 50 points. I will give midterms, worth 100 points each, on Thursday, January 29 and Tuesday, February 17.. Your 200 point final examination will be given on the University's schedule. That is, it will be in your lecture room at * _____ o'clock on _____ March ____. The total number of points possible is 500.

Your letter grade will be based on my judgment of your mastery of the subject matter and your readiness for 152. The " subject matter " includes not only the new material from 151, but all prerequisite material, including arithmetic, algebra, geometry, trigonometry, and English.

Having said all that, I would expect that 350 points would suffice for a C, 400 for a B, and 450 for an A.

*Note from FWC to Professor Popular:

Lecture Time	Exam day and time
8	Wed., 3/18, at 8
10	Mon., 3/16, at 8
12	Wed., 3/18, at 1
2	Mon., 3/16, at 1

To: The Math 151 Lecturers, Winter,1987

From: Frank Carroll

Welcome to a New Year. There will be some departmental testing in the course. I'll provide you with a multiple choice examination of between 12 and 16 questions whose purpose is to test computational accuracy and basic understanding. It will be given in the recitation classes on Thursday, March 5. I'll make up five equivalent versions, four being the actual tests and the other a practice version for you to give out on Monday, March 2. The sections covered on the the test (" the core ") will be 1.4, 1.6, 2.1-2.5, 3.1-3.9, and 4.2-4.7.

Attached is a suggested syllabus. If you wish, you may have copies made for your students. Notice that it lists additional sections, not in the core, for continuity , motivation, and completeness. You may use your best judgment in your choice of non-core topics.

Here are some suggestions:

1. Give at least two midterms in addition to the departmental test.

2. Get a new orange " How to Teach Math 151 " book as soon as they are available. Many questions that may occur to you are addressed in the orange book.

3. Either give frequent quizzes or collect homework, or both. Be ingenious in finding ways to treat your recitation teachers humanely while still having your students do the work they need to do. Example: Assign 12 problems per section, and let a roll of the dice decide which one is graded.

4. Make up and distribute a schedule, your grading policies,etc. A sample is attached.

It's my duty and pleasure to talk with any of you about 151 problems, give advice when you ask for it, etc. BUT- please, do not ever, for any reason, send your students to me. Complaints go to the Vice-Chairman. Questions about make-ups, early exams, etc. are problems to be handled by the lecturer directly with the student.

Have an inspiring quarter!

Suggested Lecture Schedule for Math 151 Winter 1987

Date	Activity
1/5	1.4
7	1.6
9	2.1, 2.2
12	2.3
14	2.4
16	2.5
19	Holiday
21	3.1
23	3.2 (First drop date)
26	3.3
28	Review
29	Lecturer's Midterm 1
30	3.4
2/2	3.5
4	3.6
6	3.7
9	3.8
11	3.9
13	3.10
2/16	Review
17	Lecturer's Midterm 2
18	4.2
20	4.3 (Second drop date)
23	4.4
25	4.5
27	4.6
3/2	4.7
4	Review
5	Departmental multiple-choice test
6	4.8 or 10.1
9	4.9 or 10.4-10.6
11	Additional material or review
13	

THE FINAL EXAMINATION WILL BE MADE UP BY THE LECTURER, AND GIVEN ON THE UNIVERSITY'S SCHEDULE.

GENERIC SYLLABUS FOR A SELF-CONTAINED
151 CLASS MEETING FIVE DAYS A WEEK

Class	Section	Class	Section
1	1.4	26	Review
2	1.6	27	Test 2
3	2.1	28	Go over test
4	2.2	29	4.2
5	Review	30	4.3
6	2.3	31	Review
7	2.4	32	4.4
8	2.5	33	4.5
9	Review	34	4.5
10	3.1	35	Review
11	3.2	36	4.6
12	3.3	37	4.7
13	Review	38	4.7
14	Review	39	Review
15	Test 1	40	Review
16	3.4	41	4.8
17	3.5	42	4.8
18	3.5	43	4.9
19	3.6	44	Review
20	3.7	45	Test 3
21	Review	46	10.1
22	3.8	47	10.4
23	3.9	48	10.5
24	3.10	49	10.6
25	Review	50	Review

1. Your role as a professor of mathematics.

   You have accepted the challenge of participating in the greatest educational enterprise in the history of the world - the education of the people of the United States. You are paid from tuition fees and taxes to bring your students into a community of scholars. You aren't expected to enjoy complete success. You are expected to use your intelligence, energy, and judgement to do the best job that can be done. "Anything that is worth doing is worth doing ..." By accepting your position you acknowledge that you consider the task worthy of your best efforts.

2. Your students.

   In almost all courses, the students will be a mixed lot. Some seem to have so little native ability that you despair of their ever learning anything. Some once had ability, but bad experiences in high school (and before) have left them inept or frightened or both. Some are cocky, think they know more than you do, and refuse to do any work. Some drink too much, some marry fraternity life, some hate the impersonal university, and some hate your personal guts. On these, as well as on the intelligent students of good will, the future of the world depends.

   Treat your students with respect. Help them to grow as much as possible within their limitations, your limitations, and the limitations due to curriculum and facilities. Don't let your students play on your sympathy to the detriment of their learning. Never say anything denigrating about any student or group of students. Do not make remarks about their previous schooling. Be, and appear to be, fair and professional in your dealings with students. Be, and appear to be, on their side as they work to master the little part of mathematics in your course. The students will forgive you many mistakes if they are convinced that you are working to help them.

3. Preparing the course.

   The quarter is considered to be twelve weeks long. The first week is for preparation, ten weeks are for instruction, and the last week is for examinations and judging final grades.

   During the preparation week you get the text and the syllabus with suggested homework, and map out your strategy for the course. Make up your own day - by - day syllabus and assignments, in view of the calendar for the quarter. Consider natural breaks in the material, the number of days of lecture and recitation, when holidays occur, etc. Should some harder material come earlier than in the suggested syllabus? How many midterms? What weight should they have? Same for homework, and quizzes. What are your expectations for the various grades?

151 EDWARDS & PENNEY NIGHT SYLLABUS

Week	Tuesday	Thursday
1	1.4, 1.6	2.1, 2.2
2	2.3, 2.4	2.5
3	3.1	3.2, 3.3
4	3.4	TEST
5	3.5, 3.6	3.7, 3.8
6	3.9, 3.10	TEST
7	4.2	4.3, 4.4
8	4.5, 4.6	4.7
9	4.8, 10.1	4.9, 10.4-10.6
10	TEST	Review

Different mathematics professors of good will have different answers. I give you mine, as advice. Consider it, and then do what seems to you to be most responsible.

Take the generic syllabus and spread it evenly over the quarter, using good sense. Give two midterms in a 3 - credit course, three midterms in a 5 - credit course so that you approximate a regular partition. Let each midterm be worth 100 points. Allow 50 points for quizzes and/or homework. Expect 60% for a D, 90% for an A. Print the syllabus, homework list, your expectations, etc., adding some words like "tentative" and "expected", and give the document to your students on the first day of class.

In my opinion, the reassurance the students get about what lies ahead for them far outweighs the negative outcomes. Moreover, students will be less inclined to try to slow you down when they see the scope of the course.

4. Preparing each class.

Very little actual learning happens during a lecture. The learning happens when the student wrestles with problems in his room that night. The main purposes of the class (lecture, or self-contained) are to alert the student to what the problems are about and to get him started on coping with them. Suppose that the topic is applications of a certain theorem. The professor might write the theorem briefly, remind the class about what it says, and draw a picture if appropriate. Some worked examples will be in the book, but the professor should proceed to some similar problem in the exercise section, one not assigned for homework. He should read it aloud, and ask whether everyone understands the problem. He should ask students about what the problem has to do with the theorem. He should solicit suggestions for ways to attack the problem. He should encourage active involvement, and discourage stenographic reflexivity. Three or four such problems, hard enough to be interesting but without obscuring complexity, will constitute a worthwhile class.

Prepare! You should pick out problems that illustrate four different aspects, rather than show one aspect four times. Besides, you need to know how to do the problems efficiently. It's all right to be caught by surprise once in a while; the students are glad to see that we're human. But if we look like fools too often, the students will lose confidence in us and in our ability to help them learn.

5. Preparing and grading tests and examinations.

A midterm should concentrate on material since the last midterm (or the start of the quarter.) It should have 7-10 problems, with different problems checking different skills or pieces of information. Avoid situations where correct work in part a is necessary to do anything with part b. At least 70% of the midterm should test skills. Any

"theoretical" questions, e.g., questions asking for definitions or statements of theorems, should be taken from a small set of such things that the students were alerted to prepare for.

When you have drafted your midterm, put it aside for a day. Then take it yourself. Write down everything you expect the students to write down. It should take you 10 ± 2 minutes. If it takes more than that, it's a bad test. Even intelligent well - prepared students do badly on tests that are too long.

Decide ahead of time how many points to allow for a correct answer and how many for perfect work. I suggest half credit for each. Deduct credit for false or irrelevant work, even if the correct solution also appears.

Prepare a solution - answer sheet, and distribute it to the students as they leave the test. This serves several purposes. First, the students are very interested in the answers at that point. They've invested thought and emotion, and are ready to learn something from the solution. Secondly, your preparing the sheet is one more safeguard against mistakes - typos, impossible problems, etc., - on your test.

Final examinations should be prepared with even more care, since few students will see them to correct oversights. The exams should be comprehensive and some problems may tie together ideas from different parts of the course. Multiple choice questions are more appropriate on final examinations than on midterms.

In designing your final examination, you should include problems that let you decide whether the student is best served by letting him take the next course.

6. The recitation professor's role.

The recitation professor is to involve the students actively in the setting of a small recitation section. He should administer quizzes assigned by the lecturer, send students to the board, ask questions, and collect homework. He is not to be an "animation of the solutions manual". Even when he is responding to a request to work a problem, he should force interaction with the students.

Lecturers are to visit the classes of their recitation teachers and see that the teachers are working the students. The lecturers are to discuss the performance with the recitation teacher, and file a report on the departmental form.

Typically a recitation teacher will have two recitation classes. His time commitment, including classes, preparation, grading, office hours, and tutor room duty, should be comfortably under twenty hours per week.

7. The grader's role.

The grader makes it possible to have much more homework corrected than could be done without his help. Homework corrections are his main duty. He may also aid in the grading of some routine parts of midterms and finals, but this should be kept to a minimum for several reasons. The most important is that the classroom professor needs to maintain mathematical contact with his students, so that his two - way radio must often be set on "receive", and not always on "send". A second reason is that the students will have difficulty finding and communicating with the grader. Their feedback should come to the classroom professor.

Typically, a grader works with two classroom professors. Each should assign grading work not to exceed ten hours a week, and, grading being what it is, nine hours a week is more humane and appropriate.

8. Office hours.

Choose 3-5 hours a week that appear to be most convenient for the students in your classes. Add "and by appointment." Keep your office hours if at all possible. In an emergency when you must be elsewhere 1) announce the change to your classes ahead of time, and 2) leave a note on your door.

Routine questions like, "I don't see how to do problem 7 on page 143" will take most of the time. As time and personalities allow, draw the work out of the student. If there are other helps available, e.g., tutor rooms, or recitation professors' office hours, make sure the student is taking advantage of them.

Some of the office hour time will be spent on more serious discussion. "After the 44 on the second test, what do I need to get a B in the course?" "I'm sorry I can't get to class, but between my CIS project and my job in the library, I need the time to sleep." "I'm thinking of dropping engineering and majoring in dramatics." For the first question, regard it as an arithmetic problem. Say something like, "I'm not making any promises or contracts. However, if 80% turns out to be the cutoff for a B, you'll need 167 out of the 200 points on the final to make it." (A problem with being too explicit is that the 167 becomes the new base from which the student can start negotiations.) For the personal crisis questions, treat the student as an adult who has taken you into his confidence, and give the best advice you can.

Occasionally a student appears who believes that his tuition payment buys 40 hours a week of your time, and the obligation for you to do his homework while he watches. I don't have a perfect cure. "Active involvement" helps. So does refusing to see him except during office hours and a small number of appointments.

There are students who wander the halls, looking for somebody to do an engineering differential equations problem for them. Be firm and polite. Tell them they have the right to the office hour time of the person who assigned the work. This holds whether that person is in engineering, mathematics, or anywhere else. Do not sneer at the subject or say, "I don't bother with that sort of stuff." Summary: You have a responsibility to your own students, but usually not to anybody else's.

9. The final grades

Establish a ranking of your students on some objective basis (e.g., total points on midterms, final, and quizzes), and don't violate the order in assigning letter grades. When you have a trial grade distribution, compare it with the historical grade distribution in the course. Don't give all A's or all E's unless 1) you're very sure of your ground and 2) you've discussed it with other wise professors.

Here's a method I've sometimes used. List your students and their numbers. Locate the worst student you're willing to pass, and let his score be L. Let the top score be P. Partition L, P into 10 equal subintervals, and label them D, D+, C, C+, B-, B, B+, A-, A. Now modify. Try not to separate with different grades students within a couple of points of each other. If you can avoid - grades, do so. Students like +'s but not -'s. Check that you're not too far off from what you told the class on the first day of the quarter.

Do not give grades over the phone. Do not post grades unless very secretly coded (not by SS number). I don't think you should post them at all.

Incompletes are a special problem. Read the University rules on incompletes, and interpret them narrowly. Allow an incomplete for a student who broke his leg on the way to the final exam. Do not allow one for a student who missed a midterm and the final, and has attended class only sporadically. If you do allow an incomplete, write out in complete detail what the student must do to remove the incomplete, give him a copy and keep one yourself. "John Smith will see me at least one week before March 1, 1985, by appointment. He will come prepared to take a two hour test equivalent to the final. He understands that a grade of 60% will allow him to pass, and that approximately 93% will be needed for him to earn an A."

If you must leave the University with such unfinished business, work out procedures in scrupulous detail with the appropriate course office.

10. Summary. How to teach mathematics at Ohio State.

Theorem. You can't. Neither can you teach swimming, basketball, or how to ride a bicycle. The student either learns by doing, or he doesn't learn at all. You can only help by introducing material at a proper pace, checking his progress regularly, and judging his degree of mastery.

Corollary The brilliance of your lectures is far less important than the quality of the work the students give back to you.

It's easy to become discouraged when students are sure that $\sin(x + y) = \sin x + \sin y$ and that the derivative of a product is the product of the derivatives. We're tempted to withdraw from the problem, give our lecture, and leave the students to sink or swim.

Resist the temptation.

RESPONSIBILITIES OF THE LECTURER

A Lecturer of a Freshman Mathematics course is responsible for up to 250 students in up to 10 sections with 3-5 Recitation Instructors *(RI)*. The Lecturer, an experienced teacher, enjoys the twin challenges of exemplary lecturing and training/management.   The RIs are generally inexperienced. With supervision and training, it is hoped that the RIs will be able to teach "solo" classes in a future semester. A generic description of RI duties is on "Responsibilities of Recitation Instructors;" you will need to add specificity.

The Course Coordinator *(CC)* passes on information concerning the administration of the course to all instructors. Sometimes these things are passed to the RIs via Lecturer, sometimes directly; in any event the Lecturer is responsible for implementation of the directions of the CC. The procedures described below have evolved from the experience of successful lecturers in meeting student needs within the structure imposed by this teaching format.

Before First Class:

1.   Be sure Recitation Instructors have text, syllabus, instructor notes, and eligibility requirements and the "Responsibilities of Recitation Instructors."
2.   Meet with RIs to discuss the first class meeting so they will be responsive during that hectic class. The departmental syllabus is given to RIs in advance.
3.   RIs should be given list of intended homework assignments, even if this list is not given to the students in advance.
4.   Prepare a *"syllabus supplement"* for students that includes information particular to the lecture mode. It should include a list of sections with recitation meeting times, rooms, and Recitation Instructor. This will be typed for you by arrangement through CC.
5.   Have RIs give you a complete schedule of their class and work commitments, including offices, phone numbers, etc. Lecturer should give RIs the same information.

Early Semester Responsibilities:

1.   Eligibility requirements must be strictly enforced. RIs do the legwork, but it is ultimately the Lecturer's reponsibility.
2.   Make it clear to students and RIs that switching sections is possible only through normal Drop/Add. RIs should not permit unregistered students to attend class.
3.   Whether ARC testing registration is carried out in lecture or recitation, follow up to ensure all students register.
4.   Check RI's gradebooks after CC have given instructions for gradebook organization.

In The Lecture Hall:

1.   Lecturer and RIs should always arrive early and plan to stay late. Be down front to handle student problems and answer questions. Have RIs bring gradebooks to class.
2.   Give RIs instructions on where to sit and how to handle disruptions. They should not sit together. At least two should sit in the balcony. Every student belongs to one RI; therefore it should not be difficult to identify a student who causes problems in a lecture.
3.   Ensure overhead transparencies and chalkboards are cleaned each lecture by the RIs. Let them divide up the duties as they like.

Observations of Recitation Instructors:

1. All RIs must be observed in recitation. It is a good idea to observe the second week of the semester and again in mid to late semester, unless you see real problems and return sooner. Fill out the appropriate forms and see they are placed in the RI's file. Avoid reobserving a section. Usually a RI has several sections.
2. If the observation is to be beneficial, schedule a meeting with RI within a week after the observation to discuss it. Be positive. Give suggestions for improvement.
3. If <u>real</u> problems are discovered, discuss with CC or Dr. Case immediately. Make recommendations to the CC and to Dr. Case for future teaching assignments of RIs.
4. Immediately put the original observation form in the Instructor's file, give a copy to the CC and to Dr. Case, keep a copy.

Quizzes and Tests:

1. A short quiz is normally given each week in recitation. The Lecturer must review quiz drafts in advance and return to RIs with changes and corrections. To save time in recitation, some Lecturers instruct RIs to run off quiz solutions on the back of graded quizzes. Show RIs how to do this.
2. Hour tests from <u>test bank</u> materials must be given in the Lecture Hall in the presence of the Lecturer to assure security; if exception must be made to this, the Lecturer or an Instructor substitute must be present for recitation section administration. It is good if the CC can be present for the first time a Lecturer runs a <u>FLH</u> test.
3. Carefully plan for the administration of in-class tests. Consider carefully number of forms, seating, number of proctors, ID and calculator checking, etc.
4. Outline grading procedures for quizzes and tests. Go over a few graded quizzes *(and all class unit tests)* with RIs to ensure correct, consistent grading and partial credit.
5. Ensure proper security handling of departmental <u>test bank</u> materials by RIs.

Problems, Questions, and Complaints:

1. Whenever possible simply refer students to the student syllabus; students may be asked to submit complaints, regarding requests, etc.. in writing.
2. RIs refer out-of-ordinary problems to Lecturer.
3. Lecturer will make decision on written excuses for missed tests, in consultation with Course Coordinator if situation warrants.

End of Semester:

1. CC checks over worked, weighted final exam drafts of all Lecturers and Instructors.
2. Instruct RIs in proper handling of exam materials, including the outline for the free response portion of the exam. CC will give outline to Lecturer.
3. Outline grading criteria for exam problems. Check graded exams for consistency.
4. Give RIs instructions on how to average grades and complete gradebooks.
5. Discuss any unusual situations about grades, with RI as well as W-F's, *See GIG).*
6. Collect all exams, <u>ARC</u> printouts, student excuses, withdrawal forms, PLATO results, etc...
7. Check gradebooks before letting RIs fill out final grade rosters. Sign and turn in required materials to administrative secretary (or CC, if so directed).
8. Keep a copy of each grade roster and gradebook since students often come to Lecturer the next semester to discuss grades.

RESPONSIBILITIES OF RECITATION INSTRUCTORS

Your immediate supervisors are your Lecturer and your Course Coordinator. Your term evaluation will be done jointly by them and Dr. Case. You will see Dr. Case about matters concerning your teaching or course work which do not relate directly to this course. You are responsible to the Lecturer; make sure you clear it with her/him (Course Coordinator for 1207) before making any changes in normal procedures. This includes attending meetings when called by Lecturer or Course Coordinator.

Final grades must be checked and approved by Lecturer. The rosters are signed by both Recitation Instructor and Lecturer.

Specifically, your responsibilities include the following; your Lecturer may make other requests.

Lecture Responsibilities

1. You must attend all of the lecture classes because it is important for you to work the problems in recitation using the methods represented in lecture.

2. Set up any films, etc.; fix the microphone and overhead for the Lecturer at the beginning of the period, and turn them off at the end; clean chalkboards and overhead transparency roll. These duties concerning equipment may be split up, but all RIs should be at all lectures.

3. Be at lecture ten minutes early and plan to remain ten minutes after the lecture. Be down front to help answer questions at the beginning and end. Implement attendance or quiz procedures in lecture as directed by Lecturer.

4. During the lecture classes, DO NOT SIT TOGETHER. Stand separately, spread around the hall, INCLUDING ONE OR TWO IN THE BALCONY. Move quickly to discipline disruptive students. ALWAYS WALK AND LOOK AND PROCTOR when quizzes or tests are given in the lecture hall.

Recitation Responsibilities

1. Make sure you work all of the homework problems before you go to your recitations. It is very embarrassing to both you and the department if you aren't sure how to work a problem. Keep a notebook of problems; the Lecturer may ask to see this. It is best to work problems for students in a manner as similar as possible to that used by the Lecturer.

2. Use ALL 50 MINUTES of recitation time each week. Generally you will conduct a 25-40 minute session of answering questions (guiding them to important concepts as necessary) and administer quiz during the remainder; follow directions of Lecturer in making quizzes; never cancel class.

3. Give quiz each week; draft of quiz with key and weighting must be checked and approved if directed by Lecturer or Course Coordinator; meet all time deadlines.

4. Keep 3 office hours per week for helping students in the course.

5. If a student complains about course policies, grades etc... during class, ask the student to see you after class. If a student is obnoxious and persistent, ask the student to put the complaint in writing and pass it on to the Lecturer. Always inform the Lecturer of problems which arise.

Record Keeping Responsibilities

1. Keep all records for the course --- ARC etc., and grade all papers and do all averaging. Never get behind! Give a copy of your gradebook when asked and in particular before the end of the term, but after recording the last hour test.

2. Keep CAREFUL attendance records at recitation meetings.

3. 1207 only: RIs must meet to group-grade exams given in class at time directed by Coordinator.

4. Grading of quizzes and exams must be consistent with departmental guidelines for partial credit and the instructions of your lecturer/coordinator. Ask for assistance if unsure.

5. At end of tenure of RI, gradebook will be turned in to Lecturer or Department.

Teaching Load Explanation

A "1/2 - time load", 20 hours per week, is 3 recitations or 6 solo class hours. i.e., "1-hour" of solo class teaching is equivalent to 3-1/3 clock hours; 1 recitation (which requires attending the lectures) counts like 2 solo hours. If you have questions about your assigned duties, please see Dr. Case.

The Torch or the Firehose?

A Guide to

Section Teaching

Undergraduate Academic Support Office
The Class of 1922 Fund

Massachusetts Institute of Technology
Cambridge, Massachusetts

Contents

Prologue 3

The Glass Wall: Encouraging Interaction 5

Before You Walk in 10

A Word about Pedagogy 14

In the Classroom 16

The First Day and the Last Day 20

Seeing is Understanding: Using the Blackboard 23

Evaluating Your Students: Assignments, Exams, Grades 25

Evaluating You: Feedback 29

Small Groups: Tutorials and Office Hours 31

Problem Students and Students with Problems 33

Epilogue: A Word about You 36

Classroom Observation Checklist 38

Getting an education at MIT is like trying to drink from a firehose.   —folk saying

Acknowledgements

This booklet grew out of an Independent Activities Period seminar on recitation teaching sponsored by the Undergraduate Academic Support Office, involving the group of people listed below. We hope it is worthy of its MIT predecessor, "You and Your Students", by Robley Evans, over 50,000 copies of which have been distributed since it first appeared in 1950. The pattern of teaching has changed somewhat since then, and our emphasis is therefore a bit different, but it was the immediate inspiration and we have drawn on its material.

Our thanks go to Linda Schaffir ('82) for the cartoons and Glen Apseloff ('81) for the "Mr. Stu" comics; to Connie Donaghey for the typing; and to Susan Colley and Frank Morgan of the Mathematics Department for helping to make the seminar take place.

Thanks also to the many T.A.'s and members of the faculty and administration (David Wiley and Ned Holt deserve special mention) who covered earlier drafts with their perceptive and amusing comments and tried in vain to correct the colloquial English and expunge the bad jokes, as well as to the anonymous T.A.'s and students whose comments on the local teaching scene are sprinkled through the pages.

Our gratitude finally to the MIT Class of 1922, whose Fund for the Advancement of Teaching has supported publication of this booklet.

Holliday Heine
Peggy Richardson
Undergraduate Academic Support Office
Office of the Dean for Student Affairs

Arthur Mattuck
Class of 1922 Professor, Mathematics

Edwin Taylor
Educational Video Resources; Senior Research Scientist, Physics

Stewart Brown
G, Physics

Allen Olsen
G, Physics

Craig Russell
'83, Mathematics

Reprinting of this booklet in 1986 was made possible with funds provided by the Office of the Dean of the Graduate School.

Prologue

You're down on the list to teach Recitation 09 of a large course, Introduction to Applied Science. The 400 students enrolled will hear—or at least sit through—three lectures a week by a big man in the field still hoping for his Nobel Prize. Those students lucky enough to have been assigned to Recitation 09 will have you as their recitation teacher for two additional classes a week; it's to you they will consign their weekly problem sets, their hour exams, and their hopes for an A.

Twice a week you walk through the classroom door to be greeted by thirty upturned faces—there were twenty to start with, but your spreading renown has attracted others—and a clamor of questions about the week's lectures and the problem set. Suavely and with unfailing good humor you deal with these, providing the needed insight with a few deft words. Now and then you delve into your experience to provide the striking example which animates what had seemed to be just dead knowledge, or a lively anecdote which places the subject in a human context. If the week's lectures have been difficult, you may spend part of a period offering to your students your own view of the material: it is penetrating and clarifying, and your students interrupt you with perceptive questions and comments. Grading their problem sets and exams is a joy as you chart their steadily growing competence. They badger you for literature references, and at the term's end crowd around to find out what you are teaching next semester and what the best sequel to the course would be. No evaluation is needed: the glow in their eyes tells all . . .

Wake up, we're back on Earth: it's the first day of classes, and in just two hours Recitation 09 will materialize as a bunch of interested but appraising faces. What happens from then on is partly up to them, but mostly up to you.

In this booklet we'll talk about some common problems in teaching and offer a variety of ideas you can try. (Some of these contradict each other, but what can you expect from a committee?) We're concerned primarily with the teaching of recitations that accompany large lectures, so you won't be reading about how to design a course, or how to lecture. Still, teaching a small class all by yourself has a lot in common with teaching a recitation, and the generals who give the large lecture classes need to be in touch with the problems faced by their lieutenants. So we hope that a large group of faculty—both experienced and inexperienced—will be able to find something of value in this booklet.

Right now, however, we'd like to say a few words to the new teachers—important things that might otherwise be submerged in the sea of suggestions we're coming to.

If you're just starting out as a teacher, you may find all this advice a bit bewildering and hard to keep in mind. A lot of it will only come alive after you meet your students and have had a few sessions with them. We suggest you read the booklet through quickly now and come back to it from time to time as the term goes on. Basically your task is to develop your own natural teaching style, choosing and adapting suggestions to fit your own personality. Don't follow advice that makes you really uncomfortable—you need to be relaxed to be effective. But shy teachers sometimes have to push themselves a little.

No two recitations are alike. Popular and experienced faculty having two recitation sections of the same subject will report that one is regularly a joy to teach, the other like pulling teeth all semester. The mix of students, the time of day, some subtle interaction between your personality and theirs—who knows? Remember this if things are sometimes difficult. Remember too that you are not solely and personally responsible for the success or failure of your students. They have many other resources—lectures, books, tutoring sessions, friends. Just do what you can.

By the same token, you have resources to help you improve as a teacher, and the most important are your colleagues. Make every effort to meet other recitation teachers and regularly share classroom experiences with them, trade material and discuss different explanations with them, exchange classroom visits with them. Talk to the lecturer too; at staff meetings you can provide important feedback from the students—how they like the course, and what troubles they are having. In this way not only you but the course itself will improve.

The Glass Wall: Encouraging Interaction

The number one problem in teaching recitations is the "glass wall"—the teacher on one side doing a passable job of explaining, talking, and writing, but rarely interacting with the students on the other side. Listen to an undergraduate describe it.

We're usually all there when he walks in. He looks sort of embarrassed, stares down at the desk and asks if we have any questions. There's an awkward silence, like at a party where nobody can think of anything to say. Then he starts to work a problem from the homework. He talks to the blackboard in a steady even way. You can hear, but you can't tell what's important and what isn't. I can't follow one of the steps, but I'm afraid to say anything. Every now and then he says, O.K.?, but it doesn't mean anything and he doesn't stop. After a while you don't really understand much and wonder why you're there. I copy the stuff into my notebook—I'll probably be able to figure it out at home—but if it weren't for the exam I know I'd never look at it.

I guess I keep going because I know that otherwise I'd just waste the hour some other way. He knows his stuff all right, but it's like he's up front and we're back there and there's a glass wall between us.

A recitation without interaction—what problem coud be more basic? Without the ease of communication that's supposed to be fostered by small groups, why have recitations at all?

The silence is bad for you: it makes it hard for you to know the difficulties your students are having. You can't tell if your explanation is opening the door, or whether you need to try another key in the lock.

It's bad for your students, who already have sat through many lectures and don't really want another one from you. They need a chance to talk and express themselves, to clarify their own thinking, to share their difficulties with each other, to experience the feeling of a group working together on mutual problems. This is what recitations are about, and getting it to happen in yours ought to be your number one priority.

5

**Breaking down
the glass wall**

Achieving real communication with your students isn't always easy. Think of all the situations in ordinary life where two people find communication difficult; add to them the extra complications that arise when one of them is an authority figure, and the other must talk with an audience of peers listening. You're going to need some tact and skill; here are some suggestions.

Get them thinking: ask questions
Your students have just arrived and are sitting there thinking about life's problems, about the class they just came from, or maybe about nothing at all.

"Any questions?"

Questions? They can't even remember what the current topic is in your course. They thumb through their notebooks, but it's hard to start out cold.

Your first task is to get their mental sap flowing. Ask some easy review questions (very easy, if you suspect they are far behind). If you are handing back a problem set, give them a typical mistake and ask what the error is. You could give them a few minutes to work by themselves or in small groups on an easy problem you give out, maybe cast as an informal quick quiz. This can be done in the middle of the period too, to break up the class hour. You'll see how much livelier things will become when everyone has thought about the same problem.

When you start on new material, try to cast it in the form of a problem to be solved, and have them think about it for a bit. Ask for suggestions and deal with them seriously: don't just dismiss all the ideas which ultimately won't work, or it will be the last time you'll get any response.
   In preparing, look for paradoxes that you can spring on your students, and be rewarded by the sight of them all sitting forward in their seats thinking. Actually, they will do the same thing if you've made a mistake you can't find right away, and some wily teachers have been known to make mistakes on purpose for that reason.

Peer-group pressure
If things still seem sticky—you are asking good questions, but get only nervous looks in return—your students are probably afraid of saying something that will make them look foolish in front of the others. Sometimes you will see students silently mouthing an answer they don't dare to say out loud.

6

Perhaps you have felt this way yourself in advanced courses or seminars. If so, you will appreciate that it is a difficult problem to deal with. The ultimate solution has to be for them to feel comfortable with each other, but this won't happen right away. Meanwhile you can help a lot by being supportive: "That's a good question, Jennifer" or "I'll bet others of you were wondering about that point" will go a long way toward relieving anxiety and convincing others that they should not be afraid to speak. Alas how many teachers respond instead with,

"Didn't we cover that last time in great detail?"

"Is he the only one who couldn't do this problem?"

Critical answers like these discourage questioners. Putting out the welcome mat instead invites interaction.

Student time-delay

Students always seem to be behind. If there's a weekly problem set, many will not start studying for it until the night before it is due; before that they will understand very little of what's been said in lecture during the week. No wonder they are silent.

You can deal with the problem three ways: ignore it, accept it, or fight it.

If you ignore it by pretending they are up-to-date, you'll find yourself doing most of the problem-choosing and talking, with class participation limited to just the few students who are prepared.

If you accept the situation, you'll probably feel it your duty to teach them what they haven't yet studied. It's a bad habit to get into, but if you do this, at least avoid straight lecturing—try to teach interactively, with questions and work they can do either together as a class, or individually or in small groups, with you going around and helping.

It's best of all to get your students to prepare for the recitation: best for the class as a whole and best for them as individuals. Some instructors give a very simple short weekly quiz which anyone can do after having looked at the week's reading. Students can grade each others' papers, and gain additional insight that way, since you will be discussing common mistakes.

Other teachers announce at the previous recitation a few problems that will be gone over at the next meeting. Some require a single problem to be handed in at each section meeting for grading. Often just a clear statement to your students of what you expect from them and what your plans are for the next section meeting can work wonders in encouraging them to prepare for class.

Eye contact

Do you look at your students? Think how hard it is to talk socially with someone who doesn't look at you. Nothing gives the "glass wall feeling" faster than to see a teacher explaining to the blackboard, the walls, the window, or a point about one foot over the students' heads.

If this describes you, one way to get started is to pick one or two students you know and address your remarks primarily to them. Once you get used to looking at them, you'll be able to branch out and look at the others. Eye contact will improve automatically as you get to know your students better—even very shy people are usually able to look at their friends when they talk with them.

Listening

To communicate, you have to be a good listener, because students often don't say exactly what they mean to say. If you give an elaborate response to an unasked question, it frustrates everyone. Try responding to a question with a question of your own, until you feel sure you know what it is that's being asked. If in doubt, enlist the aid of other students in the interpretation.

Q I couldn't get problem 22 on page 253, could you work it?
A Did you have trouble getting started, or was it something in the working out that bothered you?

Q Is the gyroscope important?
A Do you mean for your understanding of physics, or for the world of technology? Or do you mean for the exam next Wednesday?

The steamroller

It's common to see recitation teachers carefully prepare a lot of problems and a lecture review for the hour, then realize in class it's a bit too much. So they shift into high gear and deliver it all as a fast lecture.

Sir, could you explain that last step?

If you're going to interrupt me with questions we'll never be able to cover the material. —math lecture

Such tactics "cover" the material, but they stifle interaction, and the class will give up and settle back in silence. Instead, relax and don't worry if you don't cover everything you've prepared; it's better to have a less tense atmosphere.

Your voice

Let's continue talking about the overall atmosphere in your class, since it controls the interaction so strongly. The way you say things gives powerful indications to students. Put a cassette recorder on the desk while you are teaching. Most people are shocked by the playback and don't believe what they hear. Here are some common speaking problems:

- Voice dropping into inaudibility at the end of each sentence

- A monotone or sing-song voice that makes concentration difficult

- Sentences interlarded with "O.K.?", "All right?" with no response expected

Your voice will give an unhappy emotional tone to your class if it sighs frequently with apparent boredom, or seems to express underlying anger or irritation. Again, listen to a playback to hear if this is you.

Does Prof. R. _____ realize that it's very hard to take notes after he puts us to sleep?

Certain habitual ways of speaking will seem like put-downs to your students and inhibit response. Phrases such as "it's obvious that" or "I think everyone should be able to see that" are very unsettling to insecure students. Think how much courage it takes to respond, "Well, I may be dumb, but I don't see how you did that. . ."

In the same category one can put sarcasm, which is unpleasant because it also plays with the student's feeling of self-worth. A teacher having excellent rapport with the class can get away with such phrases as
    "You're all geniuses so I know I won't have to explain this to you. . ."
but it's too risky for the average class situation—it just isn't a joke to students who feel insecure.

Feedback time
If despite your best efforts interaction is still sluggish, try taking 10 minutes off at the end of a period, putting your feet up, and asking the students in a general way what they think of the recitation. A good discussion can help a lot in clearing the air and pointing to the difficulties. Maybe there's something you didn't know. If they seem reluctant to talk, ask them to write down their thoughts and the next time you meet report back to them what the sentiments were.

Knowing your students
Sometimes you'll be lucky—your recitation will have a few students who seem to spark everybody. But more typically, your class will just sit there in the beginning and won't seem too keen on talking. Don't worry about it. Communication is easier with people we know better, and as the students get to know each others' quirks and yours too, and as you start recognizing them in the halls, you'll find yourself relaxing and interaction will improve.
    The section of this booklet titled The First Day has some suggestions on how to get to know your students and make them comfortable with each other. The important thing is to be patient and keep trying—they will respect your efforts and sooner or later will start responding.

9

Before You Walk in . . .

Lack of interaction may be the number one problem, but if you run a silent recitation, you can at least console yourself with the thought that it's not all your fault—the students share some of the blame—and anyway, they still probably learn something. These comforting thoughts are not available to the teachers of recitations afflicted with the number two problem: poor preparation. Listen to another unhappy undergraduate.

Monday morning, 10 am. My recitation teacher walks into class with the homework assignment for next week. He is not sure exactly what the lecturer has covered so far, but hopes that he will be able to rely on his knowledge of the basics to get him by. He guardedly asks us if there are any questions.
After a while, someone finally speaks up, asking about a point made in lecture. The instructor, not too sure of himself, gives a hand-waving, hot-air answer, which he concludes by citing the text as a reference. Another student interjects an explanation, giving examples from the lecture. Thanking the student profusely, the teacher immediately begins to work through the next assignment, using sketchy solutions he has scrawled on the back of an envelope. Every once in a while he gets stuck, and the same student has to help. Other students are talking among themselves or napping.
Towards the end of class, he remembers the graded assignments he was supposed to return; he removes the paper clip and hands the stack to a student in the front row, who passes them around. One at a time the students shuffle through the stack to find their paper, while the instructor continues to work on a problem he is having trouble with. Finally the class is over. Someone returns the unclaimed papers as the students file out.

Wednesday morning, 9:15 am. The alarm clock rings. I roll over and reset it for 10:15.

The number two problem with recitations is that the teachers don't seem to know what they're doing—either they fumble around, or they come to class with apparently nothing special in mind and end up improvising.

What? Me unprepared? I know the subject cold.

Good, but can you explain it? That needs technique and forethought. Even if you are an experienced teacher, different lecturers have different emphases: the course may not be the same as the one you took as an undergraduate, lectured in yourself three years ago, or even taught in recitation last year.

Before the term begins

Course material
You should receive a course outline well before the term starts. Look it over to see what's emphasized and what the course aims are. Talk to the other recitation instructors, particularly those who have taught the recitations before. Get the textbook and read at least some of it for the flavor, and to judge how much explaining you'll have to do. Can the students read it by themselves?

If you are new to the course, there may be things in it you don't know so well. The earlier you start studying them the better. Later in the term when they come up you may be at a critical point in your own work and not have as much time to prepare as you would like.

Staff meetings

Before classes begin there should be a meeting with the lecturer and the other section instructors at which you can discuss the level and emphasis. In how much detail are things done, how deeply? Will old problem sets and exams be a reliable guide, or will there be changes? Administrative matters should be settled at this meeting, so that you'll have the information for your students at the first class: exams, problem sets, how it will be graded, tutorials, and so on. Sometimes the lecturer will suggest what to do the first week, to help you get started.

If the course is a new one, it may not be possible to decide everything in advance. In that case, frequent and regular course meetings during the term are essential. You should expect to give the lecturer important input on how things are going, and should expect to influence policy on exams, problem sets, and course material.

Even experienced hands have to do a little thinking before each meeting, if they don't want to risk having a recitation like the one just described. At a minimum, you should have looked at the problems students are likely to ask about—make sure you can both do them and explain them. You should also know where in the syllabus the lectures have gotten, and be ready to explain what you judge to be difficult points. Talk to other instructors, read suggestion sheets put out by the lecturers, and best of all, go to the lectures yourself (this is actually required by many large courses). All these will be helpful, and let's face it, there will be days when you just don't have the time or energy for much more.

Preparing for the section meeting

But a good section meeting demands a little more planning. Here are some things from a T.A. in physics that you might find useful:

"To prepare a presentation for the first day and indeed to map out a strategy for the entire semester, I had to decide in my own mind just what it was that I should be to the students.

First and foremost I had to hold their attention and motivate them to succeed in learning the material.

Once that was accomplished, I would teach them how to think about problem-solving, how to approach problems.

Finally, I would be a person from whom the students could get additional help with course material (both in class and privately).

Having decided that, how was I to go about realizing these objectives?

Holding attention. It seemed to me that providing motivation and holding their attention required two things.

First, I had to earn their confidence since they would neither be motivated by nor interested in what I planned to say unless they could be sure that I knew what I was talking about. That meant that my part was to know the material thoroughly and prepare my presentations with a view to anticipating where possible their questions.

11

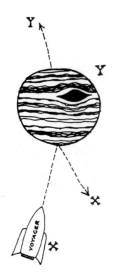

Second, choosing problems and examples that appealed to common experience or spectacular occurrence was most likely to capture their imagination and provide motivation as well as interest. For example, working out the mechanics of the Voyager/Jupiter encounter was sure to be more interesting than just about any other momentum transfer problem. That meant that I had to look around me and think just a bit to find suitable problems—not just in text books, but all around me. (Some of the things I came up with later were: the earth as an electrical machine; the auroras; and nerve conduction.)

Clearly preparation is important. However, a problem I had noticed in conducting tutorials was that repetition led to boredom—with each repetition, the explanation became shorter and less lucid. Therefore I decided to do many problems "off the top of my head" in class. In preparing and selecting examples I decided whether or not I saw the whole method of each problem. Those that were perfectly clear were done for the first time in front of the class, whereas those that required some more thought I worked out beforehand. In this way, I felt that an element of spontaneity was introduced which, if nothing else, allowed me to be more enthusiastic in class.

The last element of preparation was hitting the appropriate level. Many problems can be worked out on a variety of levels. There are mathematically elegant though abstruse solutions and brute force solutions which are long but conceptually simple. I felt that this would have to be left to experience, but decided to ask some diagnostic questions the first day, to help assess the level of the students.

Thinking about problem solving. The way to teach students to think about problems was, in my opinion, to emphasize in my explanations the logical method. For example, making clear what the input to a problem is, which input is physics and which is mathematics, using a diagram, and checking solutions against physical impossibilities (e.g. infinities) are all part of "method" and should be stressed.

Giving additional help. As for being someone who the students might turn to for help, that only requires frequent encouragement. As I got to know the students, they felt more comfortable about seeking help from me both in class and during office hours. A good practice was to be the last one to leave the classroom, as that provided a tacit encouragement for students to come up and ask questions or indeed simply talk."

Good material, presented at the right level, and so that it teaches what you want them to learn . . . it's beginning to sound as if one could spend a whole day preparing for each section meeting. Well, things aren't that bad, though preparation does take time.

Material
Look in the textbook, perhaps at some of the non-assigned problems, and in other books; consult other section instructors and the lecturer; dredge your memory and use your imagination. Look for paradoxical results, problems where obvious reasoning leads astray, real-world problems.

Level
Guess to begin with, ask other instructors, feel your way. You want something for every student—part of the time a level where everyone can follow without difficulty, part of the time a level which stretches your students just a little (nothing too complicated).

Planning the hour

Deciding how to divide up the recitation time isn't easy. Roughly in order of priority, possible activities are: solving homework problems, reviewing the week's lectures with your own commentaries and explanations, working together on new problems, giving brief quizzes, digressing to related material not in the syllabus, telling jokes. . . Different recitations do different things.

At least do <u>some</u> advance planning—have some modest objective in mind that you can head for; it will give you an internal compass that your students will sense and appreciate. Have some alternatives in mind, if you aren't sure what your class will need that day. Do what you do best and enjoy doing, what your class seems to appreciate, what they need. Keep it varied, and keep experimenting.

Remember though to start things off with some warm-up exercises. Some teachers like to start with an outline of what went on in lecture—which can serve the same purpose, if done briefly and interactively.

A Word about Pedagogy

There still remains the problem of how to present the material and explain
it clearly. Here we are at the limits of what a little booklet can say. Still,
even long-time teachers make such elementary errors of pedagogy that it
might be worth-while to talk a little about the subject from a practical
viewpoint.

Three different approaches to classical psychology offer valuable insights
that can be used in every class.

1 Define the tasks

Behavioristic psychology might urge you to think, what is it you want your
students to be able to do? Differentiate all functions of a certain type?
Check an answer by qualitative reasoning? Isolate and analyze the forces on
a rigid body? Synthesize an organic molecule on paper? Trace the themes of a
Bach fugue? These will require different approaches, but they are alike in
this: they are specific tasks that your students can aim at mastering. By contrast, the vaguely formulated tasks implied by enthusiastic lecturers and recitation teachers (understand the role of the mean value theorem in calculus
theory, appreciate how Maxwell's equations summarize electricity and
magnetism) leave students uneasy and asking about exams, to their teachers'
irritation.

Are we responsible for the laws of planetary motion?

No, Kepler was.   —from a physics class

Let your students know what the tasks will be; if there won't be any—because
you're embarked on a fascinating digression, say—tell them; if you don't
know what the tasks will be, ask around or bring it up at the next staff meeting.

2 Explain your thought process

The introspective approach to pedagogy would suggest that you look inside
yourself for clues to good teaching. Solve a problem and watch yourself do
it. What guided your thought? How did you know to try what you tried?
Tell your students. While you're at it, sit and think for a while about the recitations you attended as a student—what worked and what didn't? Use that
as a guide, because things haven't changed much. Do unto your students as
you would have liked to have been done unto.

3 Thinking starts with a problem

Located somewhere between behaviorism and introspection, the school of
gestalt psychology teaches that thinking starts with a problem, a difficulty,
a contradiction. It sounds like a truism, yet is widely ignored in practise.
Teachers say their aim is to get their students to think, yet in classroom after
classroom they violate this psychological principle by giving the solution
before there is any problem.

"Now we are going to study a new kind of integral, called the line integral.
Here is the way it is defined..."

14

Sound familiar? Of course, the motivating problem must be an interesting one—that's where it helps to try to find things in the real world to serve as the problem source.

The above ideas barely scratch the surface of what psychology has to contribute to our understanding of pedagogy. But we're after just a few basic principles that will be easy to remember and will help your teaching immediately. Here are three more that you'll find use for constantly.

Go by degrees from the easy to the hard

Things can be hard because they are complex, requiring many steps and processes (like multiple integration in calculus), or because they involve subtle ideas (like the Coriolis force in mechanics). If you think the complex problem will be too hard, first give your students practice with simpler problems involving only a few steps at a time. For subtle ideas, look for simpler analogies, allow a lot of time, and be patient.

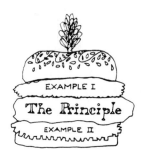

The sandwich method for general principles

To present an abstract idea—a general formula, a general law, a theorem— a good way is to present in order:

- An easy example illustrating the principle
- The general statement and explanation of the principle
- A harder example using the principle

This corresponds to the way people think. It's much easier to understand an abstract idea if you can test it out on a simple example. Then once you've understood the principle, it's fun to see its application to a more complex example that was too difficult to handle before.

The bird's-eye view and review

Often students can't follow an explanation because they can't tell where it is headed, what the ultimate goal is. You know, but they don't. So begin a complicated explanation by giving an overview of what it is you're going to do—the purpose, the general method, how long it will take, whatever. When it's all over, turn back and survey the ground you've covered together, and congratulate them on having made it.

First I tell 'em what I'm gonna tell 'em, then I tell 'em, then I tell 'em what I told 'em.

These ideas will get to be second nature with you if you use them a few times. Preparation gets easier with experience and as you get to know your class and feel comfortable with it. If you are a new teacher, it is probably better to over-prepare a little in the beginning, as it will help in overcoming some initial nervousness. Later on you can relax.

Of course some anxious teachers would prepare forever, if they could. But the bell ringeth and all preparation hath an end. We'll look for you next in the classroom.

In the Classroom

In you go, prepared to the teeth with mind-boggling stuff, determined that, prodded by your skillful questioning, your class will have an intense thinking experience and reveal a brilliance they didn't know they had in them.

"Could you go over problem 3 on page 112, I didn't get the answer in the back of the book." (General nodding agreement.)

It's a boring, stupid problem, one you hadn't planned to do. Grrr. . .!

The best laid plans of mice and recitation instructors go oft awry. Relax: you're a tour guide who has been asked to make a side trip. If the group is interested, maybe that side trip will turn out to be more interesting than you thought. But keep your original plans in mind, because your section needs a sense of purpose. Firm, yet flexible. Remember what was said before about the myth of "covering" the material—just because you don't do something doesn't mean it's lost forever. Of course it will help if you've allowed some soft time in the preparation—topics or problems to take up that you won't be too unhappy about abandoning if necessary.

Your personal schedule for the classroom

Come early. It says you're interested and available, and gives you a good chance to chat informally with a few students and find out what's going on.

Start on time. Assume those who come in late were coping with some crisis, but don't review for them.

End on time. Nothing was ever learned after the bell. If you've got two more lines to write that will set the keystone in place for the whole hour, beg them abjectly to stay for one more minute and swear you'll make it up to them somehow, but don't think they'll actually be paying any attention—it's just for your own sake you're doing this.

End ahead of time if there aren't enough minutes left to start something new and the class has run out of steam. In particular, don't start pumping them for questions to fill up the last few minutes—let them be first on the lunch line instead.

Stay late, until the stragglers leave—they might want to ask about something they didn't dare ask in front of the others.

Flub-up time

- You made a mistake at the board somewhere, but you can't find it.

- You can't answer someone's question.

- You get stuck working out problem 3 on page 112.

These things happen to everyone, but are hardest on younger teachers who often feel insecure at the beginning. The main thing is not to become defensive or hostile, and above all, don't fake an answer or b------t around: even the poorest students in your section won't be fooled, and you may lose their respect. Ask for help instead, hope you get it, and listen to the students who are trying to give it—some of them might be right. Or promise the answer for the next meeting. It's all right to make mistakes and you don't have to know it all, but you do have to be honest.

I asked a teacher a question; the answer was something like, "I'm not really sure, but I don't think it will be on the test." I'd prefer and respect an answer on the order of "I'm not really sure, but I'll get back to you on it if you like".

Q The students hissed at me yesterday, and I was shocked. Don't you think this is incredibly rude?

A It's not hissing, just letting out steam. It is triggered automatically by either of the following:

- Announcement of a problem set due the day after vacation

- Referring in class to any course other than the one you are teaching (double hissing for referring to any topic that you expect them to have learned in said course; redouble if you say the topic will be included in the next exam)

Q My students all sit in the back row.

A This is a common problem. They are hoping to be inconspicuous so you won't call on them. (I am assuming of course that you have not been eating onions.) Possible solutions are to

Adapt: use a megaphone and write big.

Fight back: mumble, write small, and call only on people in the back row.

Legislate: come early and rope off the back three rows with a sign, "Reserved for Visitors".

Use psychology: tell them that studies have shown that weaker students tend to sit nearer the back.

Offer positive reinforcement: ask them as an experiment to sit up front and that day give your most scintillating class.

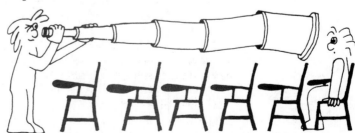

17

Q What should my approach be toward a student who mumbles his question?
A Have the student repeat the question until you can hear it. Then repeat it yourself so the whole class can hear it. This is called linear amplification.

Seriously, even experienced teachers regularly fail to make sure all the students have heard a question that's been asked. Hearing your answer without knowing the question is like listening to just one end of a telephone conversation. For many, the question may be one they have wondered about themselves; hearing it will give everyone a chance to think about it. And if you misinterpret it or get stuck, they'll be able to help you out.

Q Should I be informal or formal in my approach to the class?
A There are a lot of ways to teach—energetic or relaxed, in tweeds or jeans, the Olympian authority or the humble just-another-student. If you're not relaxed about your attitude, your students will be uncomfortable too. Just be yourself. However, if you feel you can manage it, being informal in a small section is usually more fun. It encourages interaction, and it's a bit more comfortable when error time comes around—the higher up you set yourself, the further you fall.

My instructor seemed to have a poor grasp of the material and compensated for this by being very authoritative.

Some roles teachers play (and mostly shouldn't)

We cast you a little while ago as the tour leader for your recitation. Other roles that teachers play reflect how they feel about the course and their own place in it, and some have unfortunate consequences.

The union organizer
It's tempting to ingratiate yourself with your section by telling them that it's not their fault they are having difficulties—they are victims of the System. You run down the book, the lecturer, the course in general. Of course any book or lecturer will occasionally be obscure, and it's all right to remark on it. But don't overdo this, because it can demoralize some of your students: why bother struggling if the cards are stacked against you?

18

The lifeboat captain

Do you give your section the feeling of being tossed helplessly about on the seas of your course by Angry Implacable Lecturer-Administrator Gods? That's the effect of habitually using "they":

"I heard they are going to make the next problem set really tough."

"They are leaving infinite series out this year."

"I don't know if they will give you a thermo problem on the exam."

If this is the way you really feel, then there is too much of a gap between you and the lecturer for the good of the course and your students. There should be more staff meetings and you should be having more input on the problem sets and exams. In the meantime, at least compromise by staying on neutral ground verbally: "this was a difficult problem set", rather than "they made the problem set too difficult".

The man with the white kid gloves

"The rest of this problem is just the usual algebraic garbage which I certainly don't intend to go through."

"That's just a standard F = ma problem—doesn't anyone have something more interesting they want to ask about?"

Such remarks convey to your students that routine things are beneath you. Many of them will imitate your attitude, with disastrous results later on. The most common complaints from upperclass teachers are things like, "They can't even integrate dx/x," "Doesn't anyone teach them the periodic table any more?"

So let students know by your attitude that you consider such things important and that everyone must learn them well. Brief quizzes will reinforce this. Poorer students should concentrate particularly on this type of material; finding they know something well may encourage them to go further.

Leader of the expedition

Every course has its winter of discontent—those times when most students are just struggling along and it seems as if they will never get to feel any mastery. It's up to you to cheer them on, to reassure them that others have made it and they can too, to hold up to them visions of the delights that await them on the summit. A little of your energy and enthusiasm supplied at the right moment can give them the lift that will take them the rest of the way.

In this section and the one on interaction, The Glass Wall, we've made just a few general observations about some common attitudes and problems in the classroom. In the next few sections, we turn to more specific areas where teachers often experience difficulty: classes at the beginning and end of term, and the use of the blackboard.

The First Day and the Last Day

The first day of class is important: "First impressions last". Some students seem to think the saying is "First impressions are the last"—they bolt out of their first-day recitation to the undergraduate office where they queue up to see what other section is open.

Your problem is to get over your first-day jitters and off to a good start. Instead of thinking of yourself as up for inspection, imagine yourself as the host of a party full of strangers, or the conductor of an amateur orchestra meeting for the first time: if you can make your students feel a part of something exciting and interesting, you'll have taken a big step toward ensuring the success of the next twelve weeks.

Let's hear how our physics T.A. handled the first day of his freshman section.

"I was somewhat nervous and in need of a good beginning. Introducing myself, giving my office, phone number, and office hours was the ideal way to start. General course information with an explanation of homework and grading policy was useful to the students, and served as an "ice-breaker" for me.

While doing this a paper was circulated to get names, majors, and any advance comments.

Then to help assess the level of the students, I posed some mathematical problems which were essential to success in the homework: the law of cosines, common integrals, simple differential equations, changing coordinate systems. A second purpose was to get some frequently-used formulas into their notes that would save them from having to constantly repeat these calculations. The feedback was not the best, but by watching faces, I obtained a rough idea of what was new and what was familiar.

I concluded by asking for questions and comments, and succeeded in getting a few to say what sort of section they wanted. Some wanted to go over homework, others wanted to see different problems, and still others wanted to rehash the lectures. At different times I endeavored to accommodate all of these views.

As I see it now, the first day is important in getting off to a good start. The students get their first impression of the instructor and the better that is, the easier it will be for the teacher to reach his or her students."

The first day

That was a good first day, though not the only possible one. Let's amplify the account.

Introduce yourself
Your name, office, phone, office hours for sure. If you feel comfortable doing it, why not briefly tell your class something about yourself—your schooling, research, interests, and in general what you do besides teaching their class?

Introduce the class to each other
To help students know who is in their section, make up a class list by passing around a paper on which they write their name, address, phone number, major (or possible major), and advisor's name. Circulate the list at the next couple of meetings as well, since some students try out different recitations at the beginning of the term. When it firms up, give everyone a copy.

Some students have suggested that they would like to hear the class members introduce themselves the first day. Of course this will take a while, and some may feel awkward about it. You could ask them if they'd like to do it. A relatively painless way might be to do it by living group ("Who here is from Baker House?") since students are interested in finding others who live near them to study with. (They'll tell you what the living groups are.) Instead of doing this the first day, you might wait with this until the second week when the class has stabilized, and the students are already beginning to recognize faces and voices and are getting curious about each other.

First day subject matter

The main thing is to choose material which allows interaction. If the lecture hasn't met yet, you could quiz the class orally on background material. If the lecture has met, you could in addition ask them about the lecture or start in on the homework. If the lecturer isn't going to do it, you could out-line the topics to be covered; you might give them some examples of what they'll be able to do by the end of the course, to whet their appetites. But avoid giving a long uninterrupted lecture: the message for today is that you want them to talk too.

Learning names

There's magic in a name, Knowing your students' names will tell them you are interested in them as individuals, and will help interaction. Here are ways to learn names and in general get acquainted.

ID picture

Students get some on registration day (though fall semester freshmen don't receive them until October). Ask for these pictures, and doctor them up as needed by inking in new mustaches, beards, glasses, longer hair (removing these poses more of a problem. . .)

Return homework individually

Call the students by their full name. If there isn't any homework, you could instead call the roll occasionally during the first few weeks. (Both of these let the students also see who's who.) As a further help, you can note down ob-vious physical characteristics, and where they sit, since for some obscure reason they usually stay in the same place all term. Note any unusually pronounced names; how you say their names is important to students.

Quick 5 or 10 minute quizzes.

Discuss them right afterwards; you can use the results as part of the section grades, or just use them for diagnostic purposes. Go around the room while the students take the quiz and look at their names. This a good way to see who needs help getting started—offer it if it is needed.

Use the names in class

This is the fastest way to learn them. You can ask the students to give their names whenever they speak for the first week or two. Don't be inhibited by the fear of making mistakes; you'll be forgiven.

I urge students to come in and visit during office hours in the first few weeks so that I'll get to see them as individuals instead of just as a sea of faces.
　　　　　　　　　　　　　　　　　　　　　　　　　　　—Math T.A.

In the lecture/recitation format, it's only the recitation teacher who has a chance to know the student. You'll be relied on in a variety of ways (completing evaluation forms, giving grades, talking to advisors, writing letters of recommendation, etc.) to provide information about your students. Students themselves will drop by for help and advice. The more you know about them, the more you'll be able to help. Treat getting to know them as an important business.

Don't be discouraged if the first day doesn't go as well as you had hoped. It's easy to exaggerate its importance. There's always the second day and all the others to come. Expect some quietness in the beginning as you and your students feel each other out. As you get to know each other, most of the first-day awkwardness will disappear and things will go more easily.

Just keep coming to bat—once in a while you'll hit a home run.

—Chem T.A.

The last day

Students get mired in the day-to-day details. A review on the last day gives you a chance to point out what's really important and how it all fits together —in other words, the Big Picture.

The simplest approach is to go back over the course outline, making comments—why we did this, where that topic leads, what we should have emphasized more at the time, and so on.

Or you may want to organize the material for them in a different way which sheds new light and makes it look fresh again. Pick something central to the course; if we wanted to understand just this, what else would we have had to study? Why did we study the topics in this particular order? If you had only one hour to give the essence of the course, what would you have taught? Three hours? Twelve? What's the thing you most regret having omitted? What will the students study in the sequel?

Or in a different vein, you could spend a while with them discussing the recitation or the course itself. In retrospect, what would have made the recitation more interesting? The course easier to learn?

Naturally, if there's a final, the students will want to know anything you can tell them about it. But don't spend the whole period on this. You might try to find out in advance what the lecturer will do the last day, and make your recitation complement that in some way.

It's been said that many a course is made by what happens on the last day. Don't just let your recitation stop: give it an ending.

Seeing Is Understanding: Using the Blackboard

Science and engineering teachers almost always have to write things down—
the diagrams, the formulas, the derivations. Yet their blackboard often seems
to be more a record of their stream of consciousness than anything else.

    The reason you're going to fuss over making the board look good is that
you know that at any moment

- Half your students aren't listening—they're woolgathering;
- Half your students aren't understanding—they're just taking notes.
  (The two halves might overlap some.)

You want the board to tell what went on well enough so that the day-
dreamers won't be lost when they tune in again, and so the baffled can use
their notes to figure things out later (maybe with their roommate's help).

Neatness counts

The basic rule is: don't skip around the board, tucking in formulas where-
ever there's a little space; use the board sections in an orderly way. One good
method is to start at the extreme left panel, go down, continue with the next
panel to its right, and start over at the left again when the entire board is
full. The writing itself should be clear, the right size (easily read but not
wasting a lot of space), and written level. Check these things occasionally
after class by looking at the board from the rear of the classroom.

Write it and leave it

Write down enough (including the statement of the problem for the sake of
those who didn't bring their books or notes to class) so things can be figured
out later; standard abbreviations will help to save space. Don't erase until
all the boards are filled, and don't simplify expressions by using the eraser,
as this drives note-takers up the wall. Put important things in boxes to
emphasize them visually, or use colored chalk.

**My chem T.A. draws all the structural formulas in the air with his fingers. He must think chalk is one of
the rare earths.**

**A few more
suggestions
on blackboard use**

- Practice drawing pictures or diagrams ahead of time, if you have trouble with them.

- Pull the shades if there is sunlight or glare on the board—even your best students would rather squint than tell you.

- Your chalk squeaks? Watch out for this since it is something that usually annoys the class much more than the teacher. Just break it in half, and hold it at a $45°$ angle to the board.

- Try to stay inside the squares, that is, don't write across the vertical cracks of the blackboard if you can help it. It can be rather confusing visually (also, the bears might get you).

- If you see your students' heads waving back and forth, it means they can't see what's written on the bottom third of the board. Don't use this part. Compensate by writing the top line higher up: stand on tiptoe if you have to. (It might also mean they can't see through you—standing in front of what they write is a method many teachers use to cover the material.)

Aim not to cover the material, but to uncover part of it.   —quoted in "You and Your Students"

Other visual aids

Occasionally using other aids besides the blackboard can lend some variety and excitement to the recitation—just the feeling that you've gone to a little trouble on their behalf can mean a lot to your students and to the general atmosphere.

The overhead projector can show complicated tables and diagrams, photographs of famous people in the field, and pretty commercial color transparencies with successive overlays that add information. It lets you face the class, and you can refer back to earlier points easily. On the negative side, the material disappears from view rather quickly and in the wrong hands things can be soporific.

Slides and film strips might be other possibilities. Your department may have models lying around, or working equipment that can be shown. See the course head, the departmental undergraduate office, or the lecture-preparation staff to find out what's available that can be used.

Evaluating Your Students: Assignments, Exams, Grades

Assignments

In a system where courses compete with each other for the student's time, those without regular assignments and clear expectations will lose out. Students in them get so far behind they can no longer even ask questions, let alone follow what is new.

Students tend to gauge their understanding by how well they can do the assignments, so these should be set carefully. This is usually done centrally; your job as recitation teacher will be to help your students with the homework, smoothing over some of the difficulties without robbing them of all exertion. If there's a weekly problem set, look it over as soon as it comes out and keep it in mind when planning what problems to work in class, but don't sabotage it by working essentially identical problems! (Baby chicks should not be helped out of the shell.)

Daily assignments, handed back at the next recitation, probably are best of all. They give you and the students steady feedback, your students get used to regular study and stay up-to-date, you get to know them better, the recitation is livelier, and the course more exciting. Sometimes the work is reduced by grading just one problem carefully and checking the physical existence of the others, but this can produce frustrated students. How about just giving one problem?

It would be nice if recitation instructors graded _all_ of our homework. If we didn't happen to get the one problem they graded, it was like we didn't even do the homework at all (in terms of the grade).

In most courses, homework isn't counted too heavily in the final grade, since students work with each other on it. It should certainly make a difference in borderline cases (there are often a lot of these), and in letting students who choke on exams still pass the course.

The most important thing about correcting the assignments is to do it promptly—which means handing them back at the next recitation after they come in. You're going to grade them sooner or later, after all; doing it sooner means that the students get the papers back while the work is still relatively fresh in their minds. You can then discuss common difficulties in the recitation, and the material still will be current in the course. It's also important for morale—it gives students the feeling that you are interested in their work and take them seriously, and they in turn will respond by putting effort into your recitation.

Examinations

Students take exams very seriously, which means that regardless of what you think of tests as teaching devices or as tools for evaluation, you have to take them seriously also.

Review sessions
At the last meeting before an exam, your students will hope for a review. Giving a good review session means a little work in the preparation, but you'll be rewarded by the feeling that they are hanging on your every word. You might at long last get some questions.

For a <u>quiz review</u>, usually the class hour is divided up: a brief outline of what the quiz will cover (very brief if it was done in lecture), comments and hints about those parts of the material which you feel are most likely to give them trouble, a question period, and finally working sample problems from old exams if there is time.

Exams tend to fall into routine patterns, and you or your students may be lucky and score a direct hit during the quiz review. They'll love you for it, but after the elation has worn off you may feel some qualms. Try for near misses instead. Remember that your overall purpose is to review a section of the course and help them see things in perspective, not just teach them how to pass the exam.

Grading the exam
This is often done by the entire course staff together, or at least the course head supplies detailed instructions. If not, you'll have to use your own judgment. If this is your first time as a section teacher, especially if you come from abroad, talk to some of the experienced section teachers to get a feeling for the standards—show them a few of your graded papers, or ask to see theirs.

Look over some of the papers to see what the common mistakes are, and decide in advance on the partial credit you will allow for them. It is fairer to grade one problem at a time, shuffling the papers between problems.
　　Everything said about grading homework on time applies with even greater force to the quizzes.

26

Handing back the papers

Your students will want to know at least the section averages: some teachers also give a histogram which may be so detailed that every grade is actually listed. Some give no information at all, hoping to downplay competition, but of course the students can find out other section averages, and it's silly to make them struggle for the information. Much depends on your manner and on course policy.

There also should be solution sheets. Discuss common mistakes, as everyone can learn from others' mistakes, but keep it interactive: "What's wrong with this. . . ?" It's also interesting to present unusual or elegant methods that appeared on some papers. (This goes for problem sets as well.)

The bad exam

It happens all the time. The exam was too long or too hard—your section average was 42 and your students sit there stunned. They've never gotten such low grades in their life.

Or the average was 65, but there was a trickily worded question that misled half your students, who feel they could have gotten 20 more points if it had been worded as it was in the homework.

The average was 92, and the students who have been walking on air all weekend because they felt they did so well have suddenly fallen to earth. "If everybody's somebody, then no one's anybody."

All you can do is reassure, offer sympathy, tell them it's not an uncommon occurrence. Even if you are as angry as they are, you shouldn't play union organizer: "This exam could have been better," not "They really goofed on this one".

How does it happen? The lecturer or course head may be inexperienced at making up exams, he or she may not have had enough contact with the problem sets, they may have worked so long over the exam that it seemed easy at the end, etc., etc. For sure, it wasn't pretested. Ask that future exams be taken in advance by two recitation instructors, to judge the length and difficulty and spot poor questions. Volunteer yourself (you'll have to work out the exam later anyway.) You should be able to write down the answers to an hour exam in ten to twenty minutes depending on the field.

Final grades

Lecture-recitation courses should have a uniform grading policy: how much each exam and the homework (and class quizzes and labs) will count. Student grades often are assigned in a meeting of all section instructors. You will usually have some leeway, however, and should take subjective considerations into account—giving a higher grade to the student who has steadily improved, for instance, even though the numerical total is the same as another who has steadily declined. Regularity and performance on assignments and brief quizzes can be used to raise or lower grades near the boundaries.

MIT freshmen are usually given letter grades like anyone else, but only P (pass) is recorded on their permanent transcripts. The purpose is to remove some of the pressure on them, ease the transition from high school, and give them time to fill in gaps in their academic background and to develop their work habits.

The pass-no record system can obviously be abused if they take it as a license to get by with the minimal amount of work. Few first-term freshmen seem to adopt this attitude, but it is more common in the spring term. Checking up on them frequently with problem sets and quizzes seems to be the easiest remedy and most direct approach.

Evaluating You: Feedback

You are evaluating your students steadily: the questions you ask in class, their problem sets, their exams. But it goes two ways—they are evaluating you as well. Do you want to know what they think?

Let's face it, many teachers do not. They may say it's undignified, inappropriate, students lack the understanding and judgement, etc.—but the most important reason is probably to avoid getting hurt. Shy people are particularly vulnerable. Still, feedback can be such a help to improvement that all teachers should try to overcome any natural inclinations they may have against it.

Try a bull session

Of course, when you look at your students, you'll see their expressions of perplexity, delight, or boredom. That's a form of evaluation. But you can get a lot more information by taking time out now and then to discuss with your class how they feel things are going. Listen quietly and carefully to what they say. You can tell them how you feel about things too—after all, some of the fault is theirs if the class is dull—but don't criticize them too much. Maybe they'll have some ideas on how to improve things.

If in the beginning they are too embarrassed to discuss the recitation, give them a few minutes to write down (anonymously) what they think, then discuss their comments with them during the next recitation hour. You'll have to assure them by your manner that you really want feedback and aren't going to give a D– to those who stick their necks out.

Once or twice during the term you might invite the students to discuss broader aspects of the course—its pace, the text, the assignments, grading, administration, whether they feel motivated to study it, its difficulty, and so on. Sometimes you can pick up this sort of information by coming early to the classroom and chatting with the students there.

Other evaluation processes

- Give out a course evaluation questionnaire near the end of the term and you'll learn things that will help you in your next class. Or give one out in the middle of the term, and it will help your present class.

- Bring a cassette tape recorder to class one day, and listen to the playback.

- Some departments can videotape your recitation, which lets you see yourself as others see you. You might request such a program if it doesn't exist, because it provides excellent feedback.

- Ask someone in the department whose teaching you respect—perhaps another recitation instructor—to observe your class.

At the end of this booklet is a two-page form that can be copied and used as a checklist for observing classes.

Small Groups: Tutorials and Office Hours

In some courses, in addition to having lectures and recitations, the students in a recitation section are further subdivided into several tutorial groups of 4-5 each, meeting once a week with a tutor who grades their problem sets and gives further assistance.

Group tutorials

Conducting such a tutorial is an excellent way to start out with a course—you'll be close to the students and will perceive at first hand the sort of difficulties they have. It's a good way to start out in teaching, too—problems of getting acquainted and interaction are much less troublesome. But don't think that a small group makes them automatically disappear. Some tutors seem to end up lecturing to three students as if they were teaching 300.

Running the tutorial

Assuming you've gone (as you should) to the lectures and recitations, you'll probably feel it your duty to clear up difficulties students had with poor explanations there. Do so, but don't spend all your time on this. Reinforce your explanations by working problems with them. There are different ways to do this.

One way is for everyone to contribute ideas, while you act as secretary at the board. This way they interact and get to know each other, and it's all the more social.

Or you can have students work individually, at their seats or the board, and you offer individual help. Maybe at the end one of them explains to the others. Working at their seats has the advantage that they have something to carry away with them; working at the board means they can see each other's work. Do both.

In general, try to keep the hour varied—some of this, some of that, but emphasizing whatever seems to work best with the particular group. Encourage them to keep track of difficulties during the week and bring them in. What bothers one student is likely to be a problem for others also.

Handling students of widely varying ability in a small tutorial is difficult. Your primary responsibility has to be to the ones having troubles. The others can help them, or work on more difficult things while you help them.

The tutor and the recitation teacher

The divided responsibility causes problems. If you're running a small tutorial, you will probably feel closer to the students than to the lecturer or recitation teacher. Be tactful however about blaming the lecturer or recitation teacher for the students' troubles: "This point seemed to confuse some of you. . .", not "R_____ sure did a lousy job of explaining this. . .".

The recitation teacher for a section which also has tutorials may feel somewhat distant from the students, since someone else is grading their problem sets and clearing up their confusions. Sometimes the tutor is felt to be sitting as a silent critic in the recitation. That's bad. Students should see recitation teacher and tutor working together in the classroom. The tutor should speak up in recitation if there's a suggestion to be made, and the recitation teacher should encourage this.

The teacher and tutor should meet regularly—weekly, or at least biweekly—to discuss what sort of errors students are making on the problem sets and what sort of troubles they are having with the course in general. Conveying the complaints of the students is an important responsibility, since the feedback provided can be a significant influence on the course.

Individual tutorials and office hours

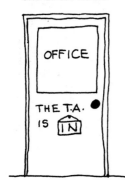

Some of the above applies also to one-on-one tutorial sessions with students, either by appointment, at a clinic where you sit in a room and help whoever comes in, or as part of your regular office hours. For this one-on-one tutoring the main points are

- Knowing your stuff
- Being a good listener: try to find out what the difficulty is—ask, or have the student try to do the work in front of you and watch what he or she does, or look at graded quizzes and problem sets.
- Being flexible about trying one explanation after another, until you find one that works
- Having patience, patience, patience: no matter how frustrated you may get, try not to show it.

It is quite possible to handle several students at once in such individual sessions: one student can be thinking about what you've just said and working, while you are helping a second on a new problem. Always introduce them to each other. If you've just explained something to one student and another comes in with the same question, ask the first student to explain to the second, while you look on or help a third. It's good for everyone, and many students actually prefer to get help from other students rather than from the staff.

Problem Students and Students with Problems

The classroom and section meeting are the business end of education. Here students and staff deal together with the stuff of the intellect. Here also idiosyncrasies in both become apparent. Sometimes unusual behavior can be overlooked, but not if it disrupts the work of the group or upsets some of its members. Some problems you can handle yourself; more serious ones need to be referred to professionals competent in handling them. In some sense you are the "antennae of the system" and need to exert your influence in making it humane. Here are a few kinds of problem students you may encounter.

The loud-mouth

There's a young man in the first row who talks too much: he blurts out answers before others have a chance, asks complicated questions off the subject being discussed, or holds forth at length on a pet topic. An occasional digression is fine, but if this behavior persists for several meetings you need to take action.

Start indirectly, trying to head off trouble by saying "Let's spread the answers around a bit," and obviously passing over the loud-mouth, or saying, "That's a fascinating topic; let's discuss it after class."

If this does not work, it is confrontation time. As class breaks up, ask him to stay for a minute. Then in the empty classroom, in an office with the door closed, or over coffee where you can talk privately, tell him you have a problem and ask for his help. (The problem is indeed yours; after all, talking too much is not causing any pain to him!) Tell him you value his participation and wish more students contributed. If his answers in class are generally good, say so. Do not criticize him but point out matter-of-factly the difficulty of involving everyone if someone dominates. If his knowledge of the subject is really advanced, should he be in another class?

When the point has been made and acknowedged, change the subject, At the end repeat your determination to involve everyone in the class.

The possible date

It's a situation that occurs in various sexual guises, but typically let's say you're a male recitation teacher of average libido, who can't help noticing that cute female with the turned-up nose who sits in the second row, asks good questions, and sometimes comes up after class with a comment. You find yourself drawn to her and wonder if you should ask her out for coffee, or dinner and a show. And see where it leads. Our advice is brief and it's official MIT policy: DON'T.

You are not equals: this is a power relationship. This means that if you've misinterpreted things and she really doesn't want to go out with you, she may feel pressured and unhappy. If she does like you and you become involved with each other, the unequal relationship and possible sense of ulterior motives on either side can make for a lot of grief or trouble. Even if the two of you can manage it, others in the section will be suspicious and cynical, and your relationship with them will deteriorate.

So don't make a move until the semester is over and the grades come out. (If it's love at first sight and the two of you are desperate, she should at a minimum change sections.)

Though the above scenario occasionally happens with the sexes reversed, the problem for a female recitation teacher is more likely to be an ambiguous invitation—say to dinner— from an attractive male in her class. It's better to ask a few discreet questions (will there be other faculty there? other students?) than to get in an awkward situation later. Again: take no first steps which could lead to a serious involvement during the semester.

The silent student

"You have the right to remain silent. Anything you say may be used against you." Students should have at least as much right to silence as those arrested for a crime. No one can or should force participation. Nevertheless, students who attend regularly but never speak up may be waiting for encouragement. Learn their names; when handing back assignments, compliment a particular solution if you can; ask them to come in during your office hour. After calling on three or four others in class, call on the silent one by name.

None of this may work. That is all right. Only gentleness is justified here.

The dependent student

It may be flattering to have a student continually asking questions after class, filling your office hours, perhaps seeking extensive personal advice. It can also be a pain in the neck. If so, you have to decide where the problem is and act.

- If the student has too weak a background because of missing prerequisites, it's not your job to supply these by personal tutoring—recommend delaying the subject a semester, or a transfer to an easier version of the course if one is available.

- If the prerequisites are all there, but the student is just very weak in the subject, you do have some responsibility to help. However, it should be shared with the other tutoring services available (see the next sections). Set firm bounds on the amount of time you can spend and let the student know.

- Some students are "dependent types"—they like to be taken care of, or are used to leaning on some one. But this doesn't have to be you. Anyway, it's not good for them. Encourage them to stand on their own feet: "I could help you with this, but I think you'll learn more doing it by yourself."

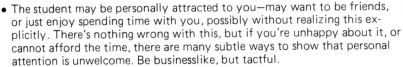

- The student may be personally attracted to you—may want to be friends, or just enjoy spending time with you, possibly without realizing this explicitly. There's nothing wrong with this, but if you're unhappy about it, or cannot afford the time, there are many subtle ways to show that personal attention is unwelcome. Be businesslike, but tactful.

The troubled student

Students may come to you with personal or academic problems, because they like and trust you. Listen as you would listen to a friend who wanted to talk, and respond as you would to a friend, offering what advice you can.

If you feel more experienced guidance is called for—this will certainly be so if the problems are serious, with possibly deep-lying causes—then be wary of offering too much advice of your own. It's better to refer the student to one of the Institute's counseling resources, usually in the Office of the Dean for Student Affairs. Normally students are referred there by their advisors, and you should find out if the student has talked to his or her advisor, or would like to. You need to be tactful. In general, for serious cases, try to have the student make the appropriate appointment by phone while still in your office.

If it seems better initially not to involve the Dean's Office, in some cases the Campus Patrol can be the right resource—it enjoys excellent relations with students, the contact can be very informal, and it is helpful if the law may be involved.

A more formal resource is the Medical Department. Psychiatric referral is usually done through the Dean's Office, but you can call the Medical Department yourself if the case seems very serious or an emergency.

All of these resources are available to you for consultation. If you're not sure what the best thing to do is, call them up and ask their advice. They want you to, and you may be able to head off serious trouble.

If a life may be at stake, no matter how unlikely you think it is, act fast. One section leader noticed razor scratches on the wrists of a student who came to his office for help with a physics problem. He said nothing, but phoned the Dean's Office immediately afterwards. Within hours the student had called in, expressing appreciation that "someone would care".

Academic problems

You should get in touch with students who do poorly on the first exam, or who miss a couple of homework assignments, to find out what the problem is. It is usually you who will have to do the seeking out, since students are embarrassed by poor grades or performance and thus feel awkward about seeing you. Many will try to pretend to themselves there is no problem, or optimistically hope that things will go better "when they get things together". Freshmen often behave this way, when after twelve years of success in school they find themselves for the first time in academic trouble and have to cope with the resulting internal and external pressures.

Your job is to confront these students gently with reality and get them to make sensible plans for their academic work. They may need suggestions on how to study and manage their time, as well as help with the course material. Encourage them to take advantage of the tutoring services available—those offered by the large freshman courses, the Black Students Union Tutoring Program, the dorm and fraternity tutors. Perhaps there is someone else in the section who lives nearby with whom they could study, or some upperclassman who can help.

If the problems seem serious, particularly if they extend to the other courses (be tactful about inquiring), you should contact the student's advisor. If the advisor doesn't seem like an appropriate first resource, it may be better to call the Undergraduate Academic Support Office (particularly for freshmen and undesignated sophomores), or where appropriate the Office of Minority Education. These offices do academic counseling and have programs designed to assist students with study skills and time management.

If you cannot get in touch with the student, at least alert the advisor.

Epilogue: A Word about You

Although we hope you've been able to pick up something useful from this booklet, it is from the actual teaching that you will learn the most, both about the subject and about yourself. It can be rewarding and a lot of fun.

For some of you, it may become too all-absorbing. A few teachers get so deeply involved that their own graduate work or research starts to suffer. Here are a few danger signs. You may be in trouble if

- You regularly hand out notes because you don't like the way the book or lectures are going and feel you can do a much better job yourself;

- You find yourself very involved with one or more students, see them a lot in your office, and feel personally responsible for seeing them improve;

- You are spending a great deal of time grading assignments, writing elaborate comments on the papers, and finding it hard to keep up with all this work;

- You find yourself worrying a lot about how things are going and spending excessive amounts of time in preparation.

These things may be great for your students, but you have to think of your-self as well. You must pace yourself—you are going to be teaching for a while, a lifetime perhaps—and you can't sprint the whole distance. Instructors who give their utmost for one semester sometimes feel "burned out" the next semester; they just can't seem to get interested in doing it again. Save part of yourself for the future, and look for a balance between your teaching and your other activities.

No two teachers are alike. The suggestions in this little booklet are meant to apply to a wide variety of teaching styles, but in the end they are just sugges-tions. You have to choose and adapt them to your own personality and style. You must feel comfortable yourself with what you are doing if your students are also to feel comfortable.

Remember that recitations also differ. If your particular section isn't going well, do all you can to improve things, but don't take it all personally and blame yourself. Even the most gifted teachers rarely have a recitation like the one we dreamt about on the first page.

We said it a few times already, but it's worth repeating here at the end. One of your most important resources is the other teachers in the course. Talk to them about the course material, and about problems you are having with the students or with pedagogy. Exchange classroom visits with them. Some courses sponsor a meeting of the recitation teachers (without the lecturer) after two or three weeks have gone by, so they can trade experiences or comments. But talk to the lecturer too – he or she needs the input from you about the course, and may in turn have valuable suggestions to offer.

If things go well or go badly, or are somewhere in the middle, we'd like to know. In future editions of this booklet, we'd like to include suggestions from you—what was misleading, what was helpful, other good ideas you had which worked out well in your classroom. Drop a line to the Undergraduate Academic Support Office (Room 7-103) and maybe you'll see yourself in print.

Good luck!

Classroom Observation Checklist

teacher	subject	day/hr	room	recit., lect., tut.?

observer	date	time: from___to___	number of students

Check or circle items deserving attention; comment in space provided.

general topics covered
types of work, percent of time spent on each
(problemsolving, lecture review, new material, etc.)

Summary _____
_____
_____

knows subject, confident, competent
knows what goes on in lecture, knows book
well-organized in class
right amount of material

Preparation and Pedagogy

Overall impressions_____
_____
_____

clear or fuzzy?   orients, summarizes
at right level   repeats just enough
goes easy → hard   stresses hard points
starts from problem   defines the tasks

Explanations _____
_____
_____
_____

is prepared with good problems and sol'ns
uses a variety of methods
explains the thought processes
hits the right level

Problem-solving _____
_____
_____
_____

interesting, relevant   offers motivation
well-prepared, clear   right level
gives examples   overview, summary
starts from problem   defines the tasks
relates material to lectures, other courses

Lecture (new material or commentary)_____
_____
_____
_____

recognized?   what were they?
how handled?   class help asked for?
admitted or covered up?

Inadequacies or errors _____
_____

Classroom technique

right size, legible   level
no squeak   no glare
clear diagrams   diagrams labeled

Board writing _____
_____
_____

gives complete story   used in orderly way
minimal erasing
gives problem number and summary
board clean at beginning of class

Board use _____
_____
_____

clear, understandable   faces class
right volume   right speed
varied, not monotone   er—OK?—alright?
energetic, not bored or sighing or droopy

Speech _____
_____
_____

starts on time   ends on time

well-apportioned   not rushed

Time _____

_____

hair-combing   beard-stroking

excessive pacing   speech mannerisms

Mannerisms _____

_____

listens carefully   repeats Q if needed

encourages questions   patient

answers well

Interaction

Questions _____

_____

_____

formulates problems   poses questions

"what's wrong with. . ?"   poses paradoxes

solicits class help

gives adequate time for response

Class participation _____

_____

_____

proper seating — not bunched in back

good eye contact

uses names

Other aids _____

_____

_____

enthusiastic, finds it exciting

neutral

seems bored by it

Attitudes and atmosphere

To material _____

_____

_____

friendly   supportive

concerned   available

distant, withdrawn   indifferent

hostile   sarcastic

To students _____

_____

_____

attentive, interested, responsive

daydreaming, doing other work, bored,

silent, hostile

are there problem students?

Student attitude _____

_____

_____

rushed/slow   exciting/dull

stiff/informal

too singleminded   too meandering

Overall atmosphere _____

_____

_____

Other comments _____

_____

_____

_____

_____

Suggestions for improvement _____

_____

_____

_____

_____

A packet of materials is sent to newly accepted TAs several months before their anticipated fall arrival date. It includes:

LETTER OF WELCOME from the Associate Chair for Graduate Studies.

A SCHEDULE of orientation activities for a Monday, Tuesday and Thursday in late August. Wednesday is left open on the schedule "in order to give you one full day to 'settle in'." It is suggested that Wednesday be used to: complete housing arrangements, open a local bank account (some banking information is given), register car or investigate transportation to campus, move into office, socialize.

AREA MAP, with directions to the campus.

CAMPUS MAP.

NAME AND ADDRESS form to send in for an advance exchange list to go to all incoming graduate students.   (Interesting idea!)

SHORT-TERM HOUSING INFORMATION. Several choices are described of "emergency" or temporary housing available while permanent housing is sought. Advance reservation information is given.

Guide to Graduate Life: 60 page university booklet with information about housing, transportation, food, income, services and organizations, academic information, state laws, the local media—newspapers, TV, radio, etc., and on- and off-campus entertainment.

APARTMENT GUIDE with directory and map to commercial housing; also university apartment and advance reservation information.

GRADUATE COURSE OFFERINGS for Fall with time of day and professor.

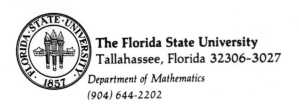

The Florida State University
Tallahassee, Florida 32306-3027
Department of Mathematics
(904) 644-2202

MAT 5941 NOTES
Internship in College Mathematics Teaching
SUMMER 1988

Welcome to the instructional staff of the Department of Mathematics!

This course will include a variety of organized and individual activities designed to improve success in and satisfaction with college mathematics teaching at the lower division level. We call it "the TA class."

Professor Bettye Anne Case, who directs the work of all TAs, instructs this Internship class. Ms. Annette Blackwelder, an Instructor and Course Coordinator in the department assists with class instruction and activities.

Dr. Case: 221 LOV   Ms. Blackwelder: 110 LOV

Dr. Case's secretary is Ms. Deborah Russ. Her office is 205-D LOV. She will try to answer any questions you may have.

Please locate your mailbox in 208 LOV and check it twice daily. See Ms. Michaels, 224 LOV, to be sure you are on the payroll.

Temporary office assignments will be made in TA class after 6/27.

Be sure to fill in a time grid with your other class(es)/obligations on the first day of TA class. This will be used to schedule your activities.

READINGS:

        GIG and these Notes
        Academic Advisement information
        CHALKING IT UP
        COLLEGE MATHEMATICS: Suggestions on How to Teach It
        Excerpted pages from "The Torch or the Firehose" (M.I.T.'s TA handbook).
        Editorial: "Intellectual Arrogance and Better Teaching Methods"
        Observations forms (general and RI)
        Student evaluation form (SIRS)
        Recitation Instructor Duties
        Grader Duties
        Student Syllabus and Instructor Notes for courses
        Texts for various FSU courses
        Problem-sheets for blackboard presentations
        Other readings to be announced

INTERNATIONAL TAs

A course in spoken English will be provided and is required during your first term. You must take the TSE at the first administration after you arrive on campus. Please see Ms. Deborah Russ, 205-D LOV, to sign up for it.

It is essential that you work to reduce any unusual pronounciations and the "accent" which students may find difficult to understand. No one can improve your speech for you. You must work on it by:

--talking with Americans
--reading aloud; using a tape recorder
--listening to TV and movies
--going shopping
--practicing presenting problems at the board to an empty room (or to friends)
--read some of the comic strips and ask someone if you don't understand the "jokes"
--discussing American food, sports, etc., with people

Also Read:

"Listening Comprehension and Accent Problems"; "How do Americans speak?"
Manual for Foreign Assistants, Althen

ATTENDANCE AND COURSE REQUIREMENTS:

TA class has a number of different types of activities, duties, and assignments designed to acquaint you with the department's teaching activities and support systems, and the Florida climate affecting lower division mathematics teaching.

(a)   Unless otherwise announced, class meets each Monday, Tuesday, Wednesday, and Thursday 3:30-4:45, in LOV 102 or   LOV 201. (On days we videotape, please plan to stay until 5 or 5:15.) You are expected to attend all classes.

(b)   You will also have some duties at other hours. Please be prompt and careful on all assigned duties.

(c)   You will be assigned one or two advanced TAs or Instructors to observe and do some grading for. (Their class time is fixed, and you must attend the assigned number of times.)

(d)   You will additionally be asked to observe a couple of class days in each 1- and 2000 level course being taught this summer.

(e)   Dr. Karsteter will tell you about help available to students, and you will be assigned Help Center hours to work each week.

(f)   You will be assisting Mr. Dodaro in the proctoring of freshman placement tests in June, July and August. The first such test is Monday, June 27. Be at Fisher Lecture Hall at 2 p.m. Thursday, June 23 for an organization meeting with Mr. Dodaro.

In general, please understand that all activities of this class are part of your TA appointment, which requires 20 hours work per week. You may not miss any classes or fail to complete any assignments without danger to your appointment.

FIRST BLACKBOARD PRESENTATION

You will do problems in a "simulated" class environment. You will assume that you "taught" the appropriate lesson at the last class meeting and that several students are asking about the problem which is a medium level problem from the homework. (You should re-explain the theory adequately but obviously cannot take too long because you have to get on to the new lesson.)

First write your name on the board as "Mr. Smith" or "Ms. Chen", and write the book page and problem number. Then proceed to work the problem as you would for a class keeping in mind the following criteria.

> Loud enough?
> Eye contact?
> Uses blackboard well? (book page and problem number written down)
> Concepts clearly presented?
> Encourages class participation?
> Needed concepts reviewed without getting "bogged down"?
> Effective presentation? (speech, confidence, mannerisms)?

ASSIGNMENTS

There are a variety of activities and assignments for TA class. Some involve presentations for you to make in class. Some are written assignments, which are numbered below, and on the following pages (**1**., **2**., etc ...) are to be handed in. Each number must begin on a separate page, and it is helpful if you write in ink (or type).

Hand in the written assignments to Ms. Blackwelder's box by Monday, August 1 or earlier in class.

1. Read materials you have been given. Prepare at least 10 questions which occur to you because of this reading. BRING THESE TO TA CLASS ON THURSDAY, JULY 7. Included is a DRAFT of the new GIG. Please "proof" it before July 7 and be ready to give us corrections and improvements.

2. Please tell us what you think about the helpfulness of the activities of the TA class. Please indicate additional helpful things we might do in the future. Comment on the dealings you had with the department before/upon arrival and make suggestions, etc...

3. Write comments on your experience with the AMP proctoring.

4. You have been videotaped several times. Please comment on you earliest tape. (Tapes may be viewed in the Math Help Center, 110 EDU.) *Did you improve*?

TA class Schedule

Thursday	June 23	2:00 p.m.	Meet in 272 FLH with Mr. Dodaro AMP proctoring organizational meeting.
Monday	June 27	7:30 a.m.	AMP proctoring 275 FLH
Monday	June 27	3:30 p.m.	First class meeting 102 LOV
Tuesday	June 28	3:30 p.m.	TA Class meets, 102 LOV

Wednesday	June 29	7:30 a.m.	AMP proctoring, 275 FLH
Wednesday	June 29	3:30 p.m.	TA Class meets in 110 EDU with Dr. Karester for a tour of the Math Help Center
Thursday	June 30	3:30 p.m.	Problem Presentations 102 LOV

ASSIGNMENTS, (cont.)

5. Note your name in the left column below; if you do not find your name, please ask which TA/Instructor(s) you are paired with. You should observe these classes a minimum of 7 or 8 time each, learn about or do some grading, and possibly present some material later in the term. Write up, to hand in, an account of your experience working with these classes and instructors. (Check carefully; you are down to attend two different types of classes.)

Experienced Instructor	His Office	Course to Observe	Class Hours/ Room
Salstrand [Sun][Rogers]	210 MH-5R	1102	MWF 11-12 107 LOV
Lederberg [Chen, J.] [Zhang] [Turner]	218 STB	1102	MTWRF 12:30-1:45 200 LOV
Encinosa [Rizk] [O'Donnell]	205 Outer EDU	1102	MTWRF 12:30-1:45 101 LOV
Youn [Gutierrez] [Ellis]	304B EDU	1102	MTWRF 9:30-10:45 103 LOV
Rhoda [Lutfi] [Bense] [Lennertz]	108B EDU	1141	MTWRF 11-12:15 114 HTL
Hung [Ellis] [Turner]	210 MH-1L	1140	MWF 2-3, 114HTL
Du [Martin] [O'Donnell]	210 MH-6R	1140	MTWRF 9:30-10:45 114 HTL
Daugherty [Zhang] [Chen]	210 MH-6R	3233	MWF 8-9, 102 LOV
Daugherty [Lutfi] [Sun] [Gutierrez]	210 MH-6R	3233	MTWRF 12:30-1:45 107 LOV
Desloge [Lennertz] [Bense] [Rizk]	210 MH-5L	3233	MWF 9:30-10:30 102 LOV
Chen, P. [Rogers] [Martin]	210 MH-5R	3233	MWF 2-3, 217 HTL

6. There are several other classes which you may observe, these are listed below. If you are assigned to 1102 above, then you should observe an 1141 class, once or twice. If you are assigned to 1141 above, then you should observe 1102. Additionally, everyone must observe once or twice in both the lecture and recitation part of the 1141 sections listed below. Observe any of the following classes that you can. Write a few sentences about your reactions to each of the classes which you observe.

Experienced Instructor	Course to Observe	Class Hours/ Room
King	1141	MTWRF 3:30-4:45, 106 LOV
LeNoir	1141 (Lecture)	MWF 8-9, 101 CAR
Choi	1141 (Recitation)	R 8-9, 101 CAR
Yang	1141 (Recitation)	R 12:30-1:30, 201 LOV
Zhuang	1141 (Recitation)	R 2-3, 200 LOV
Heydari	2311 (Calc I)	MTWR 11-12, 103 LOV
Diao	3312 (Calc II)	MTWR 11-12, 106 LOV
Dodaro	3312 (Calc II)	MTWR 12:30-1:30, 102 LOV
Wooland	1113 (Trig)	MTWR 12:30-1:45, 114 HTL

<u>7</u>.   Dr. Karsteter will conduct a tour of the Help Center Wednesday, June 29, at 3:30 p.m. Class will meet at 110 EDU. She will assign your regular working hours there each week and give you instructions. Write a critique of your experience including comments on the following assignments:

((a), (b) and (c) must be done at a time other than your Help Center shift. It is ot necessary to sign up in advance.)

(a)   View one CLAST video tape in the HC. Check out the tape from instructor working HC PLATO room. You must have your student ID. Be sure that you learn how to work the VCR.

(b)   Sign on to PLATO and complete one objective on the PLATO-CLAST review. To sign on to this review, type "student" for PLATO name and type "clasta" for your PLATO group. PLATO terminals are located in the Help Center, Strozier Library, and Stone Building.

(c)   Sign on to PLATO and complete the first module for Unit 1 of MAC 1102. Type "last name (space) first name" in lower case letters for your PLATO name and type "MA1102TA" for your PLATO group.

<u>ASSIGNMENTS</u>, (cont.)

<u>8</u>.   Comment briefly on

a)   appropriate clothing for a TA
b)   outside employment of TAs (part-time jobs)
c)   appropriate forms of address for instructors and students in the university setting
d)   personal relationships between instructors and students
e)   sexual harrassment and gender differentiation in the classroom
f)   tutoring for money
g)   departmental test security
h)   class preparation and professionalism

<u>9</u>.   <u>Seven</u> consequences of legislation strongly affect our lower division and service courses. Write a few sentences about the effects of the following:

1)   State "Common Course Numbering" System.
2)   Board of Regents ruling that universities may not teach "precollege" courses.
3)   Class size mandate: 27.5 average at the 1000 and 2000-level.
4)   What does the "Gordon Rule" require and of whom?

5)   What is the CLAST? Help Center has copies for you to see of a "practice" test which we think is pretty close to the real thing (it fits the specifications). Your OPINION OF THE ITEMS. Do you know who takes it and when?

6)   State University -- Community College Articulation Agreement.

7)   220 TSE requirement.

10.   Read the editorial "Intellectual arrogance and better teaching methods."Note the questions to think about on the back. Get out the SIRS and Classroom Observation forms you were given. Thinking about all of this, indicate some of the items on the SIRS form that you plan to try to be very careful that the students perceive "well" about you. Note: You must have SIRS administered in each section you instruct or for which you conduct recitation.

11.   Using THE TORCH OR THE FIREHOSE. Please re-read the entire excerpt you have. Briefly, in the light of your experience since the beginning of the term, discuss

a)   What is the "Sandwich Method"?

b)   What is a "birds-eye view and review"?

c)   One sentence each to describe "Some (4) roles teachers play (*and mostly shouldn't*).

d)   Comment on "the loud mouth". (*What do YOU do?*)

e)   Comment on "the potential date". (*Note the sexism in the wording of the last sentence!*) What is our department's policy on this?

12.   Write comments about anything you try to look up but don't find, or any other ideas you may have for improving the usefulness of the GIG.

13.   You have been worked with on fair and consistent grading involving partial credit. You will also be working with the instructor/TA whose class you are observing. Describe in you own words a "philosophy of the granting of partial credit" which you think summarizes these discussions.

14.   Mr. Gaunt conducted a tour of the ARC. Please comment on this tour, its benefit to you, and on "ARC testing" in general.

15.   ALL FOR WHOM AMERICAN ENGLISH WAS NOT THE LANGUAGE OF THE CHILDHOOD HOME: Please discuss this factor, widely perceived by students to be a problem, and ways you have found to cope with it. Comment on the helpfulness of the MANUAL FOR FOREIGN TEACHING ASSISTANTS and the articles from the New York Times and Higher Education Chronicle.

THOSE FOR WHOM AMERICAN ENGLISH WAS THE LANGUAGE OF THE CHILDHOOD HOME: Any suggestions for your fellow students? Have you INDIVIDUALLY done anything to help one of them or their students? *What could you do*?

16.   Discuss cheating (*"academic dishonesty"*) prevention and apprehension.

17.   Give comments on Test making using the Test Banks.

COURSE GUIDELINE FOR THE
TA WORKSHOP (MATH 300)

MANUAL FOR THE INSTRUCTOR

(i)

Preface

The TA Workshop, in its present format, has been with us for eight years and has by and large proven its usefulness. Amorphous though it was at the beginning, it has slowly taken shape through the years. By now, the point has been reached where the content of the course should be standardized and the cumulative experience of its past instructors be codified for future reference. Since I have taught this course more often than anybody else, I have (quite foolishly) taken it upon myself to perform this thankless task. My secret hope is that, with the availability of this course guideline, more of our colleagues can be persuaded to teach the workshop without undue coercion.

While I am responsible for the actual writing of these notes, the whole undertaking of compiling this Guideline must be regarded as a collaborative effort among many people. At the time of its inception in 1975, Murray Protter and I were the first instructors of the Workshop. Our joint strategy sessions to chart the unknown at that time spawned many of the ideas recorded in the succeeding pages. Moreover, Murray's line-by-line critique of a preliminary version of this guideline led to vast improvements on its readability and accuracy. The many comments and suggestions of Paul Chernoff, Leon Henkin, Moe Hirsch, Jerry Marsden and Alan Weinstein at various stages of preparation were equally essential to its completion. Whatever merit there is in what follows must therefore be shared among all these colleagues.

I would like to thank Christine Baker for a beautiful job of typing.

H. Wu
June, 1983

I have made a few corrections to bring this Guideline up-to-date as of December, 1987. Note that TA's are now referred to as GSI's (Graduate Student Instructors), but in order to minimize the clerical work, I have left all the "TA's" in the following pages unchanged.

H. Wu
December, 1987

CONTENTS

I. Departmental Rules Concerning the TA Workshop (Math 300)----------1.

II. Things to Do Before Classes Begin----------2.

III. The Orientation Meeting----------4.

IV. Structure of the Workshop: An Overview----------10.

V. General Discussion of Teaching----------11.

VI. Miscellaneous Suggestions----------16.

VII. Viewing of TAs' Videotapes----------18.

VIII. Mini-teaching Session----------20.

IX. Final Remarks----------29.

I. Departmental Rules Concerning
the TA Workshop (Math 300)   1.

A. The course consists of an Orientation Meeting before the beginning of classes and one meeting per week (1½ to 2 hours) for seven or eight weeks. The enrollment of each section of the Workshop is generically limited to 20.

B. All beginning TA's must enroll in the Workshop during their first semester of teaching. Showing up for the Workshop, like showing up for each section meeting, is part of the TA's duties.

[Exemption from the Workshop is often sought but almost never granted. In particular, extensive teaching experience in another university or in another department of the Berkeley campus is not a valid ground for exemption. On the other hand, an instance where exemption was granted was to someone who had taught two years of calculus at MIT and could produce documents of student evaluations of his teaching with a nearly perfect score; even this person was not exempted from the Orientation Meeting however.]

C. Grading policy of the Workshop: successful completion of the Workshop (a grade of "S") is determined by the student's satisfactory attendance and participation in the Workshop activities. Any absence from a meeting, except for reasons beyond the student's control, will jeopardize a student's passing the course and his/her employment as a Graduate Student Instructor. Meetings missed for legitimate reasons must be made up in a manner determined by the instructor of the Workshop.

IX. Final Remarks

(A) The preceding sections by no means exhaust all the possibilities of running the Workshop. With a little experience, each instructor would obviously develop his/her own style. For example, one past instructor asked the TA's to write essays on what they did to improve their teaching over the quarter. Another thing he tried was getting the TA's to experiment with specific techniques (e.g., "breaking students into groups to solve a problem") and report back the following week. Another instructor made it a point of visiting every TA in the Workshop in his/her TA section, and spent at least half an hour afterwards to discuss his findings with the TA. This without a doubt is what ideally all Workshop instructors should attempt to do, because it brings the most direct benefits to each TA in the Workshop. Unfortunately, this practice is also extremely time and energy consuming, and the dedication required of its execution is definitely beyond the normal call of duty.

However, when all is said and done, there is no way that every aspect of teaching can be touched on in seven or eight Workshop meetings. The main virtue of the Workshop would seem to be that of forcing each TA to think, at least once in his/her life, about teaching and possibly about improving it.

(B) One should take note of the fact that every TA in the Workshop attends it under coercion. Those that do so willingly are rare, and rarer still are those who do not consider the Workshop an outright waste of time. This being the case, the Workshop instructor must learn to cope with a hostile audience. There is no known method of dealing with this which guarantees success.

What complicates the situation even more is the fact that, in contrast with the teaching of mathematics, there is no universal criterion to judge wrong or right in teaching the "teaching of mathematics". There is no authority to fall back on save one's own common sense. If one tries to be considerate of the TA's feelings and abstain from ever giving clearcut criticisms, one runs the risk of losing the respect of the class. After all, if everything is supposed to have both good and bad points, why then should the TA's bother to come to the Workshop? On the other hand, open criticism tends to incite open rebellion, particularly from a few of the more aggressive TA's.

Teaching the Workshop has its rewarding aspects, but it is well to know the pitfalls ahead of time.

(C) An informal survey seems to indicate that since the inception of the Workshop in 1975, the general level of teaching performance among TA's has definitely gotten better. The most flagrant abuses (coming to class unprepared, not holding office hours, etc.) have been sharply reduced. Thus, there is reason to believe that the Workshop performs an important role in the overall operation of the department.

II. Things to Do Before Classes Begin

A. Since the Orientation Meeting takes place during the week preceding the first day of classes (generically Wednesday 2-3:30 p.m. in the Fall Semester and Friday 2-3:30 p.m. in the Spring Semester), be sure to coordinate with the Supervisor of the Student Services Unit (967 Evans) to inform all new TA's of the time and place of, as well as the need to attend, the Orientation Meeting.

B. Check with the Office of Student Services (967 Evans) no later than one week before the beginning of classes to make sure that no TA in the Workshop is assigned a self-paced course (1S, 16S, or PS) for that semester.

(There are two reasons for this: (1) it is important to have actual experience teaching a section in order to exchange ideas with other TA's and (2) those who happen to be TA's of a self-paced course would miss the opportunity to be videotaped. Being videotaped and analyzing the tape critically is one of the main features of the Workshop (see §IV). One of the past Workshop instructors strongly recommends that all the TA's in the Workshop be assigned sections of the same subject (Math 16A, for instance). If possible, the instructor could even arrange to be teaching the very same subject himself/herself during that semester. Because of the administrative difficulties involved, it is not possible to elevate these recommendations to the status of departmental rules. However, the benefits of such arrangements are obvious and many, and each Workshop instructor would do well to make serious inquiries into their feasibility each time.)

C. Call the Education TV Office in 9 Dwinelle (TV Office for short), no less than a week before classes:

(C1) Call 3-8637 to make tentative arrangements for videotaping of the TA's in the Workshop (cf. §VII). You need to supply the TV Office with at least the following information:

-A rough estimate of the number of TA's in the Workshop (obtainable in advance from the Office of Student Services in 967 Evans).

-The length of time each TA should be videotaped (the usual length is 15-20 minutes).

-When you can hand over the teaching schedules of the TA's in the Workshop (usually the day after the first meeting of the workshop, see §III (E)).

3.

-When you want the videotaping completed (certainly before your last videotape-viewing session, see §VII).

(The TV Office needs the TA's teaching schedule in order to set up the videotaping. If the viewing of the videotapes is expected to start on the third meeting of the Workshop (as is the common practice), then handing over this teaching schedule right after the first Workshop meeting is a must.)

(C2) Request that at least 3/4 of each tape should be devoted to the TA and his boardwork.

(Obviously 3/4 is not a canonical number. It is suggested here only because the focussing of the camera is crucial to getting the maximum value from the videotape-viewing sessions (see §VII). In the past, much footage was wasted because of random and prolonged focussing on the facial expressions of the students.)

(C3) Reserve one of the three classrooms with TA's (4 Dw or 117 Dw or B4 Dw) for three meetings of the Workshop between the 3rd and 7th (or 8th) weeks of the semester for the showing of the videotapes. As of 1987, the person in the TV Office in charge of this task is Ms. Ann Juell, 3-8633.

(The reason for this particular time period (3rd to 7th or 8th) is that probably no videotape would be ready before the third week and the Workshop is finished by the 7th or 8th. Reserving the room does not necessarily mean that you must use it all three times; cf. §IV. These three rooms are in great demand, so the earlier you get to this task the better.)

III. The Orientation Meeting

The purpose of the Orientation Meeting is to acquaint the new TA's with the most basic facts of teaching life at Berkeley before they meet their students for the first time. For this reason, this meeting takes place before the beginning of classes. Since many TA's are new to the Berkeley campus, they have to be told everything about how our system works.

The first order of business would be to announce the grading policy of the Workshop.

Among the many possible topics for discussion, those listed in the Handout below would seem to be the rock bottom minimum. One suggestion would be to pass this out in the Orientation Meeting and then elaborate on each topic verbally.

5.

HANDOUT

BASIC DO'S AND DONT'S FOR TA'S IN THE MATHEMATICS DEPT.

I. A. Always meet your classes promptly and keep them the full 50 minutes.

   B. Attend your professor's lectures and coordinate them with your own section(s).

   C. Announce your office hours at the first meeting and hold them faithfully throughout the semester.

   D. Return homework assignments regularly; be firm in supervising your reader.

II. A. Speak loudly and slowly.

    B. When speaking, maintain eye contact with students when appropriate and don't block the writing on the board.

III. A. Erase all the boards before beginning a class.

     B. Write legibly on the boards by pressing firmly on the chalk.

     C. Write systematically: start from the upper left corner of each board and go in an orderly manner down to the lower right corner.

     D. Never erase unless there is no board space left; always erase the oldest writing before the more recent ones.

     E. Better to have large board-writing than small.

     F. Don't cram everything into one board when other boards are available.

6.

The following gives a little more substance to the list in the Handout plus some essential topics not mentioned there.

A. The TA's should check their mailboxes often, particularly during the first two weeks of classes. Leaving a message in the TA's mailbox is the only way the professor or the Graduate Appointments Committee can communicate with the TA.

(It is common practice for a professor to leave notes in the TAs' mailboxes after a lecture to indicate what he/she would like to see done in the next TA section meeting. The TA must therefore be specially alert to this form of communication.)

B. Class size, change of sections, etc.

(B1) Students in each TA section are now enrolled (as of 1987) through an advanced enrollment procedure (ACE). Each TA is given a list of his or her students in each section before the first meeting. Normal enrollment is limited to at most 24 students. However, if the TA is experienced enough, or exceptionally accommodating, up to 27 students can be put in each section. This is the absolute maximum.

(B2) Up to the end of the third week of classes, adding a section to a student's study list or changing from one section to another is handled by the Head TA (one for each of 1A, 1B, 16A, 16B, 50A, 50B). The name of the Head TA is posted in the lobby of the 9th Floor of Evans, and can also be obtained from 970. Beginning with the fourth week, the TA handles such matters at his or her discretion.

(B3) Undergraduates can drop a course without any formality up to the end of the fifth week. Do not bother the Head TA with question concerning dropping.

(B4) Since students may drop a course without penalty and without consulting the TA or Professor up to the end of the 5th week, the TA has no precise idea of who his/her students are until an official list of enrollment arrives from the Registrar sometime near the midterm. Even this list may not be final. There is then no cause for panic if discrepancies crop up.

C. The formal duties of a TA.

(C1) One of the serious errors a TA can make is the failure to meet his/her class and let the students wait in vain. Each TA should arrange in advance with a fellow TA to be a replacement in case of a serious emergency or a bed-confining illness. If a replacement is not available, call up the Office of Student Services (642-2479) immediately.

7.

[This most obvious of duties appears also to be one regularly violated. One can hardly over-emphasize this point.]

(C2) The TA's serve as the Professor's deputies in the course. To this end:

(C2a) The TA should attend the professor's lectures. The idea is not that the TA would learn from the lectures, but that the TA must be aware of what goes on in the lectures in order to be able to coordinate the TA section with the lectures (e.g., complete an occasional computation left unfinished in the lecture, take up topics which are left out of the lecture for lack of time, give examples not given in the lecture, etc.) However, the TA should never repeat the lecture (see § v, (2a)).

[Traditionally the professor feels guilty or embarrassed about requiring the TA's attendance in the lectures. Experience indicates that this requirement is, by and large, essential for a successful TA section. The TA's should be impressed with this particular requirement from the start.]

(C2b) If the TA believes that the professor is making unreasonable demands on the TA's, the TA should discuss this honestly with the professor. The TA's can group together to do this if courage is lacking. In extreme cases, see the vice-chairman of instruction.

(C3) The TA's are the ones directly in charge of the readers of the course (a reader is someone who grades the homework papers for the TA section; he/she is usually an undergraduate mathematics major here.)

(C3a) The TA should be firm in supervising the reader. Make sure that the reader's work is both punctual and satisfactory; normally at most one week is allowed between the pickup and return of the homework papers by the reader. Do not hesitate to specify to the reader how the papers should be graded (consult with the professor if necessary). It is good to require that the readers provide a brief note with each homework set indicating where the students make the most mistakes. The TA should always look over the graded papers, if only briefly, before handing them back to the students.

[TA's often have a mistaken notion of being "nice" to readers without due regard for the welfare of the suffering students. An irate student storming into a professor's office to complain about failure to get back homeworks is an all too common occurrence.]

(C3b) Keep in mind that readers are allowed to spend ≤ 10 minutes per student per week. Do not make unreasonable demands on the reader. If the professor assigns too many problems per

8.

week, the TA should consult with the professor to arrive at a compromise grading policy. Be sure to inform the reader about it.

(C3c) The TA should report a delinquent reader immediately to the person in charge of hiring the readers (as of 1987, this is Harold Raffill of 968 Evans). In serious cases, the reader should be replaced promptly. This has been done on many occasions.

(C3d) The TA or the reader should keep a record of homework grades throughout the semester since most professors count the homework grade as part of the course grade.

(C4) The professor has no direct contact with the reader except in the extreme case of having to track down delinquent readers (see (C3b) above). The professor's decision on homework assignments also affects the reader's work (see (C3c) above).

(C5) The TA should never dismiss a class early simply because there is no material to cover. Think of problems to challenge a class, or simply call students to the board to do problems. In other words the TA should always have something in reserve to fill up the 50 minutes of each TA section meeting.

(C6) Hold office hours regularly and faithfully. Announce office number and office hours during the first TA section meeting. For a regular TA teaching two section, the normal number of office hours is three per week.

(C7) Participate in exam grading sessions for the midterms and final. The TA's work is not done at the end of the instructional period. He/she also has to help assign grades to the students in his/her sections.

D. The basic mechanics of being in front of a class.

These pertain to the items under headings II and III of the Handout. They are perhaps too obvious to bear any elaborate discussion except that experience indicates it is a good idea to go over these items one by one. One can of course enliven this litany with some humorous real-life anecdotes that spring from the infraction of these basic rules.

E. The workshop participants' teaching schedules.

This is not so much a discussion of the basic facts of teaching as setting up a schedule for the videotaping of the TA's in the Workshop (see §VI). The TA office (§II (c)) needs the room and time of each TA's sections. So be sure to tell the TA's during the Orientation Meeting to bring this information to the first Workshop meeting.

9.

A SUGGESTION. This may be the appropriate time to assign some reading material on teaching to the participants for the purpose of discussion in the first meeting of the Workshop. The Office of Student Service (967) has copies of:

    (i) College Mathematics: Suggestions on How to Teach It, MAA, 1979.

    (ii) Guideline for Teaching Assistants in Mathematics, by H.L. Alder.

    (iii) Helpful Hints to Instructors of Large Lecture Courses.

Item (i) has a good set of specific guidelines and rules for teaching in general. Since most TA's will eventually be teachers, they are interested in going over these guideline in class and discussing each of these for 5-10 minutes. See also §VI, (1), for another way of making use of (i).

Item (ii) is a set of notes written by Alder of UC Davis for the internal consumption of the Math. Dept. in UCD; it is reproduced here with his special permission. Item (iii) is a write-up whose specific purpose is to help the faculty members here cope with large lecture courses, but the TA's can certainly learn from it what kind of performance they should expect from their professors. Needless to say, the Workshop instructor should use any reading material on teaching that he/she considers appropriate.

10.

IV. Structure of the Workshop:
An Overview

By common consent among all past instructors, there are three basic components in this Workshop:

(A) General discussion of the techniques of teaching.

(B) Viewing of TAs' videotapes.

(C) Mini-teaching sessions or other devices for the concrete illustration of the abstract ideas.

There are pros and cons concerning what goes into each and which of these three should be emphasized over the other two. Such uncertainties notwithstanding, it seems safe to recommend that out of the seven or eight weeks allotted to the Workshop two weeks should go into each of these, with the remaining week or two to be disposed of in whatever manner the Workshop instructor sees fit. Moreover, since the first meeting of the Workshop starts from ground zero, it has to be devoted to (A) (both (B) and (c) require prior preparation, see § II (C), § VII and § VIII). However, the order of presenting (B) and (C) is up to the instructor.

The remainder of this Course Guideline will be devoted, in the main, to the discussion of these basic components. However, attention should be called to the other possible variations on this format briefly mentioned in § VI and § IX.

Note that there is a basic distinction between this Workshop in the Math. Dept. and the general kind of TA training program available elsewhere on campus. Mathematics is a formidable subject to the average student precisely because it is a highly technical subject requiring specialized skill. Therefore the basic objective of this Workshop is to equip the TA's with the basic skill to help them get this technical subject across to the students. A large part of the difficulty of teaching a TA section in Mathematics lies squarely in the technical aspect of the subject and any successful teacher must be, above all, technically proficient. Thus while this Workshop does not teach the new TA's mathematics per se, mathematics is very much in the foreground in every phase of the instruction in the Workshop. Given this highly specialized task, the Workshop instructor must address the problem of mathematics teaching before he/she takes up any related issues such as innovative approaches to teaching.

11.

V. General Discussion of Teaching

This discussion will probably take place in the first meeting of the Workshop, so here is the right place to mention the things that ought to be attended to right away in the first meeting:

(i) Inform the TA's again about the grading policy of the Workshop (§ I (C)).

(ii) Ask for the TAs' teaching schedules (time and place of each TA section). This schedule should be handed over to the TV office in 9 DW as soon as possible (§ II (C)).

(iii) Inform the TA's when the Workshop will be devoted to the viewing of the videotapes and where (§ II (c)).

Now to the discussion of teaching itself. The following lists a set of topics that could serve as a basis for discussion. Note that on the one hand abstract discussions of teaching should be avoided if at all possible. On the other, a little abstract discussion at the beginning would seem to be necessary and, for the first meeting at least, inevitable. The first five topics below may be considered basic.

(1) Preparation: always prepare well before meeting one's class.

Contrary to the notion that seems to be gaining currency, sweating vainly to do a calculus problem in front of the whole class is not necessarily an educational experience for the students, and is decidedly not the case if it happens repeatedly. Moreover, even in lower division mathematics, there are subtle points that can be explained well to a beginner only if one is well prepared.

(2) Content of a TA section.

(2a) It should not repeat any part of the professor's lecture.

[The TA should never yield to the canonical request:"I didn't understand the lecture at all. Would you be good enough to repeat it?" The TA's job is to complement the lecture, not repeat it. Occasionally, a one or two sentence remark covering some items in the lecture is allowable.]

(2b) It should not be another lecture.

[By "lecture", one means a presentation in a systematic way of a body of material. Such a presentation basically precludes TA-student interaction and is thus contrary to the purpose of a small TA section.]

(2c) It should not present a systematics improvement of the professor's lectures.

12.

[The story is told of a certain TA who undertook the systematic presentation of ε-δ proofs in the first quarter of a freshman calculus course because he felt that the professor's lectures "were not sufficiently rigorous."]

(2d) The TA may have to finish off things left unfinished in the lecture (especially if so requested by the professor); it may be devoted to answering specific questions or explaining subtle point in the material; it may be devoted to the working out of additional examples to complement the lectures.

(2e) A large bulk of the TA section should be devoted to the solution of homework problems for the students. Two points should be kept in mind in this connection though: (i) an assigned homework problem should be worked out at the board only after the homework papers have been collected; (ii) some TA's have exaggerated their role as problem solver for students to the point of devoting each hour exclusively to the working out of homework problems; in more extreme cases, some of our colleagues have been known to ask their TA's to do nothing other than calling students to the board to work out problems in each TA section meeting. Since the TA sections are supposed to serve more educational purposes than just solving problems, caution should be advised regarding the excesses in (ii).

(3) Encourage students' participation.

Student participation is impossible in the large, impersonal lectures, and is therefore reserved for the TA sections. The TA should create an atmosphere at once relaxing and informal so that students can feel comfortable enough to participate in class discussions. The creation and maintenance of this atmosphere require constant effort. Suffice it to say that no one trick is known to achieve this effect, and the many factors that enter into the picture may well be the focus of continued discussion in the Workshop. Here are a few simple hints on the pitfalls to avoid:

(3a) Insufficient waiting time: ask the class a question and immediately supply the answer. (To an instructor in front of a class, five seconds seem like a long time to wait for a response. Time it.)

(3b) Anticipate answers or excessive help: rush the student by completing his/her incomplete sentences or simply take the words out of his/her mouth. (To correct this habit, try the following trick: announce to the class that you will give 1 or 1½ minutes to seek an answer, then look at watch and follow through. In the meantime, encourage conversation among students to search for answer.)

(3c) Immediate reward: praise or criticism of a student's answer on the spot all the time.

13.

(3d) Belittling a student's mistakes.

There are two popular devices which increase student participation, but like most things in life they may not be of universal applicability. One is calling students to the board to work out homework problems. However, this should not be carried to excess (see the discussion in (2e) above). Another is getting students to work in small groups on a specific problem. In this case, the TA has to be careful about the compatibility problem within each group.

(4) Attention to boardwork.

The importance of a TA's boardwork can be easily appreciated by inspecting the students' notes --most of them are just replicas of what is on the board. Freshmen and sophomores are generally not experienced enough to learn either to copy selectively or to add the vital remarks of the oral presentation to their notes. Here are some simple suggestions for better boardwork beyond those in Part III of the Handout in §III.

(4a) Key words should always be written down, e.g. Theorem if a theorem is to follow, hypothesis and conclusion in a theorem (especially on the first year level), page number as well as the number of the problem to be worked out, proof or solution if a proof or solution is to be presented, etc.

(4b) Use underline, capitals, or boxes for emphasis, e.g. enclose both continuous and closed in boxes in the theorem that continuous functions on closed intervals are bounded.

(4c) Bring colored chalk for the many drawings that are obviously essential to any satisfactory explanation of mathematics at this level (colored chalk is available from 970 Evans).

(4d) Write out key statements in full; define all the symbols to be used. To put down only a formula or an equation and then speak about it is often disastrous. Try writing down the noteworthy features of the formula or equation instead, using whole sentences. Also, occasionally one puts down a statement (or equation) that is wrong and says that it is wrong; in such a case, be sure to write that it is wrong.

(4e) Try to be selective in writing down the details of a calculation, especially a long one. Too many details only confuse the student.

(4f) Use mimeograph (ditto) to avoid an excessive amount of copying of uninteresting (but perhaps necessary) details on the board. All TA's are entitled to the use of the mimeograph machine in the xerox room 958 Evans.

(5) How to deal with mistakes made in front of a class.

14.

All teachers make mistakes from time to time. Once the TA becomes aware of the mistake, he/she should magnify the mistake by calling special attention to it and to its correction. There is no faster way to lose the students' respect than trying to cover up a mistake.

(6) The handling of office hours.

Other than observing the most basic rules dictated by common sense (e.g., be there during the announced office hours, answer questions courteously even if they seem stupid to you, etc.), the TA may be called upon to deal with some thorny problems. For example:

(6a) Too many students show up during the designated office hour. The TA will have to make an instantaneous judgment as to whether he/she can answer all the questions sensibly within the time limit. He/she may wish to announce clearly that each student would have to be held to x minutes maximum in order to be fair to all concerned. Ask the late comers to come back near the end of the hour instead of wasting time waiting around. (However, take note of the fact that many students like to listen to the TA's answer to another student's question.) If the TA feels that the situation is impossible, he/she may begin by finding out if many of the students have identical questions (very often such is the case). If so, the TA can obviously explain this question at the board to one and all. If not, the TA can either schedule extra office hours later in the day, or ask some of the students to come back another day. If every office hour is jammed beyond reason, the TA should consult with his/her professor for a solution.

(6b) Student's exploitation of the TA. Some students have a knack of aggressively exploiting TA's by getting them to do the homework or even to digest the text for them. It takes only two or three such students to use up all the office hours each week. The TA should learn to be aware of this and prevent it from developing into a pattern. The TA can, for example, explain to the student that given the time available, the TA cannot afford to be the student's private tutor.

(6c) Personality conflict. In a one-to-one situation in the office, this is not an uncommon phenomenon. If possible, the TA should try to clear the air by honestly discussing the problem with the student.

(7) The relation between TA and undergraduates.

This relation is not a symmetric one: the TA is in a position of power. Thus the TA must be careful not to act toward any of his/her students in a way that is likely to be interpreted, correctly or otherwise, as taking advantage of this power, or as exerting pressure for favors which are not willingly granted. What may appear to a TA as friendly interest may appear to a student as sexual harassment.

15.

[This topic is brought up because there have been such cases involving our department. They are ambiguous and embarrassing, and upsetting for all concerned. It is best if they can be avoided in the first place.]

(8) The handling of homework, quizzes, exams and grades.

The professor in charge of the course usually dictates how he/she wants the homework, quizzes, midterm and final weighted in the computation of the course grade. The TA should find out such information right from the start, if indeed this information is not made explicit by the professor. In most cases, a certain amount of input from the TA concerning the homework and quiz grades is encouraged; again, check with the professor.

The grading policy for each large lecture course should be uniform for all the sections. The TA's might have to request jointly that the professor fix such a policy if one is not already available.

(9) Know the basic facts of undergraduate life:

The last day in the semester that a course may be dropped without penalty and without a petition to the dean -- look up the calendar in the Schedule of Classes.

Meaning of the "I" grade ("incomplete") -- look up the Catalogue, under "Grades of Scholarship".

Meaning of the grades D+,D-,F, and how to erase them by repeating the course -- look up the Catalogue, under "Repetition of Courses."

(10) What to do if the TA discovers that the professor in charge of the course does not seem to be aware of the precepts the TA's are being taught.

This is not a common problem but it does come up. The Workshop instructor can legitimately use it to reinforce the need for this kind of Workshop which did not exist in the 1960's.

16.

VI. Miscellaneous Suggestions

This section presents some material complementary to §V. The following items are not necessarily interrelated.

(1) Instead of the discussion in §V it is a valid alternative to assign reading material from "College Mathematics: Suggestions on How to Teach It" (see the end of §III) during the Orientation Meeting, and then base the discussions of the subsequent Workshop meetings on this material. One past Workshop instructor makes reading assignments from this pamphlet throughout the Workshop (i.e., even during the videotape viewing sessions) and uses the reading material to orient the discussion. There are obvious merits in this approach: the discussion in each meeting would be more focused and the coverage of various aspects of teaching more thorough and systematic.

(2) Arranging mutual visitations among the TA's. This can be done at the first meeting of the Workshop, when the TA's already know the schedule of their TA sections. Pair up the TA's so that the paired TA's would visit each other's TA section. The Workshop instructor may ask for a written report of each other's impression, or may simply ask the pair to discuss their findings.

Many refinements of this arrangement are possible. Instead of pairs, cyclic visits by trio's to each other's sections is another possibility. Each pair or trio may be asked to discuss before each visit what possible weaknesses to look for, and to report to the Workshop instructor only the difference in opinion between the visitor and the "visitee". Or that each pair or trio may be asked to visit each other twice, the first time to concentrate on each other's weaknesses only, and then to observe the improvements (if any) the second time around. The Workshop instructor may choose to discuss in class the more interesting among these reports.

In theory this gives each TA the opportunity to receive some honest feedback in a real-world teaching situation. For this reason, such visitations are invaluable. In practice, however, one must be careful about the following factors that can detract from the usefulness of this arrangement: (i) each participant of the Workshop, being new to teaching, is quite possibly not sensitive enough to pick up many of the strengths and weaknesses of the "visitee", (ii) the insecurity of a TA new at the job plus a false sense of camaraderie may cause the visitor to be reluctant to honestly criticize the "visitee".

If the "visitee" is not identified in the written report to the Workshop instructor, the results may well be more honest.

(3) Instead of the abstract discussion of teaching during the second meeting of the Workshop, the following alternative has been tried with great success.

Ask each TA to describe for five minutes what he/she did during the first TA section meeting. For a novice in teaching, this sharing of experience can do wonders to his/her confidence. (In a class of 20, this will take up more than one two-hour meeting since there is a discussion after each 5 minute report.)

17.

A variant of this scheme is the following. In the third or fourth meeting of the Workshop, pick some TA's at random to recount, for 5-10 minutes, the special or unusual problems he/she has thus far encountered in teaching and his/her solutions of the problems. Obviously, much animated discussion would follow the recounting of these problems.

(4) Distribute 8 to 10 old final exams from different professors (970 Evans has these exams), all from the same course, e.g. Math 1A. Then ask the students to analyze them (as homework) for (a) thoroughness, (b) fairness, (c) coverage, (d) too much or too little theory, etc. This is important not only because it forces the TA's to "come down to the level of freshmen", but also because most TA's will be involved in making up quizzes and need to learn the difficulties in composing a good exam at any level. This also furnishes good discussion topics.

18.

VII. Viewing of the TA's Videotapes

The earliest the TA's videotapes can be shown in the Workshop is the third meeting (see § II (C), and § III (E)). Recall that the viewing of these videotapes can only take place in 4 Dwinelle or 117 Dwinelle or 4B Dwinelle (as of 1983 anyway) as these are the only rooms on campus with the requisite facilities. Thus be sure to announce to the Workshop this change of room ahead of time.

Tips for the videotape-viewing sessions:

(A) The Workshop instructor should preview the available videotapes himself/herself (go to the TV Office in 9 Dwinelle) and make a selection of those tapes to be shown in class.

Since each videotape last 15 minutes or more, and there is usually at least 15 minutes of discussion per tape, each 100 minute Workshop meeting, only has time for three videotapes at most (two is more realistic). Furthermore, since there are a maximum of three viewing sessions (see § IV), only 6 to 9 tapes from those of the 15 to 20 TA's in the Workshop can be shown to the class. Some selection is necessary. There are other reasons why the Workshop instructor must pick and choose. One is that the dedication and skill of the video-camera persons vary, so that some tapes are intrinsically unacceptable for showing in front of a class. Moreover, it would be of greater educational value to the class if only tapes of the two extremes on the scale of excellence are shown; they serve better to stimulate discussion. It takes very little time for the Workshop instructor to scan through the tapes in order to make the appropriate choices.

[An alternate way to select tapes for showing in class is the following: ask each TA to select a portion of his/her tape (after previewing) about which he/she thinks discussion would be of value.]

(B) It is suggested that a TA's consent to have his/her tape shown in class be dispensed with altogether. After all, a student in a public-speaking class need not be consulted before he/she is summoned to deliver a speech to the class.

(C) One can hardly overemphasize the fact that, in addition to watching out for the technical aspects of teaching per se (in the sense of § V), the Workshop instructor should pay particular attention to the mathematical content of each tape. Is the TA's explanation of continuity even more difficult to understand than the text? Is the TA wasting time on abstractions inappropriate in a freshman class? Could a well-drawn picture have saved ten minutes of desultory explanation at the board? Did the TA fail to underscore the key points of the solution? etc., etc.

As mentioned earlier in §IV, a teacher who is equipped with all the teaching techniques known to mankind but is otherwise not in full command of the mathematics is not going to make it as a Mathematics TA. Equal stress must be placed on mathematical substance as well as on general teaching techniques.

19.

(D) In addition to viewing and discussing the TA's tapes in class, an equally important part of the Workshop is for the TA to carefully study his/her own tape in private (the TA simply inquires at the TV Office, 9 Dwinelle). It is a surprising fact that there is always a sizable portion of the TA's in the Workshop who refuse to look at their own tapes. A possible remedy is to require a brief report form each TA on his/her own videotape.

(E) For the purpose of orienting the audience to each tape, it is suggested that the TA, whose tape is going to be shown, should state clearly what the course is and what material was being taught at the time of the videotaping.

(F) Method of Soliciting the other TAs' comments on the tapes. TA's are generally reluctant to criticize each other. Of course, there will always be one or two in each Workshop who are so aggressive as to attack everybody and everything, but they are of no help in generating useful comments either. The following method has been tried with success in breaking the ice: bring along some scrap papers and pass them out to the TA's before a tape is shown. Then ask the TA's to jot down their reactions to the tape, unsigned. Collect the papers and read out the comments before attempting any discussion. This takes a little more time, but the results are usually quite startling.

Of course once the Workshop instructor senses that the ice has been broken, he/she should forget about this trick and proceed to the class discussion directly.

A variant of the above method is the following. Instruct the class to be prepared to spend 5 minutes after viewing a tape to write their critical comments, making it clear that you, the Workshop instructor, would do the same. Then after the 5 minutes are over, ask each TA to read what he/she has written (you do likewise, of course). This insures a good discussion after the reading of each set of comments.

(G) Most past Workshop instructors remarked on the difficulty of getting a meaningful discussion going concerning the tapes. The preceding methods should eliminate this difficulty. Another common observation is that after viewing ten tapes or so, the saturation point is already reached. Implicit in this observation is the practice of showing the videotapes consecutively without attempting a serious discussion of each tape individually. It is for this reason that a prior selection of the tapes should be made before showing them in class ((A) above) and that at most three tapes be shown per Workshop meeting. This line of reasoning also suggests that there be only two videotape-viewing sessions rather than three.

20.

VIII. Mini-teaching Session

Two kinds of mini-teaching sessions have been tried. We will discuss them separately.

(1) The purpose of these mini-sessions.

Since the TA's are not supposed to lecture in their sections ($\S$ V(2b)) and since they are only asked to teach elementary material in their sections, why ask them to lecture on more advanced topics now? First of all, appropriate elementary materials from calculus texts for the purpose of such mini-sessions are rather difficult to come by, so the mixing-in of more advanced topics is at times necessary. The advanced topics also serve to relieve boredom. Second, the more advanced materials are more effective in revealing the potential weaknesses of the TA's in presenting mathematics orally. Third, these mini-sessions should be thought of as the finger-exercise pieces of the pianist or the calisthenics in the spring training of the baseball players. Typically, the following problems show up in the mini-sessions:

(a) Inability to cover the required material in the allotted time; one cannot overemphasize the need to be sensitive to the time constraints.

(b) Lack of consideration for the sophistication of the audience. For example, when asked to deliver a lecture to sophomores, the TA spends time discussing the subtleties of "definition by induction."

(c) Overestimate of the audience's ability to follow intricate and unmotivated arguments at the board. For example, in trying to lecture on Example I below, the TA tries to repeat the gruesome argument verbatim.

(d) The oral presentation sounds like reading from a text: no intuitive explanation, no motivation, and no heuristic considerations.

(e) Poor mechanics: bad boardwork, lack of voice modulation, no eye contact, rapid or indistinct speech, etc.

(2) The Workshop instructor should of course solicit comments from the other TA's on each mini-lecture as well as offer his/her own comments. If by chance the other TA's fail to provide candid and useful comments, the trick of $\S$VII (F) can be again employed. As for the Workshop instructor's comments, they obviously require a lot of tact. But if the TA's performance is entirely off the mark (unfortunately very often the case), the Workshop instructor would have to be forthright in saying so. As an example, one TA, when asked to lecture on Example 2 below, spent all 15 minutes on a rigorous proof of the case r=2 (i.e., formula (1.1) in Example 2). Obviously one must look for polite ways of saying that this is unacceptable, but there is no getting around saying it.

(3) Experience indicates that the TA's would appreciate hearing the Workshop instructor's own presentation of the same material. This is risky business because it puts the instructor's authority on the line, but it does afford the instructor the welcome opportunity to underline the essential points of his/her message.

(4) Experience also indicates that it would be best to ask all the TA's (or at least half of them) to prepare the same material although only one will be asked in the next meeting to do the actual presentation. This way the TA's would appreciate much more what is good or bad about the others person's presentation.

(5) As with every phase of this Workshop, some TA's and some past Workshop instructors believe that these mini-sessions are a waste of time. However, one survey over a two year period indicates that a clear majority of the TA's really believe this is the most valuable part of the Workshop.

-VIII B-

The second kind of mini-session more closely approximates the actual kind of teaching done by the TA's in their own section meetings. It involves getting a TA to present a 10-minute talk on some material he/she happens to be currently teaching. The TA should draft 5 to 6 students from his/her TA section and bring them to the Workshop on the day of his/her mini-session (the drafting of students usually presents no difficulty). Before the mini-session begins, the TA should give a little background of what he/she hopes to accomplish in the ten-minute talk. The talk itself is of course addressed to the students drafted from his/her section. The other TA's in the Workshop are observers in this "class". After the talk, the other TA's can discuss their reactions and evaluate how successful the teaching has been vis-à-vis the stated goals of this mini-session. At all time, the drafted students are there to provide the students' reaction as a point of reference. Providing the opportunity to check with the students themselves is the real asset of this kind of mini-session.

There is time for two such mini-sessions for each Workshop meeting. Comparison of the two kinds of mini-sessions described above is not difficult. The advantage of the first kind is that it is probably more stimulating to the TA's inasmuch as honest mathematics plays a nontrivial part. The advantage of the second kind is that it is more relevant. When done well, either one is capable of being a valuable educational experience.

23.

-VIII A-

This kind of mini-teaching session involves selecting some particularly abstract or formal passages from standard texts, handing out xerox copies to the TA's, and asking them to prepare a lecture on the material for the next meeting. Specify that, in preparing the lecture, each TA should bear in mind the following boundary conditions:

(1) the maximum time limit of the lecture (about 15 minutes would be best).

(2) the level of his/her audience (which is in reality the participants of the Workshop, of course), whether it be freshmen, sophomores, etc.

(3) the precise background material assumed (this is particularly important in view of the fact that each TA is handed only an excerpt from the middle of a text).

Some examples of such handouts together with comments are provided below.

Example 1.   Time allowed: 15 minutes.
                     Level: junior year.

Background assumed: the definition of $\mathbb{R}$ as an Archimedean ordered field. In the following, "$\mathbb{N}$" refers to the natural integers 0,1,2,3,..., "(IV)" refers to the axiom of nested intervals, and "(22.16)" refers to the theorem that every open interval in $\mathbb{R}$ contains an infinite number of rational numbers.

(2.2.17) *The set of real numbers is not denumerable.*

We argue by contradiction. Suppose we had a bijection $n \mapsto x_n$ from N onto R. We define a subsequence $n \mapsto p(n)$ of integers by induction in the following way: $p(0) = 0$, $p(1)$ is the smallest value of n such that $x_n > x_0$. Suppose that $p(n)$ has been defined for $n \leq 2m - 1$, and that $x_{p(2m-2)} < x_{p(2m-1)}$; then the set $]x_{p(2m-2)}, x_{p(2m-1)}[$ is infinite by (2.2.16), and we define $p(2m)$ to be the smallest integer $k > p(2m-1)$ such that $x_{p(2m-2)} < x_k < x_{p(2m-1)}$; then we define $p(2m+1)$ as the smallest integer $k > p(2m)$ such that $x_{p(2m)} < x_k < x_{p(2m-1)}$. It is immediate that the sequence $(p(n))$ is strictly increasing, hence $p(n) \geq n$ for all n. On the other hand, from the construction it follows that the closed interval $[x_{p(2m)}, x_{p(2m+1)}]$ is contained in the open interval $]x_{p(2m-2)}, x_{p(2m-1)}[$. By (IV) there is a real number y contained in all closed intervals $[x_{p(2m)}, x_{p(2m+1)}]$ and it cannot coincide with any extremity, since the extremities of an interval do not belong to the next one. Let q be the integer such that $y = x_q$, and let n be the largest integer such that $p(n) \leq q$, hence $q < p(n+1)$. Suppose first $n = 2m$; then, the relation $x_{p(2m)} < x_q < x_{p(2m+1)}$ contradicts the definition of $p(2m+1)$. If on the contrary $n = 2m - 1$, then the relation $x_{p(2m-2)} < x_q < x_{p(2m-1)}$ contradicts the definition of $p(2m)$. This ends the proof.

Comments. This material is taken from p.22 of Dieudonné's "Foundations of Modern Analysis". TA's usually complain that it is impossible to do a good job of presenting this complicated proof within 15 minutes. But in real life, one actually has less time than that to present this proof in a junior level course.

Most common errors. (1) The oral presentation is as formal and stiff as the one in the write-up.

(2) The definition of the intervals $[x_{p(2m)}, x_{p(2m+1)}]$ is reproduced exactly as written in the oral presentation.

(3) All 15 minutes are spent on the formal proof. No heuristic discussion of the really simple idea behind the formalism.

(4) One tries to cut the Gordian knot by dismissing the given proof altogether but giving the diagonal-process proof of Cantor instead. However, no mention is made of the big gap between the definition of $\mathbb{R}$ as an Archimedean ordered field (as done here) and the possibility of expressing every real number as a decimal.

Suggestions. (1) The complicated formal definition of the intervals as given in the write-up is perhaps not lecturable. One way out would be to tell the class explicitly that only a presentation of the main idea behind the proof will be attempted.

An oral presentation might be given along the following line. Suppose there is a one-one correspondence between $\mathbb{N}$ and $\mathbb{R}$, then each real number would acquire a unique "integral name", namely, the (nonnegative) integer to which this real number corresponds. First locate the real number with integral name 0. The real numbers to the right of this 0 give rise to a sequence of positive integers. Let the smallest integer be 6(say). Locate the real number with integral name 6.

Let I_1 be the open interval with these two real numbers corresponding to 0 and 6 as endpoints. Note the key fact: all real numbers in I_1 have integral names > 6. Now the set of integral names of the real numbers in I_1 again give rise to a sequence of positive integers bounded below by 7. Let 115(say) be the smallest number in this sequence. Again the integral names of the real numbers lying between these numbers with integral names 115 and 6 form a sequence of positive integers bounded below by 116, so there is a smallest one, say 2213. Let I_2 be the open interval between these two numbers with integral names 115 and 2213. Again observe the key fact: all numbers in I_2 have integral names > 2213. Repeat this process to arrive at a sequence of nested closed intervals $I_1 \supset I_2 \supset I_3 \supset \dots$. The integral names of the endpoints of $I_1, I_2, I_3, \dots$ form a

24.

strictly increasing sequence of positive integers p_i, while the integral names of the real numbers in each I_i exceed the integral names of the endpoints of I_i. What then is the integral name of the number in $\cap I_n$? It exceeds every member of p_i, so must be ∞. Contradiction.

(2) Since the explanation in (1) takes only about 10 minutes, the instructor might wish to encourage the prospective teachers to discuss two things in the remaining 5 minutes. One is that formal proofs are usually much more formidable than the simple ideas behind them, as this example clearly shows. Thus the value of an oral presentation lies in its potential to strip away the formalism and restore the simplicity of a proof. A second item is the Cantor diagonal-process, which is the usual proof of the noncountability of $\mathbb{R}$. Explain why the Cantor proof does not fit into this axiomatic scheme (without extra work, that is). On an intuitive level, the Cantor proof is much better and is worth knowing anyway because the reasoning has universal applicability.

(3) Encourage the students to compare the disparity between the simplicity of the idea behind a proof and the usual complications in a formal proof elsewhere in mathematics.

25.

Example 2. Time allowed: 15 minutes.
Level: sophomore year.

Background assumed: basic notions of set theory (union, intersection, cardinality of a finite set, etc.), and the principle of mathematical induction.

1.5 PRINCIPLE OF INCLUSION AND EXCLUSION

We present in this section some results related to the cardinality of finite sets. We shall use the notation $|P|$ to denote the cardinality of the set P. Some simple results, the derivation of which is left to the reader, are:

$$|P \cup Q| \leq |P| + |Q|$$

$$|P \cap Q| \leq \min(|P|, |Q|)$$

$$|P \oplus Q| = |P| + |Q| - 2|P \cap Q|$$

$$|P - Q| \geq |P| - |Q|$$

We show in the following a less obvious result. Let A_1 and A_2 be two sets. We want to show that

$$|A_1 \cup A_2| = |A_1| + |A_2| - |A_1 \cap A_2| \qquad (1.1)$$

Note that the sets A_1 and A_2 might have some common elements. To be specific, the number of common elements between A_1 and A_2 is $|A_1 \cap A_2|$. Each of these elements is counted twice in $|A_1| + |A_2|$ (once in $|A_1|$ and once in $|A_2|$), although it should be counted as one element in $|A_1 \cup A_2|$. Therefore, the *double count* of these elements in $|A_1| + |A_2|$ should be adjusted by the subtraction of the term $|A_1 \cap A_2|$ in the right-hand side of (1.1). As an example, suppose that among a set of 12 books, 6 are novels, 7 were published in the year 1975, and 3 are novels published in 1975. Let A_1 denote the set of books that are novels, and A_2 denote the set of books published in 1975. We have

$$|A_1| = 6 \qquad |A_2| = 7 \qquad |A_1 \cap A_2| = 3$$

Consequently, according to (1.1),

$$|A_1 \cup A_2| = 6 + 7 - 3 = 10$$

That is, there are ten books which are either novels or 1975 publications, or both. Consequently, among the 12 books there are 2 nonnovels that were not published in 1975.

Extending the result in (1.1), we have, for three sets A_1, A_2, and A_3,

$$|A_1 \cup A_2 \cup A_3| = |A_1| + |A_2| + |A_3| - |A_1 \cap A_2| - |A_1 \cap A_3|$$
$$- |A_2 \cap A_3| + |A_1 \cap A_2 \cap A_3| \qquad (1.2)$$

As we shall prove a more general result in the following, we shall not prove the result in (1.2) here. On the other hand, we suggest that a reader check the result in (1.2) by examining the Venn diagram in Fig. 1.5.
Let us consider some illustrative examples:

27.

Comment. This is taken from pp. 15-18 of Liu's "Elements of Descrete Mathematics", First edition (text of Math 55). Again the time limit of 15 minutes for this material is entirely realistic.

Common errors. (1) Tries to give a complete proof, including the special case of r = 2; sort of ramming the whole thing down the student's throat regardless of whether or not a sophomore is capable of taking in so much within such a short time. Clearly 15 minutes are not enough for this kind of full-scaled treatment.

(2) Gives rigorous proof of the case r = 2 (i.e. formula (1.1)). This usually takes up 10 minutes, thereby leaving very little time for the discussion of the general case.

(3) Fails to bring to the students' attention the presence of the alternating signs in the right side of (1.3) or to explain intuitively why such an alternating sum is to be expected.

Suggestions. (1) One possible way to spend the 15 minutes is the following. Quickly go through the case r = 2 by the use of Venn diagrams, and spend as much time as possible on the transition from r = 2 to r = 3. Emphasize that this is really the kernel of the induction proof in general. Moreover, one sees clearly in this case why the alternating signs must enter into (1.3).

(2) Should discourage the presentation of the general induction proof for several reasons. First, those interested few can always read the text. Second, the case r = 3 is the most important in applications and when this case is clearly understood, the general case is quite easy. Third, very few sophomores (if any) sitting in a classroom can absorb the general proof when it is presented on the blackboard in a hurry. Fourth, the general argument is not as enlightening as the passage from r = 2 to r = 3 becasue the technicalities tend to overwhelm the main ideas (or at least so it seems to a sophomore).

(3) Of course, one must make the general disclaimer that, logically speaking, knowing how to prove the case r = 3 is not enough; the formal proof in the text is necessary. However, for the purpose of intuitive understanding, the case r = 3 is quite sufficient.

26.

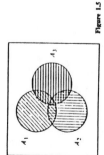

Figure 1.5

In the general case, for the sets $A_1, A_2, \ldots, A_n$ we have

$$|A_1 \cup A_2 \cup \cdots \cup A_n| = \sum |A_i| - \sum_{i<j} |A_i \cap A_j| + \sum_{i<j<k} |A_i \cap A_j \cap A_k|$$
$$+ \cdots + (-1)^{n-1} |A_1 \cap A_2 \cap \cdots \cap A_n| \qquad (1.3)$$

Although the result in (1.2) is not difficult to visualize, the result in (1.3) is not as obvious. We now prove (1.3) by induction on the number of sets r. Clearly, (1.1) can serve as the basis of induction. As the induction step, we assume that (1.3) is valid for any $r-1$ sets. We note first that, viewing $(A_1 \cup A_2 \cup \cdots \cup A_{r-1})$ and A_r as two sets, according to (1.1) we have

$$|A_1 \cup \cdots \cup A_{r-1} \cup A_r| = |A_1 \cup A_2 \cup \cdots \cup A_{r-1}| + |A_r|$$
$$- |A_r \cap (A_1 \cup A_2 \cup \cdots \cup A_{r-1})| \qquad (1.4)$$

Now

$$|A_r \cap (A_1 \cup A_2 \cup \cdots \cup A_{r-1})| = |(A_r \cap A_1) \cup (A_r \cap A_2) \cup \cdots \cup (A_r \cap A_{r-1})|$$

According to the induction hypothesis, for the $r-1$ sets $A_r \cap A_1, A_r \cap A_2, \ldots, A_r \cap A_{r-1}$, we have

$$|(A_r \cap A_1) \cup (A_r \cap A_2) \cup \cdots \cup (A_r \cap A_{r-1})|$$
$$= |A_r \cap A_1| + |A_r \cap A_2| + \cdots + |A_r \cap A_{r-1}|$$
$$- |(A_r \cap A_1) \cap (A_r \cap A_2)| - |(A_r \cap A_1) \cap (A_r \cap A_3)|$$
$$- \cdots$$
$$+ |(A_r \cap A_1) \cap (A_r \cap A_2) \cap (A_r \cap A_3)| + \cdots$$
$$- \cdots$$
$$+ (-1)^{r-2} |(A_r \cap A_1) \cap (A_r \cap A_2) \cap \cdots \cap (A_r \cap A_{r-1})|$$
$$= |A_r \cap A_1| + |A_r \cap A_2| + \cdots + |A_r \cap A_{r-1}|$$
$$- |A_r \cap A_1 \cap A_2| - |A_r \cap A_1 \cap A_3| - \cdots$$
$$+ |A_r \cap A_1 \cap A_2 \cap A_3| + \cdots$$
$$- \cdots$$
$$+ (-1)^{r-2} |A_r \cap A_1 \cap A_2 \cap \cdots \cap A_{r-1}| \qquad (1.5)$$

Also, according to the induction hypothesis, for the $r-1$ sets $A_1, A_2, \ldots, A_{r-1}$, we have

$$|A_1 \cup A_2 \cup \cdots \cup A_{r-1}| = |A_1| + |A_2| + \cdots$$
$$- |A_1 \cap A_2| - |A_1 \cap A_3| - \cdots$$
$$+ \cdots$$
$$+ (-1)^{r-2} |A_1 \cap A_2 \cap \cdots \cap A_{r-1}| \qquad (1.6)$$

Substituting (1.5) and (1.6) into (1.4), we obtain (1.3).

ETHICS AND PROFESSIONALISM:
QUESTIONS ARISING IN THE TEACHING OF COLLEGE MATHEMATICS

Many of the materials described in Chapters 3, 4, 5 and copied in this Appendix focus on the obligations, ethics and professionalism related to teaching. "Cheating" has always been a concern. Multisection courses give rise to many opportunities for academic dishonesty.

"Professionalism" includes the obvious such as preparing for classes and being on time, meeting office hours, grading papers promptly, and superficials such as observing local customs of dress, form of student address, etc...

The "privacy act" complicates previously prevalent practices such as posting grades. Media investigations into the teaching of international TAs may infringe upon the privacy rights of both undergraduates and TAs.

It is important that these issues be addressed in some manner with all instructors who are new to the particular department. Consideration of these matters by new TAs is sometimes left entirely to written materials but more often is addressed in both written materials and discussions. University- or college-wide orientation programs often include these topics.

In the description of the TA orientation program at Montana, reference is made to such discussion by the "EEO officer".

Some attention is indicated in the college-wide programs at Syracuse, Delaware, and St. Petersburg Junior College. The *Course Guideline for the TA Workshop* at Berkeley reminds the professor to consider these matters. Discussion is found in a number of the copied materials: See *For the Newest MAA 111 TA* (North Carolina State), *Helpful Hints* (Wisconsin), *How to Teach Math 151* (Ohio State), *Torch or Firehose* (MIT).

Included in Reznick's *Chalking It Up* is discussion of unruly students, cheating, examination security, grading complaints, office hours and timeliness, perceived sexual harassment, and personal relationships with students ("Dating a student is *a priori* an unprofessional act."). Some paragraphs are copied below from a university guide for nonprofessorial faculty. Reznick's section, "Coping", is incorporated by reference in this local guide, and there is then mention of some concerns perceived as important in that department's situation.

While the Department does not have a formal statement of professional ethics, you should remember to treat people with respect and to carry out your responsibilities with integrity. In particular, you have a deep obligation to your students. The teacher of a 3-hour course, for example, is guiding each student in the class through 2.5% of an undergraduate college education. It is important, therefore, to be friendly and relaxed in dealing with students, but at the same time to remember that you represent the Department, the College, the University, and the mathematical profession. Please get a copy of Chalking It Up from Ms. and read "Coping," p. 12-14.

At the risk of stating the completely obvious, it should be mentioned that it is unprofessional to date, even once, or have a personal relationship with a student enrolled in a class you are teaching, or to intrude on the personal dignity of any person. The department chair should be informed of any situation in which it seems necessary that an instructor teach a member of his or her family.

Because of the common syllabi and testing procedures, it is unprofessional to tutor a student in any section of a course you are teaching or assisting. One who has taught a 1000-level course should think carefully before agreeing to tutor a student in that course in a later semester. The burden is on the instructor who chooses to do this to be certain that there is no compromise of information not generally available. Those employed by the Department of Mathematics should not participate in a commercial tutoring service.

Responsibility for the security of the materials used to evaluate students (e.g., tests and quizzes) rests with the instructional staff. No one other than members of the instructional staff has a legitimate interest in knowing the contents of tests, quizzes, etc., and any inquiries should be referred (and reported) to Dr. If questions of an ethical nature should arise, please see Dr.   , Dr.   , or Dr.

To your students, you are the Department of Mathematics. You can enhance the Department's image by following directions so that standards are consistent, and by always explaining policies patiently to your students. Be sure you are well prepared for each class and professional in manner. Again stating the obvious, professionalism implies the following:

-- Your attire should be neat, clean, and appropriate. (Although undergraduates may wear tank tops or short shorts, these are not suitable for Instructors.) This is a hot climate; close attention to cleanliness and personal hygiene is essential.
-- You should be addressed by your students as Dr., Mr., Ms., Miss, or Mrs., and in the university situation it is most appropriate that you address your students similarly.
-- You must follow the Student Syllabus and Instructor Notes carefully. The Syllabus is a "legal contract."
-- You should work out each homework problem before assigning it. (The single complaint we have the most trouble dealing with is "My instructor couldn't do a problem assigned for homework." If you find yourself in this embarrassing situation, provide a complete dittoed solution at the next class.)

The "Privacy Act" prohibits revealing information of an "evaluative nature" to a third party without specific written permission. Do not give grades out over the telephone or post test results. Never discuss a student's performance with a parent unless the student is present and agrees to the discussion.

Do not post grades. Students wanting to know grades quickly should leave a stamped self-addressed post card. Do not tell students to see you to get their grades. (This disturbs your office mates and the staff.) The privacy act may also preclude your giving certain sorts of information to the media. See also TV, Newspaper...

T.V., Newspaper or Other Classroom Observers or Interviewers

Because of recent T.V. and newspaper coverage about teachers with foreign accents, we have asked the Dean of the faculties for clarification of the policies concerning classroom observations. Watch for future memos on this subject. Allowing media in classrooms could have serious legal ramifications. Therefore, if anyone asks to observe or film you and/or your students in your classroom, please let Dr.   , your Course Coordinator, or Dr.   know immediately and leave Dr.   a note. If you want to allow this, you will have to get a statement signed by each student that it is OK, and we will want to check these statements prior to the requested visitation.

Remember: You do not HAVE to talk to anyone other than your students, or be filmed in or out of the classroom. If you feel someone is trying to intimidate or harass you to do so, "just say no," walk away, and report the incident to Dr.

OBLIGATIONS

• • •

SYRACUSE UNIVERSITY TEACHING ASSISTANT ORIENTATION PROGRAM '88   1
International Student Portion

	Wednesday Aug 17	Thursday Aug 18	Friday Aug 19	Saturday Aug 20	Sunday Aug 21	Monday Aug 22	Tuesday Aug 23	
8 AM		Continental Breakfast 7:45-8:45	Continental Breakfast 7:45-8:45	Continental Breakfast 8:00 - 8:45	Continental Breakfast 8:00 - 8:45	Continental Breakfast 7:45-8:45	Continental Breakfast 7:45-8:45	8 AM
9 AM		Health Insurance, Employment, Social Security 9:00 - 10:00	Language Testing 1 9:00 - 9:55 / OIS-1 Taxes, Banking, Immigration 9:00 - 9:55	Oral Communications Workshop (As Assigned) 9:00 - 12:00	Oral Communications Workshop (As Assigned) 9:00 - 12:00	Getting to Know Syracuse University Undergraduates 9:00 - 10:00	Roles Of TA at SU 9:00 - 10:00 (Panel)	9 AM
10 AM	International Registration	ARENA Completing Forms 10:15 - 12:00	Language Testing 2 10:00-10:55 / OIS-2 Taxes, Banking, Immigration 10:00-10:55			Oral Communications I 10:15 - 12:00	Intensive Oral Communications 10:15 - 12:00 / Free Time	10 AM
11 AM			Language Testing 3 11:00-11:55 / OIS-3 Taxes, Banking, Immigration 11:00-11:55					11 AM
NOON		Lunch 12:15 - 1:15	Lunch 12:15 - 1:15	Lunch Buffet 12:15 - 1:15	Lunch Buffet 12:15 - 1:15	Lunch 12:15 - 1:15		NOON
1 PM								1 PM
2 PM		Culture Shock 1:30 - 2:30 (Panel)	American Slang 1:30 - 2:30			Teaching American Students 1:30 - 2:30 (Panel)		2 PM
3 PM		Housing And Banking (Forms Continued) 3:00 - 5:30	Intro. to Oral Communications (as assigned) 2:45-3:45 / Campus Tour - 1 2:45-3:45			Oral Communications II 2:45 - 5:30		3 PM
4 PM	Opening Session 3:30 - 4:30		Campus Tour - 2 4:00 - 5:00				(General Program Begins)	4 PM
5 PM	Small Group Discussions 4:45 - 5:45							5 PM
6 PM	Welcoming Dinner 6:00 - 7:30	Dinner* 6:00 - 7:00	Dinner* 6:00 - 7:00	Dinner* 6:00 - 7:00	Dinner With An American Family 5:30 - 9:00	Dinner* 6:00 - 7:00		6 PM
7 PM								7 PM
8 PM	Social Reception 7:45-10:00		GSO Movie Night 8:00 - 10:30					8 PM
9 PM								9 PM
10 PM								10 PM

[Large group sessions]   * Meal provided only to TAs residing in Sadler Hall.   CID 7/12/88

SYRACUSE UNIVERSITY TEACHING ASSISTANT ORIENTATION PROGRAM '88

2

	Tuesday Aug 23	Wednesday Aug 24	Thursday Aug 25	Friday Aug 26	Sat Aug 27	Sun Aug 28	Monday Aug 29	Tuesday Aug 30
8 AM		Continental Breakfast 7:45 - 8:45	Continental Breakfast 7:45 - 8:45	Continental Breakfast 7:45 - 8:45			Continental Breakfast 7:45 - 8:45	Continental Breakfast 7:45 - 8:45
9 AM	General Registration 8:00 - Noon (Optional Campus Tours) 9:00 10:00 11:00	Department Time or Options 9:00 - 12:00	Your First Class 9:00 - 10:00	Using Media 9:00 - 10:00			TA As Student/Teacher 9:00 - 10:30 (Panel)	Concurrent Sessions 9:15 - 10:30
10 AM								11:00 - 12:15
11 AM			Micro-Teaching II 10:15 - 12:15	Micro-Teaching III 10:15 - 12:15	Brunch* 10:00 - 12:00	Brunch* 10:00 - 12:00	Dealing With Problems I 10:45 - 12:15 (Discipline Specific)	• Grading • Laboratory Instruction • Field Trips • Review Sessions • Writing a Syllabus • Learning Theory • Tutoring • Media Production • Computer Workshops
NOON								
1 PM	Opening Luncheon 12:30 - 2:00	Departmental Lunch 12:30 - 1:30	Lunch 12:30 - 1:30	Lunch 12:30 - 1:30			Lunch 12:30 - 1:30	Lunch 12:30 - 1:30
2 PM	Small Groups 2:15 - 3:15	Lecturing 1:45 - 3:15	Leading A Discussion 1:45 - 2:15	Evaluating Students 1:45 - 2:45	State Fair (Optional)	City Tour (Optional)	Academic Honesty 1:45 - 2:45	Moving
3 PM			Discussion Exercise 2:30 - 4:00	Intro. to the Graduate School 3:00 - 4:00			Dealing With Problems II 3:00 - 5:00 (Discipline Specific)	
4 PM	Recital at Crouse College Followed by Keynote Address and Reception 3:30 - 5:45	Small Group Micro-Teaching I 3:30 - 5:30	Leading a Discussion (con't.) 4:15 - 5:00	Tours Library and Academic Computing 4:15 - 5:45				
5 PM								
6 PM	Dinner* 6:00 - 7:00	Dinner* 6:00 - 7:00	Dinner* 6:00 - 7:00	Dinner* 6:00 - 7:00			Dinner* 6:00 - 7:00	
7 PM								
8 PM		GSO Social 8:00 - 10:00	Oral Language Test (As Assigned) 7:00 - 9:00	GSO Happy Hour at the Inn Complete				
9 PM								
10 PM								

Large group sessions   * Meal provided only to TAs residing in Sadler Hall

CID 7/12/88

SYRACUSE UNIVERSITY TEACHING ASSISTANT ORIENTATION PROGRAM '88

3

DEPARTMENTAL TIME

	Wednesday Aug 31	Thursday Sept 1	Friday Sept 2		Saturday Sept 3	Sunday Sept 4	Monday Sept 5	Tuesday Sept 6	Thursday Sept 8	
7:30 AM										7:30 AM
8 AM										8 AM
9 AM										9 AM
10 AM		DEPARTMENT TIME								10 AM
11 AM		Library and OPEN	University opens for new undergraduates - Friday, September 2.							11 AM
NOON										NOON
1 PM										1 PM
2 PM								First Day Of Classes	Graduate Convocation 7:30 pm (followed by reception)	2 PM
3 PM								Graduate Registration		3 PM
4 PM		ADVANCED REGISTRATION (Next opportunity for Graduate Registration will be September 6.)								4 PM
5 PM										5 PM
6 PM										6 PM
7 PM										7 PM
8 PM										8 PM
9 PM										9 PM
10 PM										10 PM

Large group sessions

CID 7/12/88

Introduction Activity

<u>**Goals:**</u>

1. To provide TA's an opportunity to talk one-on-one to a number of other TA's while being observed by the trainers. This allows early identification of TA's with potential communication problems.
2. To develop interaction and a sharing of experiences among the TA's to begin to build a networking among them.

<u>**Procedure:**</u>   Chairs are arranged in long rows facing each other. For example, we use 6 rows (3 sets of pairs) to divide our group of 48 new TA's.

The trainer asks a question chosen from the list below. The TA's are given 3 minutes or so to share answers to the question and to note each other's name (name tags are used). Time is called and the TA's shift seats, as in musical chairs, being then paired with a new partner. Another question is asked, etc. At most we've been able to use 8 - 12 questions during this activity.

Sample Questions:

1. Describe your travels over the past 3 weeks - what was one interesting highlight?
2. Tell about one interesting or frustrating thing that happened to you in the last week.
3. What is your favorite area of mathematics? What is your least favorite area of mathematics?
4. During what time of day and under what conditions do you study best?
5. What will the most difficult part of teaching be?
6. If you could visit one country what would it be? What would you do there?
7. What hobbies next to mathematics do you enjoy?
8. What is the most important thing a teacher every did to help you learn?
9. What is the most relaxing way you could spend a week?
10. What was the last school you attended and tell one interesting thing about the city in which that school was?
11. What is your favorite pastime, next to doing mathematics?
12. What is one of the most intriguing things about teaching you expect to learn in the next year?
13. Name a game or sport you'd like to learn but never had the opportunity to.
14. If you could not go to graduate school in math, what would your second choice be?
15. If you could change one thing about the way you were taught math, what would that be?
16. How many schools have you attended in your life - what was your favorite one and why?
17. What is the most exciting thing you learned about math before you were age 15?
18. What is your favorite board game - how often do you get to play it?
19. What is one question about the University of Maryland you hope somebody answers for you in the next 2 weeks?
20. What would be the menu for your favorite meal?
21. What is your favorite color? What is the funniest thing you own in that color?
22. Describe your first visit to the state of Maryland.
23. Describe one thing you own which is in your favorite color.
24. What form of transportation do you expect to use a great deal during the next year?
25. What is your greatest expectation for the students you will be teaching?
26. What is your favorite piece of furniture in your last permanent residence - and why?
27. Describe the last time you participated in your favorite sport?
28. What is one goal you hope to accomplish by the end of September?
29. What is the last non-mathematics book you read and enjoyed? What was it about?

Department of Mathematics
Oregon State University

Orientation and Workshop
for
New Graduate Teaching Assistants and New and Visiting Faculty

September 21-23, 1987

Kidder 276 (and nearby rooms)

Monday, Sept 21

8:30-9:30   Introductions and Welcome to the department.
Dr. Frank Flaherty, Chairman—Welcome!

Dr. John Lee, assistant Chairman—Discussion of teaching responsibilities, course co-ordinator duties, grading policies and meaning of letter grades, office hours, MSLC tutoring time, teaching assignments.

9:30-9:45   Break—Coffee and Doughnuts

9:45-11:30   Dr. Mike Shaughnessy and Dr. Thomas Dick—teaching undergraduate mathematics at OSU. Description of lower division mathematics courses. Backgrounds of students in lower division mathematics courses. The Math Placement test. Handouts on teaching, and overview of the handouts. Upper division mathematics courses. The upcoming change to semesters, and curriculum revision at OSU. Tour of the main office, and computer rooms. Introduction to the office staff.

11:30-1:00   Lunch

1:00-1:45   Meet in the Math Sciences Learning Center (MSLC) in the west wing of the first floor of Kidder Hall. Dianne Hart, Director of the MSLC, will conduct a tour of the MSLC. Included will be the tutoring, testing, and computer services of the MSLC that support undergraduate mathematics instruction at OSU. Answer until correct tests (you get to take a sample test). Working as an MSLC tutor.

1:50-2:15   Back in Kidder 276. Dr. Hal Parks will talk about the graduate program at OSU, and will introduce the Graduate Committee. Discussion of graduate course requirements, program options, degree and program requirements, and advising procedures for new Graduate Students. (Advising times have been reserved for new GTA's to sign up on Tuesday and Wednesday afternoon, before registration on Thursday).

2:15-2:30   Break

2:30-4:00   Drs. Shaughnessy, Dick, and Chen. Sample minilecture and reaction and discussion. Discussion about teaching techniques and classroom conduct. Assignment of minilecture topics and practice test material for GTA's for the next two days. Textbook handouts.

4:00-4:15   "Hello" from "experienced?" TA's. GTA events throughout the year, social and otherwise.

Tuesday, Sept. 22

8:30-10:00   Drs. Shaughnessy, Dick, Musser, and Burger—your teaching committee. Minilectures with discussion from participants, experienced GTA's and faculty. Each minilecture should be about 15 minutes, with 15 minutes for feedback and discussion. The rest of us will play the role of "the class" during the minilecture.

In room 274—Dr. Burger, Elizabeth Altstadt, Gudrun Bodvarsson, Steffen Bunde, John Crow, and Sirilath De Silva.

Presenters: Bunde, Altstadt, and De Silva

In room 276—Dr. Shaughnessy Ramanathas Gowrie, Don Hickethier, Kathy Ivey, Michelle Jones, and Libby Krussel

Presenters: Jones, Hickethier, and Krussel

In room 278—Dr. Musser, Robert La Follette, Amy Owen, David Phillips, Johan Pieters, and Brian Vogt

Presenters: David Phillips, Amy Owen, and Brian Vogt

In room 280—Dr. Dick, Alison War, Andrew Warren, Peter White, Marvin Wilson, and Xueshan Yang

Presenters: White, War, and Wilson

10:00-10:20   Break

10:20-11:30   Continue minilectures

Room 274 Dr. Burger and crew, presenters: Crow and Bodvarsson

Room 276 Dr. Shaughnessy and crew, presenters: Ivey and Gowrie

Room 278 Dr. Musser and crew, presenters: La Follette and Pieters

Room 280 Dr. Dick and crew, presenters: Warren and Yang

11:30-1:00   Lunch

1:00-2:30   Kidder 276- Drs. Shaughnessy and Dick. Further discussion of teaching mathematics, making up tests, grading, dealing with problem students, cheating and academic honesty.

2:30-4:30   Individual advising and course scheduling for new graduate students. GTA's meet with members of the graduate committee to work on course selection and scheduling.

Wednesday, Sept. 23

8:30-10:00   Kidder 276 to start. Drs. Musser, Burger, Dick, and GTA crew. Discussion of sample tests composed and run off by the GTA participants. (Gowrie and Jones go with Musser; Hickethier and Ivey go with Burger, Krussel go with Dick; the rest stay with the same group as yesterday). Feedback to the new TA's provided on the difficulty, length, and appropriateness of test items.

10:00-10:15   Break

10:15-11:00   (about) Further discussion and presentation of sample tests.

11:00-11:45   Drs. Dick, Musser, and Burger. Discussion about the functions of the departmental teaching cojmmittee. Follow-up classroom visits during the academic year; mid-term evaluation forms; end of term evaluation forms; outstanding GTA awards; Carter award for undergraduate teaching; final remarks and summary of the workshop.

11:45-1:00   Lunch

1:00-3:00   Further individual advising and course scheduling with members of the graduate committee.

3:30-5:00(?)   Social at Clodfelter's (corner of Monroe and 16th, across the street from campus). Dr. Flaherty will be in attendance to lead the "ceremonies." All new GTA's, new and visiting faculty, workshop staff, and "old" GTA's and faculty are welcomed. You have been working pretty hard for a few days, time to get to know your colleagues better.

UNIVERSITY OF WISCONSIN   **MATHEMATICS DEPARTMENT**

ORIENTATION OF NEW GRADUATE STUDENTS AND
TEACHING ASSISTANTS--FALL 1987

MONDAY, AUGUST 24

I. <u>MORNING</u> (9:00 a.m., Ninth Floor Lounge, Van Vleck)
         ----- For all new Graduate Students -----
1. Welcome: Tom Kurtz, Department Chairperson.
2. Outline of Orientation, Registration Advice, Rules & Regulations for Graduate Students: Peter Ney.
3. The Employment Outlook.
         ----- Break -----
4. Faculty Advising Panel. Faculty members discuss introductory courses.

II. <u>AFTERNOON</u> (Ninth Floor)
1. (Noon to 3:30, Lounge) Individual Faculty Academic Advising.
2. (Noon to 3:30, Room 901) Advice from Coordinators.

TUESDAY, AUGUST 25

I. <u>MORNING</u>. Registration. Faculty advice available from 9:00 a.m. to 1:00 p.m. in Ninth Floor Lounge. <u>New TA Fall schedule card</u> due by 1:00. Return to Graduate Program Secretary, Sherry Lange in Room 218 Van Vleck.

II. <u>AFTERNOON</u> (1:30 p.m., B239 Van Vleck)
         ----- Orientation for New TAs -----
1. General Job Description: Steve Bauman, Coordinators, Graduate Students.
2. General discussion of teaching

WEDNESDAY, AUGUST 26

I. <u>MORNING</u> (9:00 a.m., B239 Van Vleck)
         ----- Orientation for New TAs -----
1. Teaching Assignments.
2. Discussions with Coordinators and Lecturers in Lecture Groups, and review of syllabus.

II. <u>AFTERNOON</u> (1:00 p.m.)
1. Practice Teaching in Lecture Groups.

THURSDAY, AUGUST 27

I. <u>MORNING</u>
1. Second Practice Teaching Session in Lecture Groups.
2. Evaluation of Orientation.

FRIDAY, AUGUST 28

----- Wine & Cheese Party -----
----- Ninth Floor Lounge, 3:00 -----
----- ALL DEPARTMENT INVITED -----

TO: Faculty, Advanced TAs, and Staff helping with orientation

You are involved in the New TA activities whenever your name occurs in boldface or where there are square brackets.

Fall Orientation, New TAs
1988

August 14, Sunday	2:00 p.m.	Introductory session with Dr. **Case** and Ms. **Blackwelder**, 201 LOV. Discuss Problem Presentations, divide into groups, discuss grids, packet, assign offices.
	3:00 p.m.	Getting acquainted; pictures taken. (Short break)
	4:00 p.m.	First problem presentations, all groups (**Blackwelder, Encinosa, Dodaro, Gaunt, Case**)
	5:00 p.m.	AMP organizational meeting with Mr. **Dodaro** and Mr. **Miller** and Dr. **Encinosa**, 201 LOV, new Fall TAs and Summer TAs who will do Fall AMPs.
	6:00 p.m.	Supper (Returning TAs and faculty advisors invited) (**Encinosa, Blackwelder, Dodaro, Case**)
	7:00 p.m.	Academic Advisement with Dr. **Wright** and Dr. **Loper** (discuss procedures for Touchtone Telephone Registration). Complete grids.
August 15, Monday	8:30 a.m.	Leave GRID in 208 LOV in Ms. **Russ'** mailbox. Meet with Ms. **Blackwelder** : course sequences, "totally departmental" courses vs. "ditto master syllabus" courses, details, materials etc., 201 LOV.
	9:30 a.m.	Problem Presentations, 2 groups (videotape) (TAs view videotape this afternoon in 107 EDU) (**LeNoir, Blackwelder. Gaunt:** view tapes)
	2:00 p.m.	AMP, 275 FLH (New TAs from summer also proctor AMP) (**Miller, Encinosa, Dodaro**)
	4:30 p.m.	Teaching 1102 "Nitty Gritty"; Quiz Making, 201 LOV (Ms. **Blackwelder**)
August 16, Tuesday		**TOUCHTONE TELEPHONE REGISTRATION FOR NEW GRADUATE STUDENTS 8:30 A.M. - 8:00 P.M. BE SURE TO TRY TO REGISTER TUESDAY! ASK MR. MILLER IF YOU NEED HELP!**
	7:30 a.m.	AMP, 275 FLH (New TAs from summer also proctor AMP) (Mr. **Miller**)
	10:00 a.m.	Teaching Interview sessions throughout day (one session will be videotaped). Assignments will be given Monday. Tape will be available for viewing in 107 EDU, after Noon on Wednesday. [All **instructors**]
	4:00 p.m.	Leave quiz in Ms. **Blackwelder's** mailbox

FALL ORIENTATION, NEW TAs
Page two

August 17, Wednesday		**TOUCHTONE TELEPHONE REGISTRATION FOR NEW GRADUATE STUDENTS 8:30 A.M. - 8:00 P.M.**

        8:00 a.m.　　Meet with Mr. **Miller** and staff, grids, fee waivers, eligibility, questions, etc., 201 LOV (Ms. **Russ**, Ms. **Ball**)

        10:15 a.m.　　TAs meet with Dr. **Case** in 201 LOV, SIRS, GIG, Ethics, Professionalism, Questions

        10:00 a.m.　　[**Course coordinators**: Decision meeting, 204 LOV] [**LeNoir**, lunch]

        12:00 noon　　[**Case, Dodaro, Blackwelder, LeNoir, Encinosa, Gaunt, Russ**: Scheduling]

        1:15 p.m.　　Help Center tour with Mr. **Gaunt**, Help Center Director

        3:00 p.m.　　Information session for New Graduate Students (University wide) 312 Oglesby Union

        3:00 p.m.　　[All **lecturers** meet with Dr. **Case** and 1102, 1141 **coordinators**]

August 18, Thursday　　**TOUCHTONE TELEPHONE REGISTRATION FOR NEW GRADUATE STUDENTS 8:30 A.M. - 8:00 P.M.**

        8:30 a.m.　　Problem presentations from material covered in first 2 weeks of courses (with **lecturers**) (**Blackwelder, Dodaro**)

        10:30 a.m.　　Grading Techniques/Acadmic Dishonesty Policies, Procedures, 201 LOV, (**LeNoir, Case**)

        12:30 p.m.　　AMP, 275 FLH

August 19, Friday　　10:00 a.m.　　Test Making from a Test Bank, 201 LOV (**Dodaro, LeNoir**)

                      [**Hall, Gaunt** or **Dodaro** do set up for Fall Meeting]

        3:30 p.m.　　Reception, 201 & 204 LOV

        4:00 p.m.　　Fall Organization Meeting, 101 LOV

Announce: MAT 5941, "TA Class", will meet regularly on Fridays at 2:30 p.m. (usually in 201 LOV) until Thanksgiving.

(The new TAs all register for the fall version of MAT 5941, Internship in the Teaching of College Mathematics. They meet on about ten Friday afternoons and complete the remainder of the activities described in the summer workshop notes.)

VILLANOVA   UNIVERSITY

Department of Mathematical Sciences

The Teaching Assistants' Orientation is a two day program in which University and departmental procedures are discussed the first day and a practice teaching session takes place the second day. The first day's topics are outlined below.

I. Introductions.

II. Procedural Details
   A. Distribution of academic calendars
   B. Assignment of course and supervisor
   C. Distribution of textbooks and instructor's manuals
   D. Distribution of departmental syllabi and discussion of the course coordinator concept
   E. Distribution of sample instructor syllabi
   F. Distribution of roll books and other supplies
   G. Discussion of excerpts from faculty handbook regarding class regulations
   H. Discussion of course descriptions and prerequisites
   I. Procedure for change of room requests
   J. Listing of Audio-Visual facilities
   K. Information on parking and dining privileges

III. Duties in addition to Teaching
   A. Extra help sessions for Math 1300 Calculus I
   B. Proctor exams
   C. Departmental duties

IV. Tour of Math Department supplies, facilities, computer room, and other selected areas of campus (assignment of offices, mailboxes, and keys)

V. Lunch (together) in Connelly Center

VI. In-Class Performance
   A. Course-Instructor evaluations
   B. Material organization
   C. Homework possibilities
   D. Testing and grading policies
   E. Office hours scheduling
   F. University attendance policy for freshmen
   G. Instructor dress code
   H. Teacher-student rapport
   I. Classroom atmosphere

The practice teaching session of the second day consists of 20 to 30 minute presentations by each teaching assistant on a mathematical topic inherent in their assigned Fall semester course. Constructive criticism, accompanied by fellow teaching assistants' comments, is administered by an experienced professor. The principles and techniques of instruction are thoroughly reviewed via questioning and discussion.

TEACHING ASSISTANT
EVALUATION

Name of TA _____   Course Number _____

Faculty Supervisor _____   Date _____

Number of students on current roster   _____

Number of students attending section observed _____

A.   Please respond to the following questions.

   1.   Does your TA have a comfortable understanding of the course
       and appropriate insights on the material   _____

   2.   Does your TA do the grading expected in the course?   _____

   3.   Does your TA use recitation time for problem-solving
       and discussion of the students questions?   _____

   4.   Are your TA's explanations at a level appropriate to
       the students and to the course?   _____

   5.   Would you be happy to have this TA working for you
       next quarter?   _____

B.   Please grade your TA's performance on an A, B, C, D, E, scale
    for each of the following:

   6.   The TA is well prepared for section.   _____

   7.   The TA's use of English is understood by his/her
       students.   _____

   8.   The TA understands the students questions and
       recognizes the students difficulties.   _____

   9.   The students are involved during recitation.   _____

  10.   The TA is responsible and cooperative.   _____

  11.   How would you rate the overall effectiveness of your TA.   _____

C.   Further Remarks:

CLEMSON UNIVERSITY

DEPARTMENT OF MATHEMATICAL SCIENCES
Report on One Class Visit

Teaching Assistant:_____ Date:_____

Course and Section:_____

Supervisor:_____

	Outstanding	Good	Average	Poor	Inadequate opportunity to observe	COMMENTS
Mastery of subject						
Preparation for class						
Organization of material						
Presentation of material						
Coverage of material						
Ability to motivate students						
Ability to encourage students to think independently						
Ability to answer questions and channel discussion						
Use of blackboard, visual aids, etc.						
Attitude of class toward instructor						
Attitude of instructor toward class						
Punctuality						
Appearance						
Classroom manner						
Overall performance						

Other Comments:

CLARKSON UNIVERSITY
STUDENT DESCRIPTION OF TEACHING

INSTRUCTOR: _____ DATE: _____

The following items reflect some of the ways teachers can be described. For the instructor named above, please fill in the numbered box which indicates the degree to which you feel each item is descriptive of him or her. IF YOU HAVE NO INFORMATION OR FEEL AN ITEM DOES NOT APPLY, PLEASE FILL IN THE N/A BOX.

Responses will not be returned to the instructor until after final grades have been given.

Please indicate the appropriate course number and section.

	Not At All Descriptive				Very Descriptive	N/A
1. Discusses points of view other than his/her own.	①	②	③	④	⑤	Ⓒ
2. Contrasts implications of various theories.	①	②	③	④	⑤	Ⓝ
3. Discusses recent developments in the field.	①	②	③	④	⑤	Ⓒ
4. Gives references for more interesting & involved points.	①	②	③	④	⑤	Ⓝ
5. Emphasizes conceptual understanding.	①	②	③	④	⑤	Ⓒ
6. Explains clearly.	①	②	③	④	⑤	Ⓒ
7. Is well prepared.	①	②	③	④	⑤	Ⓒ
8. Gives lectures that are easy to outline.	①	②	③	④	⑤	Ⓝ
9. Summarizes major points.	①	②	③	④	⑤	Ⓒ
10. States objectives for each class session.	①	②	③	④	⑤	Ⓒ
11. Identifies what he/she considers important.	①	②	③	④	⑤	Ⓒ
12. Encourages class discussion.	①	②	③	④	⑤	Ⓒ
13. Invites students to share their knowledge and experiences.	①	②	③	④	⑤	Ⓒ
14. Invites criticism of his/her own ideas.	①	②	③	④	⑤	Ⓝ
15. Knows if the class is understanding him/her.	①	②	③	④	⑤	Ⓒ
16 Has students apply concepts to demonstrate understanding.	①	②	③	④	⑤	Ⓝ
17. Knows when students are bored.	①	②	③	④	⑤	Ⓒ
18. Has genuine interest in students.	①	②	③	④	⑤	Ⓝ
19. Gives personal help to students having difficulties in course.	①	②	③	④	⑤	Ⓝ
20. Relates to students as individuals.	①	②	③	④	⑤	Ⓒ
21. Is accessible to students out of class.	①	②	③	④	⑤	Ⓒ
22. Has an interesting style of presentation.	①	②	③	④	⑤	Ⓝ
23. Is enthusiastic about his/her subject.	①	②	③	④	⑤	Ⓒ
24. Varies the speed and tone of his/her voice.	①	②	③	④	⑤	Ⓒ
25. Has interest in and concern for the quality of his/her teaching.	①	②	③	④	⑤	Ⓒ
26. Motivates students to do their best work.	①	②	③	④	⑤	Ⓒ
27. Gives interesting and stimulating assignments.	①	②	③	④	⑤	Ⓒ
28. Gives examinations requiring synthesis of parts of the course.	①	②	③	④	⑤	Ⓒ
29. Gives examinations permitting students to show understanding.	①	②	③	④	⑤	Ⓒ
30. Keeps students informed of their progress.	①	②	③	④	⑤	Ⓝ

COURSE NUMBER   SECTION

31. Considering both the limitations and possibilities of the subject matter and course, how would you rate the overall teaching effectiveness of this instructor?

Not At All Effective		Moderately Effective			Extremely Effective	
①	②	③	④	⑤	⑥	⑦

32. Focusing now on the course content, how worthwhile was this course in comparison with others you have taken at this University?

Not At All Worthwhile		Moderately Worthwhile			Extremely Worthwhile	
①	②	③	④	⑤	⑥	⑦

33. Are you taking this course because it is required?   Ⓨ Ⓝ

This form is a modified version of a form developed by The Center for Research and Development in Higher Education, University of California, Berkeley.

DO NOT MARK

IN SHADED AREA

Comments:

1. Please use this space to identify what you perceive as the real strengths and weaknesses of:

a) the course

b) the instructor's
teaching

What improvements
would you suggest?

TO STUDENTS: PLEASE EVALUATE <u>ONLY</u> THE PERFORMANCE OF YOUR T.A. (RECITATION INSTRUCTOR, <u>NOT</u> THE LECTURER)

DEPARTMENT OF MATHEMATICS

TEACHING EVALUATION BY STUDENTS

FORM FOR **SMALL** CLASSES

Course_____

Instructor_____

Quarter_____ Year_____

Your grade in the last Math Course you have taken_____

Rate each of the items below from A (excellent) to E (very poor). Please give your honest opinion about the teaching performance of your instructor If you feel you cannot respond to the question, write NR (no response).

1. The instructor comes to class on time. . . . . . . _____

2. The instructor was well-prepared for class . . . . . _____

3. The instructor had a thorough knowledge of the subject . _____

4. The instructor communicated his/her subject matter well. _____

5. The instructor stimulated interest in the course subject(s) _____

6. The overall teaching ability of the instructor was high. . _____

7. The instructor spoke clearly and audibly . . . . _____

8. Instructor's command of English was adequate . . . . . _____

9. Instructor's explanations were clear and concise. . . . _____

10. Instructor adequately summarized material to aid retention _____

11. Instructor used good examples and illustrations . . . . _____

12. Instructor emphasized particularly important course material _____

13. Writing and drawing at board were legible. . . . . . . _____

14. Instructor helped clarify difficult material. . . . . . _____

15. Exams were graded fairly. . . . . _____

16. Tests were too easy to be useful for the purpose of evaluating the student's knowledge--E (extremely easy), A (just right) _____

17. Tests were too hard to be useful for the purpose of evaluating the student's knowledge--E (extremely Hard), A (just right) _____

18. The instructor had a helpful attitude toward the students _____

19. The instructor was available when you sought help outside the classroom . . . . _____

20. The instructor invited questions form the students and answered questions well . . . . _____

21. Overall rating of this instructor's teaching. . . . . _____

<u>Comments</u>: (Use the back side if necessary.)

EVALUATION OF T. A. WORKSHOP

1988

I. The following is a list of the topics presented during the workshop. Please indicate the topics which you found helpful and/or how the topic might be presented more effectively.

	Helpful	Could be more helpful. How?

A. An overview of the Department of Mathematics
   a. The discussion of the graduate program by Dr. Franke
   b. The tour of the Audio-Visual Tutorial Center by Prof. Savage
   c. The discussion of the Departmental policies by Dr. Garoutte
   d. Acquaintance with the Department's administration
   e. Informal discussion with "old" T.A.'s

B. Teaching techniques
   a. Demonstration lesson by Ms. Rohrbach
   b. An overview of teaching by Dr. Burniston
   c. "For the Newest T. A. on the Harrelson Circle"
   d. "How to write a lesson plan by Ms. Barnhardt
   e. Video Tape - "How Not to Teach"
   f. "How to Tutor" by Ms. Schiermeier

C. Teaching Practicum
   a. Videotaping with self-evaluation
   b. Writing the detailed lesson plan
   c. Comparing lesson plans with peers
   d. Presenting mini-lessons to peers
   e. Observing lessons presented by peers

D. Testing
   a. Discussion of grading techniques by Prof. Petrea
   b. Grading and discussing the set of prepared tests
   c. Designing the 50-minute test by Dr. Wilson
   d. Making up and taking the 50-minute test

E. Course Content
   a. Discussions on the areas of difficulty for students
   b. Discussions on how a topic might be presented to a class

II. Please list other topics that should have been included in this workshop.

III. Please list the topics that should have been omitted from this workshop and explain why.

IV. General comments.

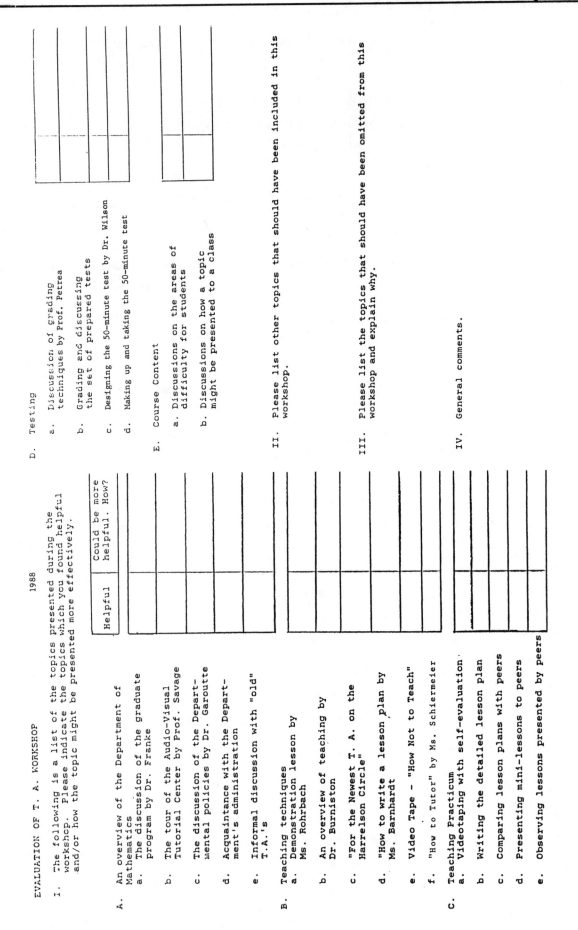

ORDERING INFORMATION

PRICE

1-9 COPIES, $2.50 EACH

10-25 COPIES, $1.75 EACH

26-50 COPIES, $1.50 EACH

51 OR MORE, $1.25 EACH

ALL PRICES INCLUDE POSTAGE (BOOK RATE) AND HANDLING.

ALL ORDERS MUST BE PREPAID.

ADDRESS ORDERS TO:

OIES
120 INTERNATIONAL CENTER
THE UNIVERSITY OF IOWA
IOWA CITY, IA 52242

MAKE CHECKS PAYABLE TO:
THE UNIVERSITY OF IOWA

❖

MANUAL FOR FOREIGN TEACHING ASSISTANTS

Revised Edition
With an Appendix for Foreign Faculty

Gary Althen
Foreign Student Adviser
Office of International Education and Services
The University of Iowa
Iowa City, Iowa 52242

TABLE OF CONTENTS

ACKNOWLEDGMENTS iv

INTRODUCTION 1
  Purpose 1
  Overview 1

INFORMATION AND SUGGESTIONS FOR FOREIGN TAs 2
  What is Expected of Teaching Assistants? 2
  What Aspects of the U.S. Higher Education System are
  Challenging to Foreign TAs? 2
    The students' level of preparation 3
    The students' level of motivation 4
    The students' general behavior 5
    The prejudice of some students 6
    The students' expectations of their teachers 7
    Language 10
    Communicating without words 13
    Guidelines concerning nonverbal behaviors 14
    The system for evaluating students 16
    Students' complaints and excuses 17
    Cheating 18
  Where can TAs Get Help? 19
    Your students 19
    Other teachers 20
    Observation of your teaching 20
    Departmental or institutional training for TAs 20
    English classes for foreign students 20
    American friends or acquaintances 20
    Readings 20

CONCLUSION 21

APPENDIX FOR FOREIGN FACULTY 22

ANNOTATED BIBLIOGRAPHY 27

ORDERING INFORMATION 28

iii

PREFACE TO THE REVISED EDITION

Since it was first published in 1981, more than 3,000 copies of the Manual for Foreign Teaching Assistants have been sold to universities throughout the United States. In addition, it has been adapted for use in universities in Canada. The book is used both as a text in training programs for foreign teaching assistants, and as an independent reference for individual foreign teaching assistants.

Since 1981 there has been considerable contention in the United States over the issue of teaching assistants from other countries. Foreign TAs have been the targets of complaints from students, faculty, and members of the public who believe that the use of foreign teaching assistants in American classrooms somehow results in educational deficiencies for U.S. students. In response to these criticisms, many universities have devised new or expanded programs for orienting foreign TAs and helping them improve their English.

The text of this revised edition of the Manual for Foreign Teaching Assistants is modified only slightly from the 1981 version. That version has proved useful to countless TAs who needed a better understanding of American culture as it is manifested in university classrooms. Also, they welcomed the specific suggestions the Manual offered for improving their performance as teachers.

The revised edition includes a new appendix for foreign faculty. In many academic fields, foreign faculty are being used in increasing numbers to teach classes to American undergraduates. With that in mind, the Manual is being enlarged to offer suggestions to foreign faculty (and to those few foreign TAs who have sole responsibility for organizing and presenting courses). A brief annotated bibliography has also been added.

The Manual's format has been improved greatly, taking advantage of desktop publishing technologies that have become available since the Manual was first produced.

ii

INTRODUCTION

Purpose

The purpose of this manual is to give new foreign teaching assistants information and suggestions that will help them (1) better understand the situation in which they are working and (2) develop ideas and practices that will make their teaching easier and more rewarding. Foreign teaching assistants come from countries where teaching may be done differently, student expectations may be different, and there may be many other contrasts with the U.S. educational system. This manual is intended to help foreign teaching assistants understand some of these differences and perform well in the U.S. system.

Many ideas in this manual come from experienced foreign teaching assistants.

While different teaching assistants (commonly called "TAs") have different experiences, most of them agree on many points about U.S. students, the U.S. educational system, and advisable ways for a TA to act toward students. This manual describes some of those points of agreement and also discusses some of the reasons why different TAs have different experiences.

Overview

This manual addresses three main questions: What is expected of teaching assistants? What aspects of the U.S. higher education system are challenging to foreign TAs? Where can TAs get help? There are many ideas and practical suggestions in these pages. You cannot expect to learn them all or put them all to use immediately. You can keep them in mind and work persistently toward adopting those suggestions that seem pertinent to you.

There are many generalizations in this manual. Not everything here will apply to every foreign TA's situation. You must make your own judgments as to what is useful for you.

1

ACKNOWLEDGMENTS

This manual was originally produced under the auspices of the Foreign Student Committee at the University of Iowa. Members of the Committee reviewed the draft manuscript, as did many others who were interested in the topic. Appreciation is hereby expressed to the following people for their encouragement and suggestions: Stephen Arum, Sally Baldus, Samir Bishara, Norlin Boyd, Barry Bratton, Maureen Burke, Angela Dalle-Vacche, Behaam Davanizadeh, Robert Ericksen, Alain Gelineau, Geoffrey Hope, John Kennan, Forrest Nelson, Soo-Byung Park, John Penick, Hector Salas, Rudolph Schulz, Stephanie Sesker, Ted Sjoerdsma, Frank Stone, and Wayne Young. Patty Meier, Karen Suarez, and Phil Carls prepared the revised edition. Kathy Thomas designed the publication.

Gary Althen
Iowa City
1988

iv

INFORMATION AND SUGGESTIONS FOR FOREIGN TAs

What is Expected of Teaching Assistants?

Different TAs have different responsibilities, depending mainly on the *academic departments* with which they are affiliated, and the particular *course* (or courses) to which they are assigned. TAs might, under the guidance and supervision of faculty members, be responsible for any or all of the following activities:

—preparing lectures
—giving lectures
—conducting laboratory sessions
—grading written assignments
—monitoring examinations
—preparing examinations
—grading examinations
—conducting drill or practice sessions

—conducting discussion or review sessions, usually in conjunction with large, lecture classes in which the instructor does not have the time or opportunity to answer all students' questions
—answering students' questions on a one-to-one basis, after class or in an office

Some TAs have full responsibility for planning and conducting a course. TAs with that responsibility will want to read the Appendix. Ask your supervising faculty member what you will be responsible for doing. You should also ask how many hours each week you are expected to devote to your TA responsibilities.

What Aspects of the U.S. Higher Education System are Challenging to Foreign TAs?

This is a complicated topic. Any educational system is the product of complex historical, cultural, economic, and political factors. Many books can be written about the history or sociology of any particular educational system, and reading such books can be interesting and helpful. The teaching assistant has a more immediate concern, however. That concern is how to teach or help a classroom full of students. What happens in that classroom? What is a TA supposed to do about it? Such questions are the concern of this part of the manual.

The students' level of preparation

Most foreign TAs are surprised at the *low level of academic preparation* of some of their U.S. students, especially the freshmen (first-year university students), and especially in the sciences and mathematics. Of course, students at more selective universities are better prepared in general than those at less selective universities. In general, U.S. students, for a variety of reasons, do not enter post-secondary studies with the high level of preparation that post-secondary students in many other countries have achieved.

It is helpful to understand that one of the basic ideas shaping the U.S. educational system is this: As many people as possible should go as far as possible in school. A very large portion of the U.S. population completes secondary school ("high school"), and a large portion of high school graduates continue to some kind of post-secondary education. There is no nation-wide school-leaving examination in the U.S., as there is in many other countries, nor are there rigid college-entrance examinations. The result of these and many other factors is that students can enter universities even though they have not learned as much as university-level students in many other countries have learned. Material covered in some freshman and sophomore (second year of university studies) courses in the U.S. is covered in secondary schools in other countries.

By the time they get to be juniors and seniors (third- and fourth-year students), U.S. university students may be more or less on a par with university students in other countries. One reason for that is that many of the people who start at a university do not continue. At large, public universities, for example, only about half of all freshmen eventually graduate. Another reason is that university-level courses are in some ways more demanding in the U.S. than they are in many other countries, once the lower level courses have been completed.

If you are teaching classes of freshmen, you should not expect them to know as much about any academic subject as first-year university students in your own country might know. The U.S. students are not necessarily any less intelligent than students at home, but they may have had a less rigorous academic experience than do primary or secondary students in your country, especially those students in your country who follow an academic or university-bound curriculum. U.S. freshmen and sophomores may also be at least as interested in their social lives as in their academic lives. For them, "college life" is not just an intellectual experience, but a phase in their lives that is intended to focus on social and cultural activities as well.

If you are teaching upper division (junior and senior) U.S. students, you are likely to find them appropriately knowledgeable about the subject matter of their courses. (Like students of a similar age in many other countries, they

2

3

might *not* be as knowledgeable about world affairs, other countries, or major political issues as you would like them to be. Nor are they as likely as students in some other countries to want to argue strongly in support of their ideas about political or social matters.)

You might find that students in professional or pre-professional courses of study are better prepared and more highly motivated than students who have not yet committed themselves to a particular career.

The students' level of motivation

University students in many countries have studied and worked very hard to get into the university. They have learned a great deal, and they are usually very interested in learning more. That is not necessarily true of all university students in the U.S. While many students here are quite interested in their studies and want to do well in their courses, *many U.S. students are not particularly interested in their studies*. Again, this is more true of freshmen and less true of seniors. Some freshmen enter a university not because they truly want to be students and learn more, but for some other reasons. For example, they may have been unable to find a job after high school, or they did not want to find a job. Perhaps their parents wanted them to go to a university, or some of their friends were going and they thought they should go too.

Those students who are least motivated are probably *freshmen* in *required courses*. If you are teaching a course that students are required to take, whether they are interested or not, you will certainly encounter students who do not want to be in the course and who do not approach it very seriously. Students in "elective" courses (ones they have chosen to take) are likely to be more highly motivated; seniors will generally be more highly motivated to study than freshmen.

At all levels, you will find some students who are *very* serious about their studies, and who work hard to get good grades.

Thus, just as your class may contain students with varying degrees of academic preparation, it is likely to contain students with varying degrees of motivation. It is more difficult to teach a class that includes this wide range of ability and interest than it is to teach a class where you can assume that all students know a certain minimum amount about the subject matter and are interested in learning more.

4

> ### WHAT TO DO
>
> There is no easy answer to the question of how a teacher ought to approach a class that includes students of widely varying ability and interest. Most teachers try to address the *interested students of average ability*, and try to gain the interest of those students who seem unconcerned about the subject matter of the course. You will probably need to *adjust your own expectations* if you are supposing that your students will generally perform well. If your expectations are inappropriately high, you will certainly become disappointed and discouraged.

The students' general behavior

The word most often used to describe U.S. students' classroom behavior and their attitudes toward their teachers is "informal." This informality shows itself in various ways: Students may dress in blue jeans or, in the summer, shorts. They may go without shoes (having been raised with the idea that it is "comfortable" to be able to "go barefoot" in warm weather, and not with the idea that being without shoes is a sign of poverty or disrespect). They may eat or drink something during a class, read a newspaper, or talk with a neighbor. They are not likely to do anything special to acknowledge the teacher's arrival or presence. They may speak to a teacher with a vocabulary and tone of voice that do not seem to convey any particular respect. They may call a teacher by his or her first (given) name. They may arrive late for class, or leave before the class is over. They may ask the teacher questions in a way that seems to challenge the teacher's knowledge or ability.

If you are from a place where students treat teachers with obvious deference and respect, you may find yourself feeling shocked or even angered by the behavior of your U.S. students. You may start believing, incorrectly, that the students are disrespectful toward you personally, or toward you as a foreigner. Even if you know intellectually that "American students are informal," you may find yourself reacting negatively when you are actually confronted by a student who comes to your class without shoes and drinks a Coke during your lecture.

5

WHAT TO DO

Realize that you will never get your U.S. students to act the way students in your country act. Realize that they treat all their TAs informally, not just you. And realize--even if it is difficult to do so--that the students' informal behavior is perfectly "normal" for them.

U.S. students behave informally toward teachers whom they like and respect very much. In fact, they may be even *more* informal, in some ways, in the presence of teachers they like and respect. Remember that U.S. students have been taught while growing up the democratic or egalitarian ideal that all people are equal, and that it is not appropriate to behave in markedly different ways around people with different social standings.

The prejudice of some students

In general, people in the U.S. are taught as they grow up that their country is superior and that they are a superior people. Along with this can go the idea that people from other countries are somehow inferior—less intelligent, less "advanced," etc. You may encounter students who assume, without really thinking about it and without necessarily being aware of what they are doing, that they are superior to you because they are Americans.

WHAT TO DO

Try to understand that many U.S. university students have had little or no interaction with people from other countries. They have not had the opportunity to know individual foreigners who are well-educated, intelligent, and sophisticated. Realize that their prejudices against foreigners are not related to you personally. Remember also that prejudices against foreigners are found everywhere, not just in the U.S. It is best if you can resolve to help your students overcome their prejudices by being an example of an intelligent, compassionate person who is not an American.

On the other hand, you are likely to encounter some students who believe that foreigners are generally superior because they come from societies with longer histories and with cultural accomplishments that have not been realized in the United States. You will probably encounter U.S. students who are very interested in people from other countries, and who want to get to know you and learn from you.

The students' expectations of their teachers

The manner in which teachers treat them through secondary school leads university students to have certain expectations about the way their teachers should behave. Those expectations differ from one system to another.

In some countries, the teacher is in an authoritarian role. The teacher can give assignments, set deadlines, formulate examinations, and give grades in whatever way he thinks is appropriate, and the students will generally accept what is done or said. This is not the case in the United States. Students here have been trained to consider it their right to be treated "fairly" and to have reasonable (to them) explanations concerning assignments, tests, and grades.

They regard the course syllabus as a binding contract that explains the terms and conditions of the course. TAs are bound by that contract, and are generally expected to be "fair" and "reasonable," rather than "arbitrary" or "authoritarian," in dealing with students.

U.S. university students are generally not as self-reliant in their academic work as are university students in many other countries. This is more true of freshmen than of seniors. They often expect their teachers to *explain everything very fully* to them, and tell *them exactly what is expected* in any particular assignment or examination, because that is what their teachers have normally done in primary and secondary school. U.S. students have generally been given the idea that they should rely on teachers to give them assignments, react to their assignments, give them frequent tests or examinations, and generally spell out in detail what they are expected to do. This is sometimes referred to as "spoon-feeding" the students, that is, giving the students step-by-step instruction in the way that an adult might feed a baby, spoonful-by-spoonful.

WHAT TO DO

Realize, again, that the students' expectations (or even demands) for explicit, step-by-step instruction and explanations are not the product of lack of intelligence, but of having been made to think over many years that this is what teachers are supposed to provide. You will probably have to follow the same practice. Tell your students specifically what assignments they are supposed to do, how long the assignments are supposed to be, and when they are due. Make all deadlines clear. It is helpful to write all this on the board or on sheets of paper you distribute. Be prepared to answer students' questions about the details of the course, and especially of examinations. (For example: "Will the questions be essay or multiple

WHAT TO DO

Memorize, practice, and use reponses that support the student's willingness to speak up in class while gently making clear that the particular contribution is inaccurate. Some examples:

— "That's an interesting idea, but I don't think it's quite right."
— "That's a reasonable suggestion, but there is probably a more accurate one. Who else has an idea?"
— "I hadn't thought of that possibility. Do any other ideas come to mind?"
— "That might be partly correct, but there's more to it. Who can elaborate on that?"

Do not make brief, direct statements that do nothing more than make clear that the student's answer is incorrect. Do not simply say things such as, "No, that's wrong." Do not ridicule the student. Even if the student's contribution is clearly incorrect, all the students in the class will feel offended if you do not offer some support along with your correction.

In general, U.S. students expect their teachers to *entertain* them, at least to some degree. This is to say that U.S. students strongly prefer teachers who do more than stand and give a "boring" lecture in a "monotonous" voice. They enjoy some humor, some interesting examples used to illustrate points, and the use of other means than just lecturing to convey information or ideas. They will be more attentive and more highly motivated if their teacher "makes the class interesting."

WHAT TO DO

If your responsibilities permit you to do so, look for ways other than lecturing in which you can convey information or ideas to your students. For example, look for appropriate films, slides, or interesting articles to use in your class. Consider using class discussion as part of your teaching. You might get ideas about these and other alternatives to lecturing by talking with experienced TAs or faculty members, and observing classes taught by teachers who are considered unusually interesting.

9

---

choice?" "Will the test cover all the material from the start of the course, or just since the previous test?" "Does spelling count?"). Grade and return assignments and examinations promptly. Meanwhile, you can try to encourage your students to take more responsibility for their own studies.

U.S. students usually want their teachers to be friendly toward them. They appreciate warmth, smiles, and some conversation that enables them to know something about the teacher as a *person*, not just as a *teacher*. They do not *demand* friendliness and openness, but they *appreciate* it and usually respond positively toward it. They usually work harder for a teacher whom they like than for one whom they do not like.

There is a danger in being *too* friendly and being viewed as just another student. It is not easy to provide guidelines on the amount of friendliness that is appropriate, since friendliness is shown in many subtle ways.

WHAT TO DO

Note the behaviors by which American teachers show a friendly interest in their students. These include learning the students' names, smiling, being willing to engage in some informal conversations outside of the class, joking, letting the students know something about themselves as individuals, and learning something about individual students. If you are comfortable in doing so, show some friendliness toward your students. Read the student newspaper so that you will be informed about the topics (especially sports) that are likely to be discussed in informal conversations among students and between students and teachers.

At the same time, remember that you are the teacher, and not just another student. You are expected to be somewhat distant, somewhat more serious, and generally concerned that your students are learning from you.

U.S. students have certain expectations concerning the manner in which teachers should respond to incorrect answers or inaccurate statements they may offer during a class. The students expect polite, encouraging responses that recognize their effort to contribute to the discussion.

8

U.S. students expect their teachers to *know their material* or to *admit they do not know something they do not know.*

WHAT TO DO

Prepare for your class or discussion group meetings as thoroughly as possible. Be familiar with the textbooks for the course and with the supervising faculty member's lecture materials. Try to anticipate the questions the students will ask, and be ready to answer them.

If students ask questions you cannot answer, say to them, "I don't know the answer to that. I will find out about it and answer your question at the next class." Or, you might see if other students have a response. It is *much better* to say you do not know something than to pretend that you know and try to "fake" an answer. U.S. students will respect you more for admitting that you do not know something than they will if they believe you are trying to cover up your ignorance of it. This is an important way in which U.S. students differ from students in many other countries. (Be sure to answer the question at the next class!)

Language

Many aspects of language cause problems between U.S. students and foreign TAs. Foreign TAs may have trouble *understanding* their students, and *being understood* by the students. Understanding can be difficult because the students use unfamiliar vocabulary (especially slang terms); speak too rapidly; refer to facts, situations, ideas, or experiences that are unfamiliar to the TA; and/or make jokes as they speak.

Being understood can be difficult because of limitations on the TA's vocabulary, pronunciation that is unfamiliar to U.S. students, differences in sense of humor, and/or lack of common experiences that can be used to make a point or illustrate an idea.

Different foreign TAs react differently to the knowledge that they have problems in communicating with their U.S. students because of language. Some TAs become very nervous at the prospect of being in front of a group of students whose language they do not fully share. (This nervousness usually reduces their proficiency in English.) Others make the assumption that it will be a good experience for the U.S. students to learn to overcome a language barrier—to learn to understand someone with an accent and to explain things to someone who does not share their language or background.

10

There are many things that you can do to reduce the problems that are caused by language differences.

WHAT TO DO

a. If your English language competence is limited, *work to improve it.* Whether through an English class, the use of a language laboratory, and/or practice with a native speaker of English, you can learn more vocabulary, improve your listening comprehension, speed up (or, if necessary, slow down) your rate of speech, and improve your pronunciation. At least 10 hours weekly should be devoted to these activities. It will be very much to your advantage to try to do all of them. Your teaching will benefit, and so will your research work, your ability to write papers and theses, and your ability to have constructive relationships with Americans. You can and should try to spend some informal time with Americans and pay special attention to their speech (especially their slang), their jokes, and the way they organize their ideas to explain something. Do not live with other students who speak your own language; to do so will very probably keep your English proficiency from improving. It might even *reduce* your proficiency.

b. Recognize that you will probably never speak English the way the natives do, without an accent and without vocabulary limitations.

c. On the first day of class, address the language issue directly: "English is not my native language, so we may have some problems understanding each other. I'll do the best I can. I'll need your help to improve. You will have to let me know when you are not sure you understand me. And I will let you know when I need you to speak more slowly or explain yourself with less slang." Write your main points on the blackboard as you go along. (But avoid speaking directly toward the blackboard.)

d. Notice that some Americans seem to erect a sort of mental wall as soon as they hear that you have an accent. They seem to say to themselves, "This person has an accent. I cannot understand her." Then they do not try to understand. You may get this response even if your accent is only a slight one. Try to retain your patience when this happens. Consider having a conversation with the person (or with an entire class) about the subject of accents in speech. You may want to do this at the first class meeting.

e. Whenever a student in your class is saying something you do not understand but that you want to understand, interrupt (perhaps with a

11

Communicating without words

Much of what people communicate to others is communicated *without words*. (Some scholars believe that *more* is communicated in "nonverbal" than in verbal ways.) As they grow up, people learn appropriate nonverbal behavior from those around them. They learn such things as what clothing is appropriate for different people in different situations, how far from another person to sit or stand during a conversation, where to direct their eyes when talking to another person, and other aspects of nonverbal behavior that are discussed briefly below.

Most people are not aware of the messages they send without words. This is not a problem when they are interacting with people who grew up in the same culture they did, and who learned the same nonverbal behavior. It is often a problem in a new culture, though. People interacting with someone from a different culture are likely to send messages they do not know they are sending, and to misinterpret incoming messages.

All this is very important for teachers. Each of us has learned certain behaviors that are considered appropriate for teachers--in our own cultures. Foreign TAs are likely to find, if they think about it, that they are behaving in the manner of teachers at home. This may cause some misunderstandings or misjudgments between them and their U.S. students.

WHAT TO DO

Observe Americans interacting: pay particular attention to the way teachers act in the classroom. (There is no *one* way in which they act, of course, but if you observe carefully enough you are likely to see common differences between their behavior and that of teachers at home.) Here are some things to watch:

— What do they wear?
— Where do they sit or stand?
— How much do they move about in the classroom?
— How much do they move their hands and arms while they talk?
— What gestures or expressions seem to indicate agreement with what is being said? Disagreement? Confusion? Friendliness? Boredom? Impatience? Interest?
— How fast do they talk? How loud?
— How far do they stand or sit from the person to whom they are talking?
— Where are their eyes directed when they are talking to an individual? To a group? When they are listening to someone?

13

---

particular gesture that you have told the students will be your sign that there is a communication problem) and ask the student to begin again, slow down, explain it in some other words, or whatever is appropriate.

f. At the beginning of the term, tell your students which English sounds are particularly difficult for you to pronounce. Teach them your pronunciation of common sounds or words that you do not pronounce the same way they do. You can try to improve your pronunciation, but at the same time they can learn your way of saying things.

g. When you are not certain that you have understood something a student said, tell the student that you are going to state what you did understand, and you want the student to tell you whether you have understood correctly. It is advisable to repeat or rephrase a student's question, to make sure you *and* the other students understand them. *Then* answer the question. If the question is not a good one, try to rephrase it in a way that makes it better.

When you are not certain that a student has understood you, ask the student to repeat or paraphrase (say in different words) what you said. Then you will know whether your idea has been conveyed.

h. Recognize that some students may use your accent as an excuse for doing poorly in your class, when their poor performance could more accurately be attributed to their own failure to pay attention and to study, or to nonverbal aspects of your behavior (see the next section).

i. It can be helpful to record or videotape your lecture (or a practice lecture) and listen to the way you sound. Could you improve?

j. In general, try to view the language difference as an interesting challenge that you and your students can approach with good humor and a cooperative spirit. Be well prepared for your classes, and present your material with confidence. If your students perceive you as interested, helpful, and friendly, they will overlook many language difficulties.

It is not unusual for one or two students in a class to assume the role of interpreter between a foreign TA and others in the class. This can be very helpful, especially at the beginning of the term, but it is unwise to allow the class to become dependent on someone who gets into that role.

12

These are often subtle matters, especially when they are complicated by the fact that certain nonverbal behaviors go along with the words that are being expressed to convey an overall message. If the meaning of the words does not coincide with the meaning of the nonverbal behavior that accompanies the words, confusion and misunderstanding result.

WHAT TO DO

Try to learn typical American nonverbal behaviors, and use them. This will help you understand Americans better, and it will certainly help them understand you better. The following are some *general guidelines* that will help you if you keep them in mind.

Guidelines concerning nonverbal behaviors

—Clothing   You will see some teachers dressed rather formally, and others dressed quite informally. TAs usually dress informally, like their students. A TA who consistently dressed formally might be considered "too formal" or "up-tight," but you can probably wear whatever you like while teaching. It is not usually clothing that distinguishes students from teachers in the U.S.

—Location   There is no particular place in the classroom where U.S. teachers are expected to remain. They may be at the blackboard, behind a desk, in front of a desk, behind a lectern, or sitting on a stool or a desk. They may sit or stand. They are likely to *move around* more than teachers in many other countries will do. A teacher who always stays in the same place in the room will probably be viewed as nervous, lacking in self-confidence, and excessively formal.

Teachers conducting small classes or discussion sessions may arrange students' chairs in a U or semi-circular way, with the teacher sitting in the middle. This is considered an appropriate and fruitful way to encourage discussion among the students.

—Gestures   Americans typically use a moderate amount of hand and arm motion when talking. They use less than people from the Mediterranean countries usually do, but more than most East Asians. Of course, some individuals use more than others. A person who uses her hands and arms "too much" is likely to be considered "emotional" or "excitable." A person who used "too little" hand and arm motion would probably be considered too formal and/or nervous.

—Facial expressions and head movements   Among Americans, a quick smile usually indicates warmth, friendliness, and interest. A prolonged smile *might* hide annoyance. Narrowed eyes usually mean concentration. Tightened jaw muscles usually reflect tension or anger. A furrowed brow might mean concentration or failure to understand.

Nodding the head up and down usually means "yes" or agreement or "I am understanding what you are saying." Shaking the head from side to side means "no," it does *not* mean "I am understanding what you are saying," as it does in parts of India.

—Volume and speed of speech   The ideal teacher will speak loudly enough to be heard at the back of the room. Speaking too softly is taken as a sign of lack of self-confidence or lack of consideration for the people who want to hear. Speaking too loudly is taken as a sign of nervousness or excessive emotion. The ideal teacher speaks at moderate rate, slowly enough to be clearly understood by all students in the room, but not so slowly that students become bored. It is advisable to vary the volume of the voice, not speaking in a monotone or in a sing-song way.

—Space between people   When two American individuals are standing and talking, they usually stand about an arm's length apart, or perhaps a little closer. This is closer together than Asians typically stand, but farther apart than Latins or Arabs typically position themselves.

Americans are likely to touch each other more than Asians might do, but much less than Latins or Arabs are likely to do. A person of lower status is unlikely to touch one of higher status. For example, a student would rarely touch a teacher. A teacher might touch a student on the lower arm or the shoulder to convey support or sympathy. Otherwise, teachers and students are not likely to touch each other.

—Eye movements   Americans are taught to believe that *direct eye contact* between two people who are talking conveys the positive characteristics of self-confidence, assuredness, attentiveness, and honesty. Thus, they are most comfortable when they look directly at each other's eyes, *intermittently*, while talking.

A person who avoids looking into the eyes of the person he is talking to is likely to be considered nervous, afraid, or even dishonest.

When speaking to a group, Americans are often advised to look at the eyes of three or four different people who are seated midway back in the audience.

Students' complaints and excuses

If they have performed poorly on an assignment or examination, or are generally doing poorly in a course, students will sometimes offer complaints or excuses that seem intended to justify their poor performance, blame it on the teacher, or perhaps to induce the teacher to behave differently. Here are some examples of complaints that a TA might hear from students:

— "You are giving us too many assignments."
— "These assignments are too hard."
— "We've already had two tests this semester."
— "I have other classes to study for, too, and I just can't do this much work for your class."
— "The questions on your test were not fair."
— "The questions on your test were not clear."
— "The test was too long. We couldn't possibly finish it."
— "You should have given me a higher grade. This answer is not wrong."
— "You should have given me at least some points for this answer because I have part of it right."
— "I answered this almost exactly the way my friend did, but she got more points for it than I did. I think I deserve a higher grade."
— "I just can't understand your explanations of things."

If a student has not met some requirement, such as turning in an assignment on time or taking a test on a designated day, that student might approach the teacher to offer an excuse for what had not been done, and request special consideration. Here are some examples of excuses that a TA might hear from students:

— "I got sick the night before the test and couldn't study."
— "My roommate got sick the night before the test and kept me awake all night with his coughing. I just was not in any condition to take the test."
— "When I started to study for this test, I realized that I didn't understand this very important part, and I couldn't find anyone to help me with it."
— "My boyfriend (or girlfriend) decided to leave town (or end our relationship), and I've been very upset. I haven't been able to study."
— "My family doesn't have much money so I have to work a lot of hours at my job, and I don't have enough time to study. I'm tired all the time."
— "My mother (or other relative) got sick (or died) and I had to go home. That's why I missed the exam."

17

A teacher whose eyes are consistently directed to the floor, the ceiling, and/or the blackboard, and not toward the members of the audience, will be considered an ineffective speaker who is too nervous, insecure, and/or lacking in knowledge to look directly at his listeners.

The system for evaluating students

This is another complicated subject, one that causes considerable concern for many foreign TAs. It causes concern because U.S. students attach great importance to grades, because TAs are often responsible for giving grades, and because the ways in which grades are determined are likely to be unfamiliar to people who have not been students in the U.S.

WHAT TO DO

Realize that many teachers are unhappy about the fact that they have to give grades to their students. Giving grades is not an activity that can be done mechanically, with fixed rules and procedures. There is nearly always an element of subjectivity (individual judgment) in assigning grades. You will need to develop a system of determining grades that makes sense to you and that you can defend if you need to. It is helpful to discuss this matter with experienced teachers in your department and to learn any guidelines your department furnishes to people who give grades.

Several different grading systems are used in the United States. Your department might instruct you which one to use, or you might be left to choose your own. These systems involve assigning numerical values to different answers on an assignment or examination, or assigning a letter grade. Students' final grades may be determined by calculating the number of points they have accumulated on a scale (so that, for example, students with at least 450 out of a possible 500 points would get an A). Or a certain number of students with the highest point totals would get a grade of A, a certain number would get B's, and so on, regardless of what their total number of points was. This approach is called "grading on the curve."

When an assignment or test is being graded, the assumption is generally made at the outset that the paper is perfect (with a score of 100, say). Then the grader *subtracts points for things that are wrong.* Thus, the grader needs to be able to explain why points were subtracted from the highest possible score. This approach contrasts to one whereby the grader assumes at the beginning that the paper has a score of zero, and then looks *to see how many points can be added for answers that are correct.*

16

WHAT TO DO

Learn your department's policies concerning assignments that are submitted late, examinations that are missed, and grades that are protested. Make sure that you understand these policies clearly and that your students are informed about them at the beginning of the course.

Talk to experienced teachers about the ways in which they respond to students' complaints and excuses. This will give you some ideas about how to respond yourself, and it will also give you some perspective on the seriousness of the students' viewpoints.

Do not be intimidated by threats from students, or tempted by offers they might make in order to get special treatment. Evaluate what they say, in light of the pertinent departmental policy, and act accordingly. For example you might say to a student that "Late assignments are lowered by one grade unless you have a note from a physician stating that you were sick and under a doctor's care. You turned your assignment in late, so I will have to lower your grade unless you bring me a note from the doctor." You might go on to suggest that this student start earlier on the next assignment, in order to allow time for unexpected interruptions.

You can almost always tell a student that you will have to consult with your supervising faculty member about his or her complaint or excuse. You can ask the student to submit the complaint in writing, so you will be able to consider it carefully.

Cheating

Some students will "cheat" on assignments or examinations. That is, they will submit written work that is represented as their own, but is in fact someone else's, or is based on access to notes or papers that were not supposed to be used during an examination. Teachers are normally expected to watch for evidence of cheating and to take certain measures in cases where cheating is suspected.

18

WHAT TO DO

Learn your department's policies concerning cheating. Learn how cheating is defined, and what a teacher is supposed to do when a particular student is suspected of cheating. Talk with experienced teachers in your department to find out how frequently they have encountered cheating, how they discovered it, and how they responded.

Make sure your students know from the beginning of the course what the consequences of cheating might be.

Where Can TAs Get Help?

Two sources of help for foreign TAs have already been mentioned several times. One is your supervising faculty member. That person is usually your best source of information about departmental policies and the particular course or courses in which you will be assisting. Do not hesitate to take questions on these topics to your supervisor.

You should realize that some faculty members have more experience than others in working with foreign TAs. Those with more experience are usually more easily able to understand your situation and your questions.

The other source of assistance that has been mentioned several times is experienced TAs. Try to arrange regular meetings or informal gatherings with TAs, both foreign and American, who have already had the teaching experience you are about to begin. Ask them specific questions. Ask them for general advice. Listen carefully to what they have to say. This will have two advantages for you. First, it will give you some specific ideas or information you can use in meeting your TA responsibilities. Second, it will help you see that some of the problems you are having are problems for all teachers. It can help you to know that some of your difficulties are inherent to teaching, and are not attributable to deficiencies of your own.

There are other sources of assistance:

Your students

U.S. students will normally view their teachers as their allies, at least in part. If you can establish reasonably friendly relationships with your students, they will help you, especially with language problems. They can explain slang terms, define words, and help you improve your pronunciation of English.

19

They can sometimes explain aspects of their own behavior that are curious to you.

Other teachers

Identify teachers (faculty members or TAs) who students consider to be good teachers, and arrange to visit their classes. See how they present their material and how they interact with their students. If you do not understand what you see happening, ask the teacher to explain it to you afterwards. It can be especially helpful to visit the classes of other TAs who are teaching courses similar to yours.

Observation of your teaching

Ask an experienced teacher visit a class you are conducting to watch what you do and offer suggestions for improvement.

Departmental or institutional training for TAs

Some academic departments and educational institutions have training or orientation sessions for new TAs, both foreign and American. Attend them and participate actively, so that you can derive as much as possible from them. The most helpful TA training programs are ones in which you can *view yourself on a videotape* and see what improvements you might make.

English classes for foreign students

There may be classes in pronunciation or listening comprehension that would help you improve your English proficiency.

American friends or acquaintances

Human behavior, whether in a classroom or anywhere else, is heavily influenced by culture. People who grow up in different cultures are taught (implicitly) assumptions about such matters as how one treats older people, or people of a higher status; what it means to "get an education," and, thus, what students are supposed to do; how and when one participates in a discussion; and how one presents an argument or point of view in a paper or a discussion. The more time you spend interacting informally with Americans, the better you will understand how they think about things, and why they act the way they do. The better you understand these things, the better teacher you will be.

Readings

There are many publications you could read to improve your understanding of U.S. culture in general, the U.S. educational system, and teaching methods that are considered desirable. The annotated bibliography at the end of this book lists a few such publications.

CONCLUSION

It is difficult to be a good teacher. You can improve with practice and experience. It is even more difficult to be a good teacher when your students do not share the values and assumptions that you learned while growing up in your own culture. Both you and your students stand to benefit if you can work together to make your teaching experience a positive one. It is hoped that this booklet provides some useful suggestions toward that end.

If you have comments or suggestions concerning this booklet, you are encouraged to send them to the author at the Office of International Education and Services, 120 International Center, University of Iowa, Iowa City, Iowa 52242.

APPENDIX FOR FOREIGN FACULTY

Teaching assistants are not normally given full responsibility for planning and conducting courses. Faculty visitors from overseas, though, commonly have that responsibility. If you should be asked to prepare a course, it is wise to remember that American students expect a great deal more than brilliant lectures—and that, regardless of your brilliance, your teaching will be found wanting if you fail to attend conscientiously to certain related duties. Here is what you will be expected to do:

Prepare a Syllabus

American students (and, in many schools, academic departments as well) expect to receive, on the first day of a course, a handout that will tell them everything about the course. Students look first at the "Schedule of Assignments" and the "Requirements," though the syllabus includes other information as well. Typically, a syllabus should include:

Purpose of the course
State what material will be covered and what students are expected to learn in the course.

Readings
List, with full bibliographic information, all materials that the students will be expected to read during the term, as well as information about *where the readings can be found*—whether purchased at the book store, borrowed from the reserve desk at the library, purchased in photocopied form, or any other way. Do not assign any material unless you see to it that the material is accessible to the students. That is *your* responsibility.

Requirements
List all tests and/or papers you plan to require in the course, the date each will be given, and the percentage of the final grade each will represent. Thus you might assign two quizzes worth 25 per cent each and one final examination worth 50 per cent, or any other combination of quizzes, examinations, and papers. You *must* inform the students what will be required of them, when it will be required, and how it will be evaluated.

It is also useful to inform the students about your expectations concerning the format required for any written assignments—that they must be typed, that a particular footnoting style must be followed, and other requirements you may wish to make.

Schedule for the course
Your syllabus must specify the *exact* reading assignments, problems, or writing assignments students should prepare for each session of the course. American students respond with panic and anger to vague assignments such as "Read up on Husserl." In preparing your syllabus, list specific titles and pages to be read for each day the course is scheduled to meet.

Your personal schedule, where your office is located, and when you can be found in your office
If you do not have this information at the time of preparing the syllabus, tell the students that you will announce your office hours at the first session—and remember to do so. Be certain you scrupulously keep all your office hours and appointments.

You can ask your department head to show you examples of well-done course syllabi.

Order Books for Your Course

It is *your* responsibility to make sure that an appropriate book store has on hand a sufficient number of copies of all texts you expect your students to buy. (In some departments, a departmental secretary will do the actual ordering once you have made clear what books are needed.) The book store will supply you (or the secretary) with the proper order form on request, but it is your responsibility to take the initiative and provide the book store with precise ordering information, including ISBN (International Standard Book Number) numbers, for every book you expect to use. This cannot be stressed strongly enough. American students react most negatively if the books are not available. They will blame you for being irresponsible, and so will your department.

Meet ALL Your Classes and Appointments

In some other countries, meeting classes is something professors do when they have nothing better to do. This is *absolutely unacceptable* in an American college or university. You are expected to meet all your classes, your office hours, and your appointments punctually and as scheduled, unless there is a very serious emergency. Cancelling classes will count heavily against you in the eyes of your students and of your department, even if you hold make-up sessions. If a serious emergency makes it absolutely impossible for you to appear for your class, you should make an effort to find a qualified replacement. Only as a last resort should you cancel class and schedule a make-up session.

Administer Quizzes and Return Them Promptly

Your American students may not care about the subject matter, but they care intensely about their grades. Especially in first- and second-year courses it is crucial to administer at least one written quiz before the semester is half over, and at least one additional one before the end of the semester. You can substitute papers for quizzes.

Your students expect you not only to grade their examinations and quizzes but *to comment on them* in detail, explaining why they received the grade they did. The more detailed your comments, the less discontent you will encounter.

Finally, it is crucial that you return the quizzes, graded and with comments, as soon as possible, preferably at the first class session after the quiz is given, and certainly no later than one week after the quiz. If you have to stay up all night to grade the papers, do so. Your job depends in part on your students' evaluation, and that in turn depends in part on the speed and care with which you grade written work.

Administer Student Evaluations, If Required

Some academic departments require all instructors to obtain student evaluations of their courses. (Many departments that do not require the evaluations strongly encourage them.) At some institutions, evaluation forms are made available to instructors; at other institutions, instructors make up their own. The outcome of the student evaluation may or may not play a heavy role in decisions about a visiting instructor's reappointment. Policies vary from department to department. You will want to find out how much weight student evaluations receive in your department.

Be Accessible

Make a point of being accessible to your students and be sure they know when and where they can find you. You will waste a great deal of time explaining material you explained in a class the student did not bother to attend. However, teacher evaluation forms nearly always ask, "Was the professor available to the student?" So, be available!

24

Finally, remember that not only students (through the teacher evaluation forms) but also your colleagues, no matter how humble and junior, are likely to vote on the question of your reappointment. In America, it is not enough to be in the good graces of the chairman or the dean. Your peers and students count heavily. Make sure that your colleagues come to think of you as one of them. It is not enough to impress with your intellect. It is important also to have mutually respectful relationships with all of your colleagues.

ANNOTATED BIBLIOGRAPHY

Althen, Gary. American Ways: A Guide for Foreigners in the United States. (Yarmouth, ME: Intercultural Press, Inc., 1988.) An introduction to American culture, with a section about cultures and cultural differences, a section about specific aspects of American culture—including education—and a section listing activities that can help visitors from abroad better understand U.S. society.

Barnes, Gregory. The American University: A World Guide. (Philadelphia: ISI Press, 1984.) An interesting, readable discussion of the American higher education system as it compares to higher education systems elsewhere.

Barnes, Gregory. Communication Skills for the Foreign-Born Professional. (Philadelphia: ISI Press, 1982.) Particularly recommended for foreign scholars, this book offers pointers about nonverbal communication among Americans and explains some important points about English usage.

Cahn, Steven M. Saints and Scamps: Ethics in Academia. (Totowa, NJ: Rowman & Littlefield, 1986.) A philosopher's discussion of the ethical obligations of college and university faculty in the United States.

Gullette, Margaret M. (Ed.). The Art and Craft of Teaching. (Cambridge: Harvard University Press, 1984.) A collection of useful essays on various aspects of being a teacher--lecturing, test questions, grading, etc. Filled with sound, practical advice.

25

REPORT ON MATHEMATICAL EXPOSITION SEMINAR
Lawrence C. Moore
June 17, 1987

This is a report of a seminar on mathematical
exposition run by me during the period May 18 - June 17,
1987, for the foreign graduate students in Mathematics
whose names are listed below:

The goals of the seminar were to improve both the
English proficiency and the oral presentation skills in
mathematics. Each participant gave three presentations of
30-45 minutes. In addition, Mr. An and Mr. Yu were asked to
present a fourth presentation and to participate in an
informal conversation session.

The material for the presentations was drawn primarily
from the books Mathematical Gems I, II and III by Ross
Honsberger, published by the Mathematical Association of
America. This material is accessible to most good
undergraduates, so that it was not necessary to present
elaborate proofs, but was so elegant that it was enjoyed by
all. Before the first presentation each student was
required to meet with me and discuss an outline. In
particular, all material to be written on the board was to
be distinguished by color. Before subsequent lectures no
preliminary meeting was required, but an outline was
prepared each time. The topics for the first two lectures
were assigned, but the students were free to select their
own for the third and fourth lectures.

Various rules were enforced during the presentations.
For example: Notes could be brought to the class, but not
held during the presentation. Lecturers were required to
establish eye contact with the audience and ask questions of
the audience. The use of colored chalk was encouraged.
During each presentation, I filled out an evaluation form
and then discussed it with the lecturer after his
presentation. (A copy of this form is attached.)

The seminar was an enjoyable mathematical experience.
The ideas presented really are "gems" and a pleasure to
consider. The students were enthusiastic and cheerfully
complied (or tried to) with the rules of presentation. Not
only did their oral presentations improve, but they gained

confidence in their ability to lecture in English. If the
need arises, such a seminar should be run again. Indeed,
the mathematical benefits were such that we might consider
running such a seminar as a "get acquainted" exercise for
the first-year students irrespective of their native
language.

<u>Seminar Evaluation Form</u>

Summer 1987

Student Name:

Date:

Topic:

<u>Evaluation</u>

Use of language (grammar, vocabulary, pronunciation):

Organization of Material:

Use of chalkboard:

Interaction with audience (eye contact, response to
questions)

Other:

Contributors

Bonnie Burns is an instructor in the English Language Program of the University Extension.

Oruc Cakmakli is a graduate student in Architecture and a cartoonist for *The Daily Californian*. He is from Turkey.

Robby Cohen is coordinator of the Graduate Assembly's TA Training Project.

Alan Cox is a teaching assistant in Business Administration from Canada.

Barbara Davis is Director of the Office of Educational Development.

Ted Goode is an adviser in the Advisers to Foreign Students and Scholars Office.

Rajesh Gupta is a teaching assistant in EECS from India.

Jacqueline Hoeppner-Freitas is an instructor in the English Language Program of the University Extension.

Dong-Gyom Kim is a teaching assistant in Mathematics from South Korea.

Eyal Naveh is a teaching assistant in History from Israel.

David Pickell is Editor-in-Chief of *The Daily Californian*.

Prabhakar Raghavan is a teaching assistant in EECS from India.

Ron Robin is a teaching assistant in History from Israel.

Stephen Small is a teaching assistant in Sociology from Great Britain.

Jyoti Sanghera is a teaching assistant in Sociology from India.

Jia-Yush Yen is a teaching assistant in Mechanical Engineering from Taiwan.

John Guansheng Zhao is a teaching assistant in Political Science from the People's Republic of China.

Teaching at Berkeley:

A Guide for Foreign Teaching Assistants

Edited by Robby Cohen and Ron Robin

Graduate Assembly
U.C. Berkeley
© 1985

opinion of some observers) to tackle those problems. One important contribution was the publication, in 1984, of the handbook *Learning to Teach*, which addresses many of the issues and concerns of TAs on this campus.

The publication of an additional instructional manual, focusing specifically on foreign TAs, is a response to the awareness, by themselves and by others, of the special problems that confront those who are teaching in an alien academic environment, in a language that is not native to them, and working with students whose cultural backgrounds and interests are quite different from their own. This handbook has two important merits. First, it recognizes and discusses those problems which are peculiarly those of the foreign instructor in an American university, and which pertain as much to experienced academics coming from abroad as to young instructors who may be teaching here for the first time in their lives. Secondly, this manual identifies the resources available on this campus to foreign TAs, to provide them with information and support, to assist them in developing their teaching skills, and in understanding Berkeley's academic milieu. I know that I speak for the campus administration and for the faculty in commending the editors and the authors of this handbook for their most valuable service to our common teaching enterprise. We welcome the new foreign TAs to the ranks of the University's teaching corps.

Gene Brucker, Chair
Academic Senate, Berkeley Division

Preface

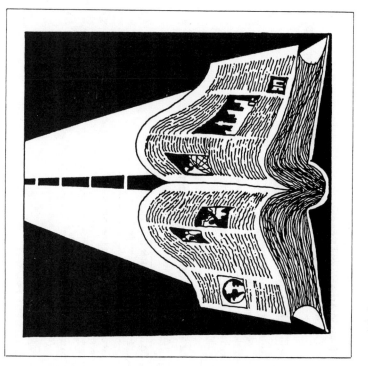

Teaching is one of the most difficult and demanding of professions, because its effectiveness requires not only the mastery of a body of knowledge, but the ability to communicate that information. These "communication skills" are not simple, nor easy to learn; their acquisition requires concentration and commitment, and an understanding, on the teachers' part, of themselves and their students. Experience in teaching, so crucial to the mastery of its skills and mysteries, begins early at Berkeley for large numbers of graduate students. This experience, this apprenticeship in the art of teaching, is an integral part of their academic training. In the past decade or so, the campus has become more aware of, and sensitive to, the problems of the fledgling instructor, and has taken some steps (not enough, in the

Table of Contents

Introduction 7

Language Skills

English and the Foreign-born TA 10
Jacqueline Hoeppner-Freitas

Overcoming the Language Barrier 17
Jia-Yush Yen

Listening Comprehension and Accent Problems 22
Dong-Gyom Kim

How Do Americans Speak? 26
Bonnie Burns

Teaching Strategies

Getting to Know the System 30
Stephen Small

How To Teach Without Talking 35
Eyal Naveh

Leading Sections in the Sciences 40
Rajesh Gupta

Grading Humanities Students 43
Eyal Naveh

Grading Options in the Sciences 47
Prabhakar Raghavan

Student Evaluations of TAs 51
Ron Robin

Cultural Issues

Foreign TAs from the Undergraduate Perspective ... 57
David Pickell

A Woman, a Foreigner, a TA 62
Jyoti Sanghera

Advantages of Foreign TAs 65
John Quansheng Zhao

Resources

Oral Communication Resources 69
Barbara Gross Davis

Peer Counseling and Support for Foreign TAs 74
Ron Robin and Alan Cox

Advisers to Foreign Students and Scholars Office
(AFSSO) 77
Ted Goode

Contributors 79

Introduction

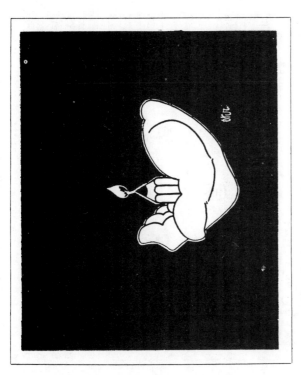

The University of California at Berkeley is known for its ability to attract outstanding graduate students from all over the world. As is the case with most graduate students at U.C. Berkeley, a significant number of these foreign scholars are active participants in the university's instructional network. Approximately thirteen percent of U.C. Berkeley's teaching assistants are from non-English speaking countries; an additional two to three percent are foreigners who list English as their native language.

Foreign TAs and their American colleagues encounter similar teaching, employment, and educational problems. Inexperience, the lack of formal pedagogical training, and a demanding study load are common obstacles shared by all TAs. Yet, the average foreign TA must overcome many unique barriers, too. Lacking familiarity with American culture, and often hampered by an imperfect command of American English, foreign TAs soon discover the need to re-evaluate their most fundamental concepts of teaching. Identifying the idiosyncracies of colloquial English, unearthing the un-written laws of student-teacher relations, comprehending local grading concepts, and even recognizing such subtle features of American life as dress codes are only some of the problems that foreign TAs confront. This handbook is designed to provide advice and information for a speedy resolution of these problems.

The foreign teaching assistant handbook is written by foreign graduate students with teaching experience and other educators who have worked closely with them. The handbook is divided into four sections: Language Skills, Teaching Strategies, Cultural Issues, and Resources. These sections offer advice and describe campus resources which can help foreign TAs overcome the cultural, pedagogical and language barriers inherent in the American college environment.

This publication should be read and used in conjunction with *Learning To Teach*, the general handbook for teaching assistants at U.C. Berkeley published by the Graduate Assembly. All too often common teaching problems are attributed erroneously to the linguistic or cultural deficiencies of non-Americans. In actual fact the lack of systematic programs for instilling TAs—both American and foreign—with the fundamentals of teaching is often the root cause of many of these complaints. By using the two handbooks together, foreign TAs should be as ready as their American colleagues to face the demanding task of teaching themselves to teach.

Funding for this handbook has been generously provided by the Committee on Graduate Student Instructors. However, this is not an official university publication. The demand for this handbook has come from the ranks of foreign TAs. The following suggestions and guidelines reflect the concerns of foreign graduate students and their sincere desire to improve undergraduate education at U.C. Berkeley. In publishing this book, the Graduate Assembly hopes to help foreign TAs overcome their teaching problems. But publication of this book is not meant to imply that foreign TAs are a "problem." Indeed, the Graduate Assembly and all the contributors to this handbook are convinced that the presence of foreign TAs at U.C. Berkeley enriches the academic community— helping to provide the cosmopolitan and intellectually charged atmosphere befitting a great university.

Robby Cohen and Ron Robin

Language Skills

Foreign TAs often identify language issues as the most unsettling aspect of their teaching experiences at U.C. Berkeley. In addition to making cultural adjustments, non-American instructors face the complex task of communicating in a foreign language. This section of the handbook is designed to help TAs overcome language barriers. The articles in this section discuss strategies for enhancing English comprehension and classroom communication.

English and the Foreign-born TA
Jacqueline Hoeppner-Freitas

You, the foreign-born TA at U.C. Berkeley, have special problems which this *Handbook* seeks to address. The English language is only one of the difficulties you face, but easy access to it can reduce the many other strains that come with living, studying and teaching in a foreign land. If you feel uncomfortable speaking English, every day will be a struggle. I know this because for ten years I have taught English as a second language to many hundreds of students, some of whom have been TAs. I wish in this article to pass on to you some of the collective wisdom about learning English that my students have taught me. I also wish to advise you of the English language training resources available at U.C. Berkeley.

The Listening Comprehension Factor

If you are newly arrived in the United States and have heard very little native-speaker English, you are in for a huge surprise. When you begin your first conversation, you may expect to hear some polite questions such as "How are you?," "Where are you living?," "Where are you from?," "What's your name?," and "Did you get your classes yet?" Instead, you hear "How are ya?," and "Where ya livin'?," "Where ya from?," "Whatcher name?," and "Didja getcher classes yet?" Is this English? Yes, it's perfectly normal spoken English, but your ear has to have a chance to get used to it. You may think that you can't understand because your vocabulary is not large enough, but you will be surprised to find out that much of the language you are hearing is well-known to you in books. You

Language Skills

simply have to "tune into" the sounds of American English.

The best remedy for this "language shock" is to immerse yourself in English. DON'T avoid speaking English. DON'T stick to speakers of your native language. DO cultivate American friends. DO go to welcome parties. DO ask people in your department for help. DO ask questions of Americans on the street. DO get a TV and watch it when you can. DO get a radio and listen to it as much as possible. DO go to an English language movie whenever possible. DO read newspapers like *The Daily Californian* (the free on-campus newspaper), the *Oakland Tribune* (which covers the East Bay), or the *San Francisco Chronicle.*

This contact with English is important for more than just language learning reasons. If you know what's going on around the area in entertainment, politics and sports, you will be able to understand the references people make. Many incidents of misunderstanding are simply a matter of not knowing the cultural and political information that everyone else takes for granted. For this reason, the local news may not be the best program for you to watch at the beginning. The many local references will make you feel that your English is not as good as it really is. Remember, an American moving to Berkeley from another part of the U.S. would also be confused by local references. If, on the other hand, you have a chance first to read the local and/or national news and then watch it on TV, this could improve your listening comprehension. For language learning purposes, it might also be wise in the beginning to choose programs that have easily understandable content, such as game shows and soap operas. (See Note below.) Summaries of soap opera plots are available in magazines and newspapers (see Saturday's *Oakland Tribune*). Those who don't have a TV or who wish to have more control over what they see may go to the Dwinelle Hall Language Laboratory or the Moffitt Library Audio-Visual Media Center (see Resources below). Here, in very pleasant surroundings, you can watch and/or listen to videotapes, audio tapes, and/or records on every imaginable subject. You can, for example, buy a copy of *The Miracle Worker,* a play about a famous blind and deaf American, and then go to Moffitt Library and watch the performance on videotape. Being able to read it as well as watch and listen to it will help your aural comprehension. In the process, you will learn about Helen Keller, a person well-known to almost every

Language Skills

American. You may or may not admire famous Americans, may or may not wish to emulate American customs, but being informed about the United States and its culture will help your listening comprehension.

The Pronunciation Factor

There is probably no other area of language difficulty that causes you as a teaching assistant more concern than your pronunciation. You may feel exposed and vulnerable to criticism from your students. This fear is very real since students sometimes try to blame their difficulties in a course on their TA's pronunciation. And, unfortunately, the American English pronunciation system is a very difficult one to master. Many languages have a perfectly adequate five-vowel system. English has at least thirteen. In addition, the stress and intonation patterns of your native language may differ radically from those of English. So, how should you go about improving your pronunciation?

Pronunciation is, of course, best improved with the aid of a knowledgeable English teacher. But the single most important factor in improving pronunciation, experts say, is the motivation of the speaker. If you are not willing to attend to the elements that make American English pronunciation so different from your native language, and to spend time practicing your pronunciation, you will not be able to improve with all the most expert help in the world. It's like trying to become a good tennis player or guitar player without practicing. It takes practice time and a certain quality of attention to improve. Through repetition you must train your tongue and lips to the point that the sounds will be produced automatically, without thinking. I always tell my students to practice the sounds of English in front of a mirror whenever possible in order to watch the movements of the mouth. In addition, you can find many little blocks of time during the day when you can repeat words and sentences for practice. When you are walking to and from class, you can be rehearsing some sentences, either in your mind or out loud. If you don't want to be heard talking to yourself, you can practice in the shower. With repetition you will lose your self-consciousness at producing such unusual sounds. But it all begins with motivation. A student of mine from Korea said

Language Skills

that everywhere he went, on buses and on the street, he listened to tapes and repeated what he heard. He said he didn't care what people thought. And his English is now native-speaker quality, an amazing fact given that he did not grow up with any native speakers of English around him. The factor that most sets people like him apart is that he cared tremendously how his English sounded.

You can turn to the following resources to improve your pronunciation. The Language Laboratory in the basement of Dwinelle Hall has pronunciation tapes (see Resources below). You can also make your own tapes at home by getting an American friend to read a short passage or dialogue on to a tape with pauses long enough for you to repeat. You must ask your American friends to read at natural speed, however, and tell them not to be tempted to slow down and enunciate "more correctly." Most Americans, untrained in linguistics, mistakenly think that rapid spoken English is somehow "incorrect." They don't realize that that's how *everybody* speaks, professors and professionals included. Correct spoken English is delivered rapidly with many elisions and reductions. Whole sounds and words are reduced to schwa (ə) or eliminated completely. So the trick is to get your American helper to speak in to the tape recorder just as she or he would on the phone to an American friend.

Of course, before you begin practicing, it is wise to get a pronunciation assessment from a trained teacher of English as a second language. On a limited basis and by appointment, such an assessment is offered by the U.C. Berkeley Extension English Language Program, exclusively for foreign-born teaching assistants. We will provide a list of some of your most serious pronunciation problems with sentences to practice on a tape recorder. We can also refer you to materials available in the language laboratory. Call 642-9833 and ask for Ellen Rosenfield if you wish such help.

The Cross-Cultural Factor

Sometimes other factors besides language difficulties prevent effective communication. When a TA and a student experience confusion and possibly even hostility towards each other, a TA might easily jump to the conclusion that his/her language ability is at fault. While this may be the case, often other factors are involved. For one thing, as with native speakers, personality differences can cause misunderstand-

Language Skills

ing between TAs and students. The most important factor, however, and one very easily overlooked, may be cultural differences. These differences can account for much of the failed communication between the foreign-born TA and the American student. Some students, out of their own ethnocentricity, may make little attempt, if any, to understand their TA's personality, attitudes and culture. It should come as no surprise, then, that this kind of student may be irritated by TA's foreign accent and "strange ways." If, for example, the TA speaks softly and avoids eye contact while teaching, the student may assume the TA is insecure, unprepared, or, even worse, ignorant. The student may not have the sophistication or the desire to look beyond the surface at the cultural differences. Studies have shown that the single most important factor influencing communication success between foreign-born TAs and American students may be the extent of the *students'* ethnocentricity.

Fortunately, not all American students lack the ability to appreciate and understand other cultures. But the foreign-born TA must be aware that such students exist and must take steps to reduce the negative effect that the inevitable cross-cultural misunderstandings will have on the class. It is therefore advisable for the foreign-born TA to learn as much as possible about American culture. Again, one need not turn to books alone for information. By watching people you can attune yourself, for example, to American-style gestures, posture and eye contact. You might notice that it is common for Americans to listen politely by looking (not staring) at the speaker and nodding and saying, "Uh-huh" once in a while to demonstrate interest. On the other hand, speakers look at their listeners briefly, look away, continue talking and then establish eye contact again. It's as if the speaker must "check in" with the listener every so often to make sure the listener is attentive. These and many other conventions comprise a large body of culturally bound behavior that needs to be understood to bring about effective communication in the classroom.

Of course, the TA must also deal with a natural inclination to find American ways strange and even undesirable. For example, the loud voices that Americans often use in conversation may irritate those from other cultures who appreciate nuance and subtlety more than volume.

Language Skills

We recommend an excellent videotape, *Take Two*, which illustrates cross-cultural misunderstandings in one-on-one situations. You may make an appointment at the English Language Program to see this video. We can also recommend readings on this subject which may help you to become more sensitive to the very important factor of cross-cultural communication. But perhaps the best way to reduce your anxiety over this factor and at the same time reduce the negative impact of cultural differences is to establish a personal, though professional, relationship with each of your students as soon as possible. This can be done by monitoring each student's progress and helping each one individually to the extent feasible.

Summary

Don't hide from English. Seek it out in every way possible. Again and again I have seen foreign students retreat from contact with English, spending time only with friends from their own language group. Though tempting, this is the single most serious mistake you can make. You will be here for several years. To make it an enjoyable and rewarding experience rather than one of isolation and fear, *force* yourself from the beginning to seek out English in every possible context, even if it's only the English of street posters or throw-away advertisements. English is not just in books. In fact, some people argue that you *can't* find the "real" English in books. If you think of every native speaker of English around you as a resource, you are fighting the English battle in the right way.

Resources

Books for Purchase: English as a second language texts and English-English dictionaries for the foreign student are available at Campus Textbook Exchange, 2470 Bancroft Way, or may be ordered from Alta California Book Center, 14 Adrian Court, Burlingame, CA. Phone: 692-1285.

Tapes for Purchase: English as a second language audio tapes can be purchased from Alta California Books (see Books above), but are rather expensive. A very limited selection of tapes can be purchased inexpensively through the Language Laboratory, Dwinelle Hall (see Audio-Visual Aids below).

Audio-Visual Aids: When school is in session, the Audio-Visual Media Center, 150 Moffitt (undergraduate) Library, 642-8197, is open Fri. 9am-5pm; Sat. 10am-5pm; and Sun.

Language Skills

1pm-10pm. Call the Center for vacation schedules.

The Language Laboratory (LL) is in the Dwinelle Hall basement (south side of the building). Information and tapes are available at B-50. Some tapes are available on loan from B-1. The video cassette viewing room is B-22. The LL will also produce at-cost copies of some taped materials in 33 Dwinelle. Call 642-4067. The hours vary depending on the service, but in general the LL is open Mon.-Thurs. 8am-6pm; Fri. 8am-5pm; Sat. 10am-4pm. During exams and vacation, the schedule changes.

The Television Office, 9 Dwinelle Hall, also has videotapes available. Hours: Mon.-Fri. 9am-noon, 1pm-5pm.

See also Barbara Davis' article, "Oral Communication Resources," on page 69 of this handbook.

Note

If you don't know what a "soap opera" is, ask an American. I give my students assignments to stop people on the street and ask questions. I recommend this method of practicing English. For example, you could say, "Excuse me, I'm a foreign student here and my English teacher gave me an assignment to find out what a 'soap opera' is. Can you help me?" They have found that many Americans are happy to help. If somebody refuses to help, say, "Thanks anyway." Choose people who do not look busy and who don't seem strange. Beware of people who are members of religious groups such as the Unification Church ("the Moonies"), who are always looking for new members, especially among foreign students.

Overcoming the Language Barrier
Jia-Yush Yen

Using a foreign language is often a struggle. Many new international students at Berkeley avoid this problem by staying with colleagues from their native country and not speaking English at all. This method of retreating from the alien American environment is understandable because it cuts down on culture shock and does seem to save one a lot of trouble, but foreign graduate students who are going to be teaching at U.C. Berkeley will have to find a better solution. If you wait until you begin teaching to start adjusting to American language and culture, you are likely to heighten the difficulties inherent in foreign TA instruction at Berkeley.

Although Berkeley is a cosmopolitan campus, undergraduates here are young and may have parochial tendencies. They do not expect to be taught by a teaching assistant who comes from a foreign country, and may assume that a foreign accent means that their TA is not first rate (a common American assumption is that all intelligent people must speak flawless English). Such negative assumptions about foreign TAs may be compounded if you are overly nervous, since this may encourage students to lose confidence in you. It will take a lot of effort to gain their respect. Therefore it is wise to prepare to meet the problems as early as possible.

The following advice is designed to help you recognize and overcome language problems which foreign TAs often face at U.C. Berkeley.

Use Every Chance to Practice Your English

The only way to make your English an effective tool for your teaching is to make it your daily means of communication. Students appreciate the TA who can react effortlessly to their questions, and who can understand their questions the first time they are raised. If students have to explain their questions several times to the TA, frustration leaves the students unwilling to ask questions—a situation particularly damaging in discussion sections, which are designed specifically to address student questions. The most effective way to sharpen the language skills you will need for teaching is to find yourself an English-speaking roommate or friend, preferably one who does not speak your native language. You can also improve your English speaking and comprehension by watching television and listening to the radio. Reading out loud can also help improve your pronunciation.

Try to Think in English

The trick in learning to communicate effectively with your students is to think in English. If you have to go to your class without any time to practice, switching into English thinking mode can improve your English a great deal. This is especially true for TAs who come from Oriental countries that have language structures entirely different from English. If they think in their native language and then translate into English, they will have great difficulty making themselves understood. The sequence of expressing thoughts differs from language to language. Having Chinese as my native language means that the last thing to be said in English is always the first thing that occurs in my mind. Consequently, I always have to break one long sentence into several shorter sentences.

Another confusing point for me was using "Yes-No" sentences. When the question is "Isn't it?," your "No" still means no, and your "Yes" still means yes. Some languages do not have tenses and genders. Non-native speakers often confuse their listeners by using the wrong tense and gender. The definite article "the" can stress the key point you want to emphasize, but it makes an awkward sentence without proper usage. The use of the plural form needs extensive attention. Careless speakers confuse students by saying things in singular form, while actually referring to the plural. If you are a more experienced English speaker, you might want to pay some attention to the prepositions.

Now, the question arises as to how one can actually consider so many problems and features of the English

Language Skills

language before speaking. This will have to be built up gradually by practicing constantly. Each time you practice a sentence, you build up a sentence structure in your mind. When the time comes to use such structures, they will automatically occur to you.

When using a language like English it is very important to pronounce each word clearly in order to transmit information accurately. Most foreigners have already adopted their native pronunciation style. It is very difficult for one to correct accent problems in a short time. Since we have to live with our accents, it is better not to agonize over this. To help your students overcome the potential barrier to comprehension that your accent may pose, it is crucial that you speak as slowly and clearly as possible.

Have Confidence in Yourself

Familiarizing yourself with class materials is the best way to build up your own confidence. If you are well prepared and fully understand all the concepts and reading for the class, this will help you speak more smoothly. The new concepts you are teaching rather than your language barrier should grab the students' attention. When you are really on top of the class material and have developed interesting ways to present it to the students, your accent, sentences, grammar and phrases —even if imperfect—will be less noticeable to students.

The degree of self-confidence you display influences the students' image of you. If students see that you have both a good grasp of the subject matter and faith in your ability to teach, they are more likely to respect you. Once you've gotten that respect, you are unlikely to encounter unjust complaints from the class. Students will take notes, ask questions, make appointments for individual instruction, and do anything else they can to understand you if they have confidence in you. Most of this confidence must be built up by the TA through conscientious efforts at effective teaching.

Before you step into your first class, it is valuable to consult your supervising professor and an experienced TA about what you should expect from the students, and how best to facilitate discussions, lead lab sections, or offer suitable lectures (TA responsibilities vary from department to department, so such consultation is critical). If you are able to learn about the students' needs and backgrounds prior to the start of the semester, this will reduce your worries before the class. It is critical that you inquire as to the level of knowledge of

Language Skills

your students. Even if your English is perfect, your students will be unable to understand you if you assume too much and teach on a higher level than they can comprehend.

Carefully Prepare Your Lecture or Questions

A good instructor always comes to section with at least a tentative plan for conducting the class. Before you enter the class you should already have thought about what you want the lecture or class discussion to cover, and how this material can most effectively be taught. If you are insecure about your English you may want to write down everything you want to say in advance. But if at all possible, try to break away from complete reliance on notes, so that there is more give and take between yourself and your students. American students like this and often learn more efficiently if they are drawn into spontaneous intellectual interactions with their classmates and TAs.

Make Extensive Use of the Blackboard

If the TA writes clearly and in a well organized fashion on the blackboard, it can serve as one of the most useful teaching tools. Many students find that a blackboard filled with key concepts expedites their comprehension of class materials. Since most foreign TAs write well in English (or at least better than they speak it), extensive use of the blackboard is in effect a way of playing to their strength. A well organized blackboard not only assists your students, but can also help you proceed through the lecture, discussion, or lab section by section. Write down as much as you can on the blackboard when explaining complex ideas and problems. If you feel that your spoken English is weak, you may want to write on the board almost everything you say. By using the blackboard you double the students' chances of understanding the material because they will be simultaneously hearing and seeing what they need to learn. This is particularly valuable with accent problems, because if they have difficulty understanding a word or phrase you are using, they will be able to look on the board and pick up that word or phrase—thereby overcoming a potential language problem.

Use Visual Aides Wherever Possible

It is much easier to show a ball than to draw one and then

try to explain it. If the model is available do not hesitate to use it. In lab courses it is much easier to operate the experiment once in front of the students than to explain all the complex procedures without visual aides. Audio-visual aids such as slides, photographs, movies, and music are often helpful in the social sciences and humanities; they are extremely effective ways to illustrate hard-to-explain points. As twentieth century America is in essence a visual culture, you will find that your students appreciate these methods.

The University provides videotaping services to record one of your class meetings. You should take advantage of this service to analyze your teaching performance. You can learn a lot from such taping, and will see, for instance, that effective teaching depends not only on speaking correctly but also on an overall sensitivity and responsiveness to your students as well as a good grasp of the material you are teaching. If you prepare well, work hard, and care about your students, you will be a good TA—even if your accent isn't as perfect as Walter Cronkite's.

Summary

- Improve your communication by thinking in English, using short sentences, paying attention to the basics (gender, tenses, etc.).
- Speak slowly instead of assuming an American accent.
- Present a confident image of a TA well prepared for the course.
- Use the blackboard and visual aids extensively.
- Be sure the students understand the material as well as your English.
- Consult more experienced TAs and faculty so that you have realistic expectations regarding the students' level of preparation.

Listening Comprehension and Accent Problems

Dong-Gyom Kim

"Sure."

I almost felt insulted when I heard this blunt reply to my heartfelt, "Thank you very much." It seemed to me to say, "Yes, I sure deserve your thanks." What else could I have made of it? After all, it was only a couple of days after I arrived in this country. It was a long time before I realized that "sure" is such a versatile word, a word that one can use to answer not only an are-you-sure question, but also a would-you request and and even a thank-you.

Studying abroad is certainly an exciting experience. Unless you are already accustomed to the new language, however, you must face annoying language problems, especially if you have to TA. What a language English is: Virtually anything can be used as a verb! I was really nervous in my first section meeting, where I had to speak for a whole hour in a language which I had hardly ever spoken for more than thirty seconds at a time. I had to ask my students constantly to repeat their questions, and then it took me what seemed like hours and hours to think out the right words for what I wanted to say.

Practicing English

Something had to be done about my difficulties. So I started watching TV a lot, even Sesame Street and Bugs Bunny cartoons. I tried hard to understand the dialogue from movies and also conversations over dinners, and sometimes I even tried to overhear people sitting next to me on BART trains. I apologize to those I overheard. And it paid off; all those

Language Skills

incredibly fast speakers appeared to be slowing down and their voices got louder and clearer. Among my happiest moments were those in which I found myself almost completely understanding people whose speech a semester earlier I could follow only fifty percent.

Learning by Listening

In many ways, making oneself understood can be more difficult than understanding others. Even if you say something that makes perfect sense, it may not be the way others would say it. For instance, none of my students would "substitute" A for B in a mathematical expression; they would just "plug" the one in for the other. It took me quite a long time to learn to plug something in instead of substituting it. I didn't want to sound like a recording from their textbook. So I listened carefully to what people said, and tried to imitate.

I have a friend, who, I am sure, thinks I am stupid because I don't understand what he says and keep asking him silly questions. Every day he teases me with a different way of saying hello: "How are you doing?," "What's up?," "How's life?," "How's it going?," "Howdy," and so forth. He seems to take pleasure in seeing my confused face, but he is such a good friend. For now I know the chances are that in the U.S. people don't always greet each other by saying "How are You?"—"I'm fine, thank you. And you?"—"Fine, too, thanks." as I had learned through books. Being made a fool of can sometimes also be a good way to learn a language.

Overcoming Accent Problems

Students complain a good deal about the accents of their foreign TAs. For many people, getting rid of an accent is indeed the most difficult thing. There are so many sounds in English, like those of *th* and *z*, which I used to have a lot of trouble pronouncing. Such sounds don't exist in Korean, my native language. When it came to a *th* right after a *z*, it seemed simply impossible. So I asked myself, *"Is this possible?"* I must have repeated this ridiculous question a million times while walking along streets, before I finally decided it *was* possible for me to pronounce those sounds.

Adjusting to American Speech Patterns

One finds that the proverbial doing as the Romans do is not

Language Skills

as easy as it sounds. It requires a lot of effort. It is of course even more difficult if one fails to notice the difference between the Romans' and one's own ways. I had expected such a different world when I first came to this country that I was struck by the similarities rather than the differences between the American way of life and my own country's. But now and then peculiar and interesting things come to my notice. For instance, in Korea we don't usually call others' names unless it is necessary to avoid confusion. But here in America, you may want to say, "Have you finished your homework yet, Tom?" even if there is no one else around besides you and Tom. The efforts the American TAs make to memorize their students' names are impressive, and I think it is indeed worthwhile to remember other people's names.

It would certainly be nice to be able to speak English like a native speaker. At the moment, I am satisfied with being able to say "sure" myself when somebody thanks me. It will take much more practice and imitation on my part to get better, but even if I am never able to speak English flawlessly, I know that things like goodwill, kindness, and willingness to help can offset some of my language handicap.

And moreover, there are some good things about being a non-native speaker: Unlike words in many other languages, English words often have quite unexpected pronunciations, and naturally, when I come across a new word, I tend to pronounce it just as it is spelled, however un-English it may sound. This made a student of mine happy: "Wow, you are the only one around here who could pronounce my name correctly at first sight!"

To Get Rid of an Accent

"Play the ape." Compare your pronunciation with native speakers', and imitate theirs.

Recording your voice and listening to it can be helpful. It may happen, for instance, that people cannot really tell whether you said "merry," "marry," or "Mary," although you think you have pronounced whatever you meant with the correct vowel.

Do not twist your lips and tongue excessively just because you think the sound is exotic enough to deserve such a strain. The outcome is likely to be a strange sound which does not exist in English at all.

Learn your favorite English songs and sing them. In some

How Do Americans Speak?
Bonnie Burns

An ability to identify the phonetic patterns of spoken English is often as important as mastering a comprehensive vocabulary or other formal linguistic techniques. Adopting the stress patterns of native speakers promises to facilitate positive responses from your students. A recognizable tone of voice will encourage students to concentrate on your remarks rather than deciphering what they may feel is a hidden message behind your utterances.

Typical American phonetic patterns and their most common foreign counterparts are best illustrated graphically. Compare these stress patterns.

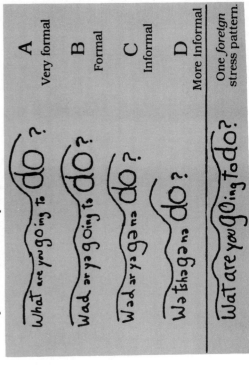

what are you going to do?	**A** Very formal
Wad ar ya going ta do?	**B** Formal
Wad ar ya gana do?	**C** Informal
We tha gana do?	**D** More Informal
Wat are you going to do?	*One foreign stress pattern.*

languages, most syllables end with a vowel, or some consonants are never used at the end of a syllable. In other languages there are no compound consonants or vowels. If your native language has such features, you may tend unwittingly to insert extra vowels in English words, changing the number of syllables in them, and thus making them sound more natural to you but strange to native English speakers. When you sing, however, you cannot help but stick to the correct number of syllables to fit the words into the melody.

If the pitch is more important in your language than the strength with which a syllable is uttered, it may give your English a conspicuous accent, because you will probably tend to emphasize the stressed syllables with pitch rather than with strength of sound. Analyzing and becoming more aware of the various factors which make up your accent is very important in getting rid of it.

To Improve Listening Comprehension

Listen a lot. It is a good idea to record, say, a radio newscast and listen to the difficult parts of it repeatedly until you can understand them. In my opinion, taking down what you hear and trying to reconstruct the text yourself is more helpful than merely listening to a tape for a text you already have.

It is important to note that quite often words will not retain their original sound values when used in fast speech. Can you tell the difference in the way *the* is pronounced in "want the money," and in "won the money"? How about *is* in "This is mine," and in "This is yours"? More striking is the way "I *want to* study," and "I'm *going to* study" are pronounced. There are hundreds of such examples.

It is very easy to get lost in a fast conversation and find yourself thinking of something else in your own native language. Try not to do that. Try to think in English.

Don't be afraid to ask people to repeat what they have said or to explain to you unfamiliar expressions. After all, you are a non-native speaker; you are not expected to be able to understand everything perfectly.

As soon as you become able to use an expression yourself you will never miss it in other people's speech. Whenever you learn a new expression, therefore, try to use it right away.

Correct Stress: The Big Three

There are exceptions, and special stresses, but these three rules usually work.

1. In a sentence, put the hardest stress on the last "important" word. By "important," I mean noun, adjective, adverb.

Examples:

Want t'e SEE e MOvie with me?
I c'n PAY fr't.
Wane Want to see a movie with me?
I can pay for it.

2. In compound nouns (two or three nouns together), stress the first noun.

Examples:

the CLASSroom   the FIREmen

He's an ENGlish teacher.   (teacher of English)

He's an English TEAcher.   (teacher from England)

In fixed adjective-noun expressions, as in compound nouns, the first word is stressed.

Examples:

Give me a HOT dog. And FRENCH fries.

It's a BLACKboard.   (fixed adjective-noun)

It's a black BOARD.   (adjective)   (noun)

3. In two-word verbs, stress the second word (i.e., the adverb). This rule follows the Rule 1, which is the basic rule of English stress.

Examples:

Bring OUT the CAKE. Bring it OUT.

Think Over my iDEA. Think it Over.

But, note:

THINK about it.

This example has a verb and a preposition (about), not an adverb. Don't stress prepositions or pronouns.

28

American Pronunciation Habits

1. Notice that at all levels, from A to D, there is a big difference between strong and weak syllables (bigger, linguists say, than in any other language).

2. Notice that stresses (what go DO) are the same from A to D. The basic stress pattern is the same at every level of English. But as pronunciation becomes less formal, weak stresses get even weaker: weak vowels change to 2 (B, C), or they drop (C, D).

3. Foreign speakers often make strong syllables too weak and weak syllables too strong and clear (see above). Because of their clear vowels in weak syllables, they sound very formal in English. And when a foreigner is misunderstood, or misunderstands someone, the trouble is usually stress, not sounds.

4. The United States is informal (especially California, *especially* Berkeley). It is uncommon for anyone to speak as formally as in A, except for special clarity (as in talking to foreign speakers) or for dramatic effect. We may believe Americans speak quite clearly (weak vowels clear too), but if we listen carefully to ourselves, we realize we don't.

Informal stress patterns are more common among academic and professional people here than in some other countries. It is common for a doctor or professor to say C or even D. The informal pronunciation has a friendly feeling.

5. C and D are spoken quickly. Quick speech is what made their weak syllables so weak in the first place.

How Do You Want To Speak?

How you learn to speak is a personal choice. You may want to practice all the levels and shift from level to level depending on the situation, like Americans do. If your personal style is informal, or if you will need English for informal situations, C level may suit you. If you are more comfortable with a formal style, or if you will need English for formal situations, maybe you will focus on B level.

Even if you prefer to cultivate a formal style, it is useful to study the informal. Why?

1. Being able to speak informally helps your listening comprehension.

2. Making weak syllables very weak (C,D) helps you stop making them too strong (foreign stress pattern) in formal speech.

27

Teaching Strategies

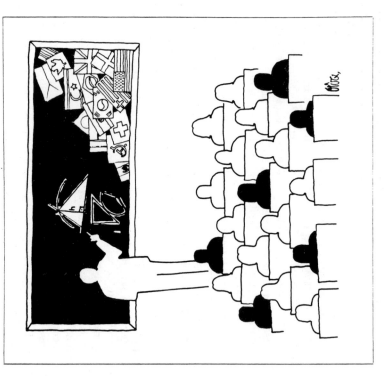

Although foreign TAs are usually extremely well grounded in their subject matter, most have no preparation for the pedagogical problems and responsibilities which await them in the classroom. Articles in this section provide vital information on undergraduates' grading expectations and the role of section leaders, as well as a discussion on how to evaluate your teaching performance.

Of course, the problems of the sciences differ from those of the humanities, and tips are presented accordingly. Nevertheless, all articles stress a common theme: the need to recognize and accommodate local educational standards.

Leading Sections in the Sciences
Rajesh Gupta

In contrast to our colleagues in the Humanities and the Social Sciences, the TA in the Sciences has the advantage of working and conversing in the universal language of numbers and equations. Since vocabulary is limited in the Sciences, basic teaching skills are of extreme importance in opening up the lines of communication between you and your students. Of course, language and cultural divergencies often pose a problem, but first and foremost the foreign TA must make judicious use of fundamental teaching props.

Preparation and Classroom Props

Skillful use of the blackboard is crucial. Neat, legible and pre-planned use of the board will endow your students with a sense of direction, while keeping you from straying to issues that do not warrant discussion with the entire section. If you have planned the section well, you should be able to anticipate all the major issues to be discussed. These problems should be set out on the blackboard before section begins. There is nothing more disheartening than attempting to write on the blackboard while carrying out a conversation in a foreign language. Among other things, it is important that throughout the lesson you be in constant touch with the students rather than the board—another reason for preparing handouts or writing up issues on the board before class.

Anticipating problems is the key to all good teaching, irrespective of language or nationality. If you have not taught American undergraduates before, you should seek guidance from the course instructor. Some instructors are happy to

Teaching Strategies

deliver their lectures and leave the rest to their TAs. Thus, TAs have to bear most of the brunt of the course. In this case you should establish a good working relationship with a veteran TA, preferably one who is sensitive to the problems you may be encountering.

Language Barriers

Language is an issue that needs to be addressed. Even though your command of English may be excellent, chances are that at least some problems will pop up when conversing with your students. Your accent may pose difficulties; your students may fail to appreciate your system of emphasizing points, or they might take supposedly humorous points seriously (what can be worse than that!). The situation is complicated by the fact that in a discussion section many students shy away from pointing out the ambiguous issues. Therefore, you need to make an effort to encourage them to express whatever might be bothering them.

Advise your students to speak clearly, and constantly probe to see if they are following your train of thought. If you are consistent and friendly, you will find that many hesitant students will build up the courage to ask questions. Try not to ignore any unclear sentences that you may utter, even though you may consider it insignificant to repeat yourself. Any incomprehension on the part of the students tends to distract even the most serious students. Listen carefully to your students, because ignored students lose interest very fast. Pay attention to facial expressions. Blank expressions mean that you need to repeat whatever you're saying, preferably in different words.

Cultural Issues

Foreign TAs will also encounter some cultural obstacles. Indeed, many of us who have just arrived in this country experience a state of cultural disorientation. Not being part of the student body is one of the problems encountered by the foreign graduate student. This personal aspect eventually affects teaching abilities, too. Some of the foreign TAs tend to feel so separated from their students that they are unaware of the general level of the class. It is best to take active interest not only in the subject matter that you are teaching, but also in the background and interests of your students. It is important to keep in mind that a typical class consists of students with a variety of backgrounds. Most foreign TAs are

41

Teaching Strategies

surprised by the wide variety of knowledge—or ignorance—that exists in a large and prestigious school such as Berkeley. Not all students can be expected to have the same level of understanding.

Often it is good practice to ask your students to write on a slip of paper the relevant courses they have already taken. You might also want to get to know their interests outside of the classroom (sports, movies, etc.) so as to improve the general rapport.

Another cultural mismatch is the level of informality expected in the classroom. Many foreign students find themselves at a loss about how formal the classroom atmosphere should be. A friendly, easy-going atmosphere is definitely helpful. However, caution is advised. Some of the students, though not very formal, are shy. Watch out, or else you might be in for a "cold shoulder" should you extend your "informality" too far. It is best to be serious without being stuffy and to work on gradually getting to know your students better.

You should expect that some of the students, being aware of your fleeting acquaintance with the American college system, will attempt to persuade you to do the work that they are supposed to grapple with alone. Make sure that you do not spend the section solving individual problems that are an integral part of the student's homework. Instead, spend as much time as possible suggesting methods for problem solving.

Leading sections at Berkeley is definitely rewarding. With a little caution, a lot of common sense, and much preparation, you will probably enjoy the interaction with your students at a time when you find yourself in the difficult situation of being far from home, in a strange society.

Summary

- Skillful use of the blackboard and other teaching props will eliminate many potential misunderstandings.
- Language barriers may be overcome if you try to anticipate likely questions and prepare answers beforehand.
- Get to know your students as early as possible, particularly their academic backgrounds.
- Learn to adapt the informal yet firm attitude expected from TAs at Berkeley.

42

Foreign TAs from
the Undergraduate Perspective
David Pickell

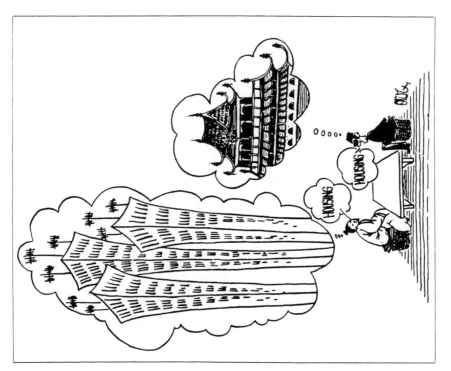

Although most undergraduates have grumbled to friends about a teaching assistant with an accent so thick that it makes a difficult subject almost impenetrable, and others with more of a sense of humor have joked about a non-native TA's language gaffe ("He thought he said problem sex!"). In general undergraduates at U.C. Berkeley find that foreign TAs add a welcome dash of spice to their undergraduate experience.

The most positive undergraduate response to foreign TAs comes from students in international disciplines, such as Latin American studies, comparative literature and geography. As one student of agricultural economy said, foreign TAs can speak not only from knowledge but also from experience on foreign political and cultural questions. "The foreign TAs that I've had would use examples that were real, and were lived," she said. "They brought a breadth and a wealth of knowledge that American TAs don't have, or have gotten only from books."

One student studying organization theory, in particular comparing bureaucracies in different countries, found the course especially rewarding because it was taught by a French TA:

"Having been born and raised in Paris, and even having worked for a year pushing papers in the French bureaucracy, the TA was able to paint a vivid picture—complete with descriptions of the standard two and a half hour lunches and job security guarantees that make the civil service here look like a temp help agency—of the notoriously moribund French system."

Communication Problems

Not all undergraduates, however, found being in a class with a foreign instructor such a rosy learning situation. Many complained, sometimes bitterly, of an instructor's difficulty with English:

Cultural Issues

The basis for a good working relationship between foreign TAs and their students is mutual understanding. Like their American counterparts, foreign TAs need to establish a comfortable learning atmosphere based on mutual respect. By identifying the cultural barriers that may divide foreign TAs from their students and offering advice on how to overcome these obstacles, this section should help you learn how to create a fertile learning environment.

whether intentional or not, can present a roadblock to participation and hands-on learning. Awe and other strange forms of academic hero-worship do not as often dampen discussion in classes led by TAs.

Undergraduates were put off by TAs—and many said this was more common among foreign TAs—who were excessively formal, who were uncomfortable being called by their first names or who keep office-hour discussions strictly to the academic topic at hand. One student in the sciences complained that her foreign TAs were less jocular, "less willing to put their feet up on the table during office hours." She added, however, that these same TAs were very strict about keeping regular office hours, in contrast to their more casual peers, who would say "just make an appointment," or "just give me a call."

Although most undergraduates objected to the formality of foreign TAs, one student said going to the office of her TA, who was from India, was a memorable taste of very non-Occidental sensibilities.

"He would always treat us as if we were walking into his house. He would offer us tea, and he was always scrupulously polite. He was very formal, but that is the custom. As an American, it seemed a little pretentious. I felt like I was being treated like royalty. But to him it was a sincere show of friendship."

The Problem of TA Intolerance

Another undergraduate told a story of a TA from Tunisia whose educational experience did not prepare him for the intellectual irreverence of American undergraduates. According to the student, the TA insisted on one interpretation of the material (and to this particular student, at least, not a very interesting one) and interpreted any student's questioning of his approach as a personal attack on his authority. At one point, the student said, the TA stormed from the room after misinterpreting a question as an insult.

"At first you weren't sure what the situation was, because he would listen to questions and just let us talk on and on. But as the class got further and further along, he increasingly insisted that we understand his approach. It was just a misunderstanding between the students and him. He just thought all the other approaches that we brought into the class, all the other analytical tools, were things that we were going to forget about when we learned his way, and we were eventually going to be converted to his way. For the papers, you got an A if you regurgitated what he said, and you got a C or D if you didn't."

"This guy was teaching Hegel. He was reading the Italian translation, which he would have to translate again into English. And his English was so horrible he would have to pause, trying to stammer out the last word of a sentence while people in the class shouted suggestions. When he heard the right word he picked it up and kept going. Needless to say, the fine points of Hegel's theory got lost somewhere in all this. It was a nightmare."

Another student found his TA's accent intelligible, and his grasp of English satisfactory, during the lecture, but when points were brought up during discussion, he would be lost.

"Precisely when you got to a very complicated counterpoint, that was when the communication would break down because he didn't understand English well enough."

But what many students found distracting was not an instructor's accent, or failure to grasp the rudiments of the King's grammar (a rare occurrence according to most undergraduates), but rather the formality of a foreign TA's use of the language. Students often depend on their TA to translate the staid and effete academic tomes they have been assigned to read into street language and American idioms that make the intellectual content of the readings accessible.

One undergraduate commented, "In this (philosophy) class, I missed the quick kind of pedestrian examples my American TAs could give. She couldn't argue a philosophical point with a McDonald's hamburger, and I've had American TAs who could."

Idiom and American slang are things that a TA can learn only by living in this country. But foreign TAs' "highbrow" use of language was mentioned by undergraduates in connection with a perceived attitude on the part of foreign TAs to define the student/teacher relationship more sharply than many would have liked.

The TA-Student Relationship

For undergraduates, TAs serve to personalize the large institution in which students can quickly seem to represent no more than numbers on computer sheets. Students have said they find classes taught by TAs more rewarding than those the professor teaches because the TA—younger, less established and perhaps less jaded than a professor—is not as threatening a figure. The authority that shrouds a professor.

"He thought the purpose of the class was to read these epics, and to understand them the way he understands them, and that was the way we were to learn. And on one hand that's an interesting approach, that we were to absorb his viewpoint. But on the other hand, that's really a very authoritarian way to teach.... He just had no conception of the Berkeleyan way of education."

Although the student said the experience was "jarring," he also said it forced him to take a fresh look at "the American approach to education," a re-evaluation he found rewarding. As he put it, "His approach to education was educational itself."

Bringing Cosmopolitan Perspectives to Berkeley

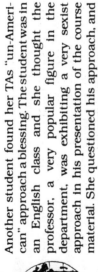

Another student found her TAs "un-American" approach a blessing. The student was in an English class and she thought the professor, a very popular figure in the department, was exhibiting a very sexist approach in his presentation of the course material. She questioned his approach, and wrote her papers using interpretive tools borrowed from some of the latest French feminist critics. The professor found her complaints distracting and had little to say about her writing. But the TA in the class, who was Swiss and had studied the new French criticism, which is popular and respected in Europe, became the student's ally and teacher. "She was very well versed in feminist theory," the student said: "she had that background which you just cannot get in the United States."

In general, undergraduates find that foreign TAs add welcome variety to their experiences at the university. Their manner and style are different: their accents, examples and phrasing, though sometimes confounding, are interesting and "exotic"; and their European and Eastern educations can shed a new light on academic questions. Non-native TAs can give students a taste of another culture, which can itself be educational.

As one undergraduate observed, "I had a statistics class where the TA was from Brazil. The way I look at it you learn more from a class than just the material. Primarily you learn the course material, and secondarily about life in general. And this TA—in a statistics class for non-mathematical types—was excellent because he would use as examples for story problems his experiences from Brazil. It made it fascinating. You're learning about another culture at the same time you're learning statistics."

A Woman, a Foreigner, a TA
Jyoti Sanghera

I paused with my hand on the door knob, ostensibly to quell my thumping heartbeat and to collect myself. In that brief moment all the events of the past ten days hurriedly stumbled over each other in dizzying succession. Only ten days ago, I had landed in this country, this alien land, from New Delhi. And now, here I was on the threshold, ready to make my entrance as a TA assigned to young Berkeley undergraduates. "It is only normal that I am nervous," I tried to console myself. Squaring my shoulders and pushing my chin up, I attempted to draw upon my previous experience as a teacher. I suppose it worked, for I managed to enter the room with an apparent equanimity I was still far from genuinely feeling.

An Apprehensive Introduction

As I introduced myself and turned to the blackboard to inscribe my strange-sounding name, I saw myself mirrored through the eyes of these students I was supposed to teach. The look I encountered from all of them was an unmistakable combination of apprehension, curiosity and nervousness. Well, to that extent we were all kindred souls. But added to this, in the case of a few young men particularly, were distinct glances of amusement and even suspicion. I suppose I stood out very conspicuously in my distinct Indian attire, as a colored woman from the Third World who spoke English with a strange accent. This was a situation that merited getting used to on both my part and on the part of the students.

The introductory process was somewhat prolonged on account of my being a foreign TA and a woman. I found myself responding to a host of queries ranging from general information on India to fairly incisive autobiographical details. For instance, I delved into India's colonial past and legacy in order to comment on their interrogative statement "How come as an Indian I spoke such fluent English?" I also had to embark on a brief lecture on the status of women in India, to break the stereotypical image of the downtrodden, silent suffering woman created by the media in the West. However, in the long run this rather arduous and protracted introduction went a long way in dispelling the initial doubts and creating an atmosphere of ease, acceptance and congeniality. My students were eager to brief me on the ways of American life and the problems faced by the university students.

Encountering Prejudice

This unique process of mutual learning marked my entire experience as a TA. My term as a TA happened to coincide with the occurrence of some political upheavals in India from which I was able to draw meaningful comparisons with the social structure in the West and generalize on the socio-political fiber of developing societies. While most of the students participated eagerly in such discussions, there were a few who were affronted and upset. Unfortunately, these dissenting voices were never aired and I became aware of their existence only after an informal evaluation when one of the students, presumably a male, ventured to assert that "This woman, this dumbo from a Third World country, from a society far less developed than the U.S.A., has the gall to stand up and criticize our country and say that she is going to put things in perspective."

I was frankly taken aback by this caustic criticism, especially because in none of the discussion sessions had I gotten wind of such hostility. Even though I am aware that this was a minority opinion, and by no means representative of the general attitude of the students, I would still advocate caution while expounding on the ills and evils afflicting American society. It may be advisable partially to couch one's criticism in softer language so as not to offend the sensibilities of some students. On the other hand, one must be wary of being intimidated by the presence of obnoxious, racist characters, a sprinkling of which may be found in each section.

Apart from the above reaction, as a woman and as a foreigner, I did not encounter any overt problems. In fact, I

found the women and minority students, including blacks, drawing much closer to me. Their sense of ease was decidedly reflected in their interaction with me during my office hours when they would come up to me without any inhibitions and often discuss their personal problems, apart from clarifying doubts on the course material.

While these gestures did create a friendly environment and made me feel good, I sensed the desirability of distancing myself somewhat from my students, especially for the expediency of grading. However, the minority students and women would make it a point to inform me of any protest meetings, conferences or events of public or political interest, and solicited my opinion on many of these issues.

It cannot be denied that a foreign woman TA has to be doubly cautious in dealing with the disadvantages she is burdened with by virtue of her status. I often wondered whether I was putting much more effort in preparing my sections than did the other TAs so as to prove myself. And to this day, I have not been able to ascertain whether I had relatively fewer students in my sections on account of my nationality and gender.

I would conclude by saying that the problems one has to cope with as a foreign woman TA are by no means insurmountable. As in every other field, in order to do well, one has to be open, receptive, human, sensitive and capable.

December 1987

Letter of Recommendation for

To whom it may concern:

          has been a <u>very solid</u> Teaching Assistant
in the Mathematics Department of the University of
Wisconsin-Madison for the past few years.   He has had
considerable experience with both Calculus and Pre-Calculus
where his student evaluations have always been among the very
highest of our approximately 150 T.A.'s.

     He has also involved himself with aspects of the
department's program outside of usual classroom teaching.   He
has been a coordinator in a Calculus lecture and in this role
was concerned with the training of new T.A.'s.   This past
summer he and I developed a summer orientation program for
new foreign T.A.'s.   We brought in seven foreign T.A.'s for
the entire summer where they studied English, practiced in a
classroom situation and "gently" assisted in the teaching of
Calculus.

          approached all these tasks with great care and
sensitivity.   During his tenure here he demonstrated all the
interpersonal, organizational and mathematical skills which
would bode well for his success as a teacher.   I recommend
him to you without qualification.

                         Sincerely,

                         Professor of Mathematics

SB/mm